ELEMENTARY PARTIAL DIFFERENTIAL EQUATIONS

HOLDEN-DAY SERIES IN MATHEMATICS

Earl E. Coddington and Andrew M. Gleason, Editors

ELEMENTARY PARTIAL DIFFERENTIAL EQUATIONS

Paul W. Berg
James L. McGregor

Department of Mathematics, Stanford University

Holden-Day

Oakland, California

Library of Congress Catalog Card Number: 66–28845

Printed in the United States of America

15 16 17 MP 90 89 88

To our families

Judy, David, and Jeremy
Alice, James, and Elaine

Preface

Partial differential equations are important in the problems of many branches of science and engineering, and as a stimulus and area of application for abstract mathematical analysis. As instruction in these fields has become more prevalent and has moved down in the university curriculum, it has become increasingly desirable for many students to begin the study of partial differential equations early in their careers.

This book is based on lectures given by us over the past ten years in a two-quarter sequence course in Elementary Partial Differential Equations. Students in the course include majors in mathematics, the physical sciences, and engineering, at differing stages in their education, varying from the sophomore to the first-year graduate level.

The book has two principal objectives: to provide students with the techniques necessary for the formulation and solution of problems involving partial differential equations in courses in other disciplines; and to prepare students for further study in partial differential equations and linear analysis by furnishing a basis for "intuition" in these subjects, namely, a thorough familiarity with their simplest typical problems.

Chapters 1–3 of the book constitute an introduction—to partial differential equations in general, to the classification of linear partial differential equations with constant coefficients, and to the formulation of a large class of typical problems of mathematical physics. In Chapters 4–7 there is developed, at first formally and finally rigorously, a unified method of solution of the problems we consider. The method is that of spectral representation, with separation of variables serving as the procedure for determining the appropriate one. In the remaining chapters the scope of this procedure is extended to a variety of other problems. Although we treat the various equations of mathematical physics with a unified method of solution, we discuss these equations individually and exhibit their different qualities in as much depth as is feasible.

The mathematical background that we require of the reader is knowledge of the contents of standard courses in the calculus and elementary ordinary differential equations. Where it has seemed appropriate we have briefly recapitulated some of this background material.

As a general pedagogical view, we have held that the understanding of general ideas is best achieved by the majority of students when they are first presented with the application of these ideas to special cases, and are then required to treat similar but novel problems, either in exercises or abbreviated text discussions, with these special cases as models. In this statement we refer not only to mathematical methods, but also to the mathematical description of physical phenomena.

The exercises play an important role in the text. In addition to the function referred to above, they serve to introduce other general ideas and procedures, and to point out that the solutions to problems which we obtain provide much more than just a means of calculating values of the solutions.

Although the book was written for use in a two-quarter course, it can be used as a text for a one-quarter or one-semester course. We have tried to organize the book to provide a maximum of opportunity for the selection of topics. Many sections, especially the later ones in a chapter, can be omitted without affecting the continuity of the discussion. Our experience has been that a course can be planned with Chapters 1–5 and Chapter 6 through Section 6.7 as a core, with the selection of additional topics depending on the interests and objectives of the instructor.

For five years the course for which this book was written had as its text multilithed lecture notes, which were successively revised and extended. A final version of these notes was published as a preliminary edition of this book and has been used as the text for our course and for courses at other institutions for the past three years. The book now being presented is a revision of this preliminary edition, with the addition of chapters based on the later lectures in our course.

Throughout the period of development of our book we have had the benefit of the comments and suggestions of numerous students and of many of our colleagues who have taught the course with us at Stanford. For their considerable assistance we express our deep gratitude. We are also indebted to a number of colleagues at other institutions for their valuable help.

Finally, we extend our thanks to Miss Gail Lemmond and Miss Jeanne Wray for their excellent typing of the manuscript, to Mr. Michael Reed for the preparation of the index, to Mr. Forrester Johnson and Mr. Thomas McReady for their assistance in writing the answers to problems, and to the staff of our publisher for their cooperation and patience in the design and production of this book.

Paul W. Berg
James L. McGregor

Stanford, California
June, 1966

Table of contents

I

Introduction

1.1. PARTIAL DIFFERENTIAL EQUATIONS

A **partial differential equation** is an equation

$$F(x, y, \ldots, u, u_x, u_y, \ldots, u_{xx}, u_{xy}, \ldots) = 0$$

involving several independent variables $x, y, \ldots$, a function u of these variables, and the partial derivatives $u_x, u_y, \ldots, u_{xx}, u_{xy}, \ldots$ of the function. The **order** of a partial differential equation is the order of the partial derivative of highest order appearing in the equation. A function $u(x, y, \ldots)$ is called a **solution** of the equation if the equation becomes an identity in the independent variables when u and its partial derivatives are substituted in the equation.

An example is the equation

$$u_{xx} + u_{yy} = 0 .$$

The order of this equation is two. The functions

$$u_1(x,y) = x^2 - y^2 ,$$
$$u_2(x,y) = e^x \cos y$$

are solutions of this equation.

We shall be mostly concerned with linear partial differential equations, that is, equations in which the unknown function and its partial derivatives appear to the first degree, at most. For example, the most general second-order linear partial differential equation in two independent variables is

$$au_{xx} + bu_{xy} + cu_{yy} + hu_x + ku_y + eu + f = 0$$

where a, b, c, h, k, e, f are any functions of x and y. Linear partial differential equations are distinguished by their wide occurrence in applications and the relative completeness and simplicity of the theory concerning them.

There are some similarities between the theory of ordinary differential equations and that of partial differential equations, but there are also profound differences. In this and the following chapter we shall examine some of these similarities and differences.

1.2. *GENERAL SOLUTIONS*

The **general solution** of a differential equation is the collection of all solutions of the equation.

The general solution of an nth order *ordinary* differential equation is a family of functions depending on n independent *arbitrary constants*. To get some idea of the nature of the general solution for *partial* differential equations we will consider some simple examples.

To solve the first-order equation

$$u_x = y \sin x \tag{1.2.1}$$

we integrate with respect to x, holding y fixed. We obtain the family of functions

$$u(x,y) = -y \cos x + f(y) \tag{1.2.2}$$

where f is an arbitrary function. By differentiation we verify that for each f the corresponding function (1.2.2) is a solution of (1.2.1), so that the family (1.2.2) is the general solution of the equation. Here we find that the general solution depends on an arbitrary element, but on an *arbitrary function* rather than an arbitrary constant.

As an example of a simple second-order equation consider

$$u_{xx} = 1\ .$$

If we integrate this equation with respect to x, holding y fixed, we obtain first

$$u_x(x,y) = x + f(y)$$

and then, upon integrating again,

$$u(x,y) = \tfrac{1}{2}x^2 + xf(y) + g(y)\ .$$

This family of functions depending on two arbitrary functions is the general solution of the equation.

Again consider the differential equation

$$u_{xx} = 1$$

but now suppose that u is a function of three variables, $u = u(x,y,z)$. Proceeding as above, we find the general solution

$$u(x,y,z) = \tfrac{1}{2}x^2 + xf(y,z) + g(y,z)$$

where f, g are arbitrary functions of two variables.

From the consideration of these and similar examples, one might conjecture that the general solution of an nth-order partial differential equation in m independent variables is a family of functions depending on n independent arbitrary functions of $m - 1$ variables. This statement is analogous to the statement concerning the general solutions of ordinary differential equations, which can be regarded as the special case in which $m = 1$. It is roughly correct, but very difficult to make precise, and impossible to prove generally. Indeed this is so even for the special case of ordinary differential equations.

Conversely, a family of functions depending on arbitrary functions can often be shown to be the family of all solutions of a certain partial differential equation, that is, the family can be described by the partial differential equation (see Exercises 1a).

In practice, one is usually called on to find, not the general solution of a differential equation, but a particular solution satisfying certain *auxiliary conditions*. In the study of ordinary differential equations the procedure for the solution of such a problem is, first to find the general solution of the differential equation, then to determine the values of the arbitrary constants in this solution required to satisfy the auxiliary conditions. Here the difficult problem is that of finding the general solution; the determination of the particular solution which satisfies the auxiliary conditions is then in principle a simple algebraic problem.

The corresponding statements for partial differential equations are not true. Even when the general solution of a partial differential equation is known, the problem of finding a particular solution satisfying given auxiliary conditions may be difficult. It may be as difficult, or even more difficult, than the problem of finding the general solution. This is so because the specification of a solution of a partial differential equation requires the determination of arbitrary functions, rather than, as in the case of ordinary differential equations, arbitrary constants. In consequence, much of our effort in this study will have to be devoted to the problem of satisfying auxiliary conditions. Indeed, in most cases we shall not even try to find the general solution, but instead will immediately seek solutions which satisfy at least some of the auxiliary conditions.

EXERCISES 1a

1. Find the general solution $u(x,y)$ of

$$u_{xxy} = 1 .$$

2. Find the general solution $u(x,y)$ of

$$u_{xx} - u = 0 .$$

3. Find the solution $u(x,y)$ of the differential equation

$$u_{xx} - u = 0$$

which satisfies the auxiliary conditions

$$u(0,y) = \varphi(y) , \quad u_x(0,y) = \psi(y) .$$

4. Find the general solution $u(x,y,z)$ of

$$u_{xxx} + u_x = 0 .$$

5. Find the general solution $u(x,y)$ of

$$uu_{xy} - u_x u_y = 0 .$$

6. Show that every function u given by

$$u = f(x^2 + y^2) ,$$

where f is an arbitrary function of one variable having a continuous derivative, is a solution of $yu_x - xu_y = 0$.

7. By introducing polar coordinates in the equation

$$yu_x - xu_y = 0 ,$$

find its general solution.

8. Show that if g is an arbitrary differentiable function of one variable, then

$$u(x,y) = -\tfrac{1}{2} + e^{y/2}g(x - \tfrac{3}{4}y)$$

is a solution of

$$3u_x + 4u_y - 2u = 1 .$$

9. Find a partial differential equation satisfied by all functions of the form
(a) $u = f(x^2 - y^2)$
(b) $u = f(x^2 + y^2) + g(x)$.
(Assume f and g to have as many derivatives as you need for your argument.)

1.3. *LINEAR OPERATORS AND LINEAR EQUATIONS*

The study of an ordinary differential equation is greatly simplified if the equation is *linear*, because for such equations all solutions can be obtained by combining particular solutions using the *principle of superposition*. We shall define and discuss these notions in this section.

We say an **operator** A is defined if there is a rule which assigns to each function u of one given class another function $v = Au$ of a second class. For example, the formulas

$$Au = \frac{\partial^2 u}{\partial x^2} + \frac{\partial^2 u}{\partial y^2},$$

$$(1.3.1) \qquad Bu = 3x\frac{\partial u}{\partial x} - 4y\frac{\partial u}{\partial y} + (x^2 + y^2)u,$$

$$Cu = \left(\frac{\partial u}{\partial x}\right)^2 + \left(\frac{\partial u}{\partial y}\right)^2$$

define three differential operators A, B, C.

An operator A is called a **linear operator** if for any functions u_1, u_2 and any constants c_1, c_2

$$(1.3.2) \qquad A(c_1u_1 + c_2u_2) = c_1Au_1 + c_2Au_2 .$$

In the examples (1.3.1), A and B are linear operators but C is not linear. By repeated application of (1.3.2) it is easily shown that if $u_1, u_2, \ldots, u_n$ are n functions and $c_1, c_2, \ldots, c_n$ are n constants, then

$$A(c_1u_1 + c_2u_2 + \cdots + c_nu_n) = c_1Au_1 + c_2Au_2 + \cdots + c_nAu_n .$$

The function $u = c_1u_1 + c_2u_2 + \cdots + c_nu_n$ is called a **linear combination** of $u_1, u_2, \ldots, u_n$.

We give three more examples of linear operators. The first transforms a function $u(x,y)$ of two variables into a function $v(x,y,z)$ of three variables, the second transforms $u(x,y)$ into a function $v(x)$ of one variable, and the third, called a linear integral operator, transforms $u(x,y)$ into a function $v(x,y)$ of two variables:

$$Ru = u(x,y) + u(y,z) + u(z,x),$$

$$Su = u(x,0) + u_y(x,0),$$

$$Tu = \int_0^1\int_0^1 u(\xi,\eta)\sqrt{(x-\xi)^2 + (y-\eta)^2}\,d\xi\,d\eta .$$

The **sum** $A + B$ of two operators A and B is the operator defined by

$$(1.3.3) \qquad (A + B)u = Au + Bu$$

for all functions u for which both Au and Bu are defined. For example, if A and B are defined by formulas (1.3.1), then

$$(A + B)u = \frac{\partial^2 u}{\partial x^2} + \frac{\partial^2 u}{\partial y^2} + 3x\frac{\partial u}{\partial x} - 4y\frac{\partial u}{\partial y} + (x^2 + y^2)u .$$

If A and B are linear operators, then

$$\begin{aligned}(A + B)(c_1u_1 + c_2u_2) &= A(c_1u_1 + c_2u_2) + B(c_1u_1 + c_2u_2)\\ &= (c_1Au_1 + c_2Au_2) + (c_1Bu_1 + c_2Bu_2)\\ &= c_1(Au_1 + Bu_1) + c_2(Au_2 + Bu_2)\\ &= c_1(A + B)u_1 + c_2(A + B)u_2\,,\end{aligned}$$

which establishes that *the sum of linear operators is a linear operator.* This statement is useful in verifying that an operator is linear.

An equation of the form

$$Au = f\,, \tag{1.3.4}$$

where A is a linear operator and f a given function, is called a **linear equation.** If f is identically zero, the equation is called **homogeneous,** otherwise it is **inhomogeneous.** If f in (1.3.4) is not identically zero, so that the equation is inhomogeneous, then the equation

$$Au = 0 \tag{1.3.5}$$

is called the **related homogeneous equation.**

A simple but useful statement is: an equation that is already known to be linear is homogeneous if and only if the function $u = 0$ is a solution. This follows immediately from the fact that if A is linear, then $A(0) = A(0\cdot u) = 0\cdot A(u) = 0$.

Let us denote the operators of partial differentiation $\partial/\partial x$, $\partial/\partial y$, . . . , by D_1, D_2, . . . and correspondingly write, if a_{11}, a_{12}, . . . are any given functions of x, y, . . . ,

$$a_{11}D_1^2u = a_{11}\frac{\partial^2 u}{\partial x^2}, \quad a_{12}D_1D_2u = a_{12}\frac{\partial^2 u}{\partial x\,\partial y}, \ldots. \tag{1.3.6}$$

Each operator in (1.3.6) is linear, as is easily verified. Hence an operator such as

$$A = a_{11}D_1^2 + a_{12}D_1D_2 + a_{22}D_2^2 + a_1D_1 + a_2D_2 + a_0\,, \tag{1.3.7}$$

where a_{11}, a_{12}, . . . , a_0 are any given functions of x and y, is linear, being a sum of linear operators. When A is an operator such as (1.3.7), the equation (1.3.4) becomes

$$a_{11}u_{xx} + a_{12}u_{xy} + a_{22}u_{yy} + a_1u_x + a_2u_y + a_0u = f \tag{1.3.8}$$

which is a linear partial differential equation. Equation (1.3.8) is the most general **linear partial differential equation of the second order in two independent variables.** In the special case in which a_{11}, a_{12}, . . . , a_0 are constant the equation is said to have **constant coefficients.**

The partial differential equations we will meet will all be linear equations. The typical problem is to find a solution of the differential equation which satisfies certain auxiliary conditions; these auxiliary conditions will be linear conditions, that is, expressible in the form

$$Bu = g \tag{1.3.9}$$

where B is a linear operator and g is a given function. The **linear auxiliary condition** (1.3.9) is called **homogeneous** or **inhomogeneous** according to whether g is or is not identically zero. As an example, consider the problem

$$\begin{cases} u_t = u_{xx}\,, & 0 < x < 1\,, \quad 0 < t \\ u_x(0,t) = 0\,, & 0 < t \\ u_x(1,t) = 0\,, & 0 < t \\ u(x,0) = x^2(1-x)^2\,, & 0 < x < 1\,, \end{cases}$$

which we will discuss in Chapter 4. The first equation is the differential equation; the remaining equations are linear auxiliary conditions.

The important property of linear equations is stated in the **principle of superposition:** *Let $f_1, f_2, \ldots, f_n$ be any functions and $c_1, c_2, \ldots, c_n$ any constants. If A is a linear operator, and if $u_1, u_2, \ldots, u_n$ are, respectively, solutions of the equations $Au_1 = f_1$, $Au_2 = f_2$, $\ldots$, $Au_n = f_n$, then $u = c_1u_1 + c_2u_2 + \cdots + c_nu_n$ is a solution of the equation $Au = c_1f_1 + c_2f_2 + \cdots + c_nf_n$.* For, by (1.3.2),

$$\begin{aligned} Au &= A(c_1u_1 + \cdots + c_nu_n) = c_1Au_1 + \cdots + c_nAu_n \\ &= c_1f_1 + \cdots + c_nf_n\,. \end{aligned}$$

Two special cases of the principle of superposition are of particular importance:

If $u_1, u_2, \ldots, u_n$ are solutions of the homogeneous linear equation $Au = 0$, and if $c_1, c_2, \ldots, c_n$ are any constants, then $c_1u_1 + c_2u_2 + \cdots + c_nu_n$ is also a solution of the equation.

If u is a solution of $Au = f$ and v is a solution of $Av = 0$, then $w = u + v$ is a solution of $Aw = f$.

The first of these is the special case in which $f_1 = f_2 = \cdots = f_n = 0$, the second that in which $n = 2$, $u_1 = u$, $f_1 = f$, $u_2 = v$, $f_2 = 0$, $c_1 = c_2 = 1$.

Sometimes we are able to find infinitely many solutions of a homogeneous linear equation. For example, the equation

$$u_x + u_y = 0$$

has the solutions

$$u_n(x,y) = (x - y)^n, \quad n = 0, 1, 2, \ldots .$$

This suggests that "infinite linear combinations"

$$u(x,y) = \sum_{n=0}^{\infty} c_n u_n(x,y)$$

will also be solutions of the homogeneous linear equation. In particular, if we take $c_n = 1/n!$, then

$$u(x,y) = \sum_{n=0}^{\infty} \frac{(x-y)^n}{n!} = e^{x-y},$$

and it is easily verified that this function is indeed a solution of $u_x + u_y = 0$.

Suppose that we have infinitely many solutions $u_1, u_2, \ldots, u_n, \ldots$ of a homogeneous linear equation $Au = 0$. Can we say that every "infinite linear combination" $c_1u_1 + c_2u_2 + \cdots + c_nu_n + \cdots$ of these solutions is again a solution of the equation? Of course, by an "infinite linear combination" we mean an infinite series

$$c_1u_1 + c_2u_2 + \cdots + c_nu_n + \cdots = \sum_{k=0}^{\infty} c_ku_k = \lim_{n\to\infty} \sum_{k=0}^{n} c_ku_k,$$

and we must require that this series be convergent. Granted that, however, it would seem we can make the statement. If we try to prove it, we write

$$A \sum_{k=0}^{\infty} c_ku_k = A \lim_{n\to\infty} \sum_{k=0}^{n} c_ku_k$$

$$A \lim_{n\to\infty} \sum_{k=0}^{n} c_ku_k = \lim_{n\to\infty} A \sum_{k=0}^{n} c_ku_k$$

$$\lim_{n\to\infty} A \sum_{k=0}^{n} c_ku_k = \lim_{n\to\infty} \sum_{k=0}^{n} c_kA(u_k) = 0 .$$

The first and last of these equations are certainly correct. In the second we have changed the order in which the operator A is applied and the limit is taken. It is by no means evident that this interchange is correct and, in fact, it is not correct in general. Nevertheless, for the operators with which we will be concerned we can state: *If $u_1, u_2, \ldots, u_n, \ldots$ is an infinite sequence of solutions of a linear homogeneous equation, and if the infinite series $\sum_{k=0}^{\infty} c_ku_k$ converges suitably, then $\sum_{k=0}^{\infty} c_ku_k$ is also a solution of the homogeneous equation.* For the present we shall assume that all infinite series "converge suitably" and put off until later the investigation of "suitable convergence."

There is another kind of "infinite linear combination" with which we will be concerned. Suppose that we have a family $u(x,y;\lambda)$ of solutions of a linear homogeneous equation, where λ may take on any real value, not just the values 1, 2, 3, ... Let $c_\lambda = c(\lambda)$ be any function of the index λ, and let a and b be any numbers, $a < b$. We subdivide the interval $a \leq \lambda \leq b$ into n equal parts of length $\Delta\lambda$. Let $a = \lambda_0 < \lambda_1 < \cdots < \lambda_n = b$ be the points of subdivision. Then

$$\sum_{k=1}^{n} c(\lambda_k)u(x,y;\lambda_k)\Delta\lambda$$

is again a solution of the homogeneous equation. If, as above, we assume that we can interchange the order in which the operator A is applied and the limit as $n \to \infty$ is taken, then we may say that

$$\lim_{n\to\infty} \sum_{k=1}^{n} c(\lambda_k)u(x,y;\lambda_k)\Delta\lambda = \int_a^b c(\lambda)u(x,y;\lambda)\, d\lambda$$

is again a solution of the homogeneous equation. Under a similar assumption, we may let $a \to -\infty$ and $b \to \infty$, so that

$$\lim_{\substack{a\to-\infty \\ b\to+\infty}} \int_a^b c(\lambda)u(x,y;\lambda)\, d\lambda = \int_{-\infty}^{\infty} c(\lambda)u(x,y;\lambda)\, d\lambda$$

is again a solution of the homogeneous equation. Under suitable conditions, which we will for the present assume satisfied and the statement of which we put off until later, the necessary interchanges can be justified. We shall state: *If* u_λ, $-\infty < \lambda < \infty$, *is a family of solutions of a homogeneous linear equation, and if* $c(\lambda)$ *is any function such that* $\int_{-\infty}^{\infty} c(\lambda)u_\lambda\, d\lambda$ *is convergent, then under suitable conditions this integral is again a solution of the homogeneous equation.*

To illustrate these ideas, consider the equation

$$Au = 3u_x + u_y = 0\ .$$

By the methods of Section 5 we find that for every fixed real λ the function

$$u(x,y;\lambda) = e^{\lambda(x-3y)}$$

is a solution (as one can easily verify). If we multiply by $c(\lambda) = e^{-\lambda}$ and integrate with respect to λ over $-1 \leq \lambda \leq 1$, we obtain

$$\begin{aligned} u(x,y) &= \int_{-1}^{1} e^{-\lambda}e^{\lambda(x-3y)}\, d\lambda \\ &= 2\,\frac{\sinh\,(x-3y-1)}{(x-3y-1)}\,, \end{aligned}$$

and it can be verified that this latter function is also a solution of $Au = 0$.

The statements we have made about "infinite linear combinations" hold for ordinary differential equations as well as partial differential equations. For ordinary differential equations they are unimportant, whereas, as we shall see in Section 5, they are essential for the study of partial differential equations.

EXERCISES 1b

1. For each of the following equations, state whether the equation is linear or nonlinear, and if it is linear, whether it is homogeneous or inhomogeneous:

(a) $u_x u_y - 3u = x$

(b) $y u_{xxy} - e^x u_x + 3 = 0$

(c) $y^3 u(0,y) + 5u_x(0,y) = 0$

(d) $u_x(x,y) - u_y(x+1,y) = xy$

(e) $u u_{xy} - u_x u_y = 0$

(f) $u_{xx} + u_{yy} + \log u = \log x$

(g) $u(x,y) + \int_0^1 u(x,t) \sin^2 t\, y\, dt = 0$

(h) $u(x,y) + \int_0^1 u^2(x,t) \sin ty\, dt = 0.$

2. Suppose that u_1 and u_2 are different solutions of the same inhomogeneous linear equation $Au = f$. Find, in terms of u_1 and u_2, a family of solutions of this equation depending on an arbitrary constant.

1.4. *COMPLEX-VALUED FUNCTIONS*

We have spoken so far of real-valued functions of real variables, but, as in the case of ordinary differential equations, it is convenient and not difficult to extend our consideration to complex-valued functions of real variables. A complex-valued function of one real variable is a function of the form

$$f(x) = f_1(x) + i f_2(x)$$

where $f_1(x)$ and $f_2(x)$ are real-valued functions. The functions $f_1(x)$ and $f_2(x)$ are called, respectively, the *real* and *imaginary* parts of $f(x)$, and we write

$$f_1(x) = \mathrm{Re}\{f(x)\}\,, \quad f_2(x) = \mathrm{Im}\{f(x)\}\,.$$

Two complex-valued functions are equal if and only if their respective real and imaginary parts are equal. Every real-valued function may be considered as a complex-valued function whose imaginary part is identically zero.

The familiar operations of algebra and the calculus extend to complex-valued functions in the natural way. For example,

$$\frac{d}{dx}\,[f_1(x) + if_2(x)] = \frac{d}{dx}\,f_1(x) + i\,\frac{d}{dx}\,f_2(x)$$

$$\int_a^b [f_1(x) + if_2(x)]\,dx = \int_a^b f_1(x)\,dx + i\int_a^b f_2(x)\,dx\,.$$

A complex-valued function will be called *continuous*, *differentiable*, etc., if both its real and imaginary parts have the corresponding property. Similar definitions and statements hold for complex-valued functions of several real variables.

Of particular importance is the complex exponential function defined by

$$e^{(\alpha+i\beta)x} = e^{\alpha x}(\cos\beta x + i\sin\beta x) \tag{1.4.1}$$

where α and β are real constants. All the familiar rules of algebra and the calculus for operating with exponentials continue to hold for this exponential with complex exponent. From (1.4.1) follow immediately

$$\cos\beta x = \frac{e^{i\beta x} + e^{-i\beta x}}{2}\,, \tag{1.4.2}$$

$$\sin\beta x = \frac{e^{i\beta x} - e^{-i\beta x}}{2i} = -\frac{i(e^{i\beta x} - e^{-i\beta x})}{2}\,, \tag{1.4.3}$$

since

$$e^{i\beta x} = \cos\beta x + i\sin\beta x\,,\quad e^{-i\beta x} = \cos\beta x - i\sin\beta x\,. \tag{1.4.4}$$

The reader should recall the use of these functions and formulas in the solution of ordinary differential equations with constant coefficients.

We also note that to solve an equation of the Euler homogeneous type

$$ax^2\varphi'' + bx\varphi' + c\varphi = 0\,,$$

where a, b, c are constants, we set $\varphi(x) = x^r$, and find that r must be a root of the equation

$$ar(r-1) + br + c = 0\,.$$

If the roots are complex, we get solutions of the form $\varphi = x^{\alpha+i\beta}$. Since $x = e^{\log x}$, this solution can be written more conveniently as

$$\begin{aligned} x^{\alpha+i\beta} &= e^{(\alpha+i\beta)\log x} \\ &= e^{\alpha \log x} e^{i\beta \log x} \\ &= e^{\alpha \log x}[\cos(\beta \log x) + i \sin(\beta \log x)] . \end{aligned}$$

In the next section we will see how one is led naturally to consider complex-valued solutions of certain partial differential equations.

1.5. *PARTICULAR SOLUTIONS*

When a linear *homogeneous* partial differential equation has *constant coefficients*, the problem of determining particular solutions of the equation can be reduced to an algebraic problem. Let us first recall how this is achieved for ordinary differential equations.

Every homogeneous linear ordinary differential equation with constant coefficients can be written in the form

$$P(D)u = 0 , \tag{1.5.1}$$

where

$$P(r) = a_n r^n + a_{n-1} r^{n-1} + \cdots + a_1 r + a_0$$

is a polynomial in r and D is the differentiation operator, $D = d/dx$. The exponential function e^{rx} has the property $De^{rx} = re^{rx}$ from which follows

$$D^n e^{rx} = r^n e^{rx} .$$

Hence if we substitute $u = e^{rx}$ in (1.5.1), we obtain

$$P(r)e^{rx} = 0$$

and therefore if r is a solution of the algebraic equation

$$P(r) = 0 ,$$

then $u = e^{rx}$ is a solution of the differential equation.

An analogous statement holds for linear homogeneous partial differential equations with constant coefficients. In the case of two independent variables, every such equation may be written in the form

$$P(D_1,D_2)u = 0 , \tag{1.5.2}$$

where

$$P(r,s) = a_{n,m} r^n s^m + a_{n,m-1} r^n s^{m-1} + \cdots + a_{1,1} rs + a_{1,0} r + a_{0,1} s + a_{0,0}$$

is a polynomial in r and s, and $D_1 = \partial/\partial x$, $D_2 = \partial/\partial y$. The exponential function e^{rx+sy} has the properties

$$D_1 e^{rx+sy} = re^{rx+sy}, \quad D_2 e^{rx+sy} = se^{rx+sy}$$

so that if we substitute $u = e^{rx+sy}$ in (1.5.2) we obtain

$$P(r,s)e^{rx+sy} = 0 .$$

Thus, if r, s is a solution of the algebraic equation

$$P(r,s) = 0 ,$$

then e^{rx+sy} is a solution of the differential equation (1.5.2).

As an example, consider

$$3u_x - 2u_y = 0 . \tag{1.5.3}$$

Substituting $u = e^{rx+sy}$ we obtain

$$(3r - 2s)e^{rx+sy} = 0 ,$$

so that e^{rx+sy} is a solution if

$$3r - 2s = 0 .$$

In this equation the value s may be chosen arbitrarily, and r is then determined. Thus we obtain a family of solutions of (1.5.3):

$$u(x,y;s) = \exp\left(\tfrac{2}{3}sx + sy\right) .$$

Here s may be *any* complex number, so that the equation (1.5.3) has infinitely many solutions of exponential form. By superposing these solutions in various ways, we can construct additional solutions of the equation.

As another example, we take

$$u_{xx} + u_{yy} = 0 . \tag{1.5.4}$$

Substitution of e^{rx+sy} leads to the algebraic equation

$$r^2 + s^2 = 0 .$$

Again we choose s arbitrarily, but for each value of s we obtain two values of r, $r = \pm is$, and correspondingly two families of solutions

$$u = e^{isx+sy}, \quad v = e^{-isx+sy} .$$

These are complex-valued solutions, but we may combine them to obtain real-valued solutions. By the principle of superposition

$$\frac{e^{isx+sy} + e^{-isx+sy}}{2} = e^{sy}\,\frac{e^{isx} + e^{-isx}}{2} = e^{sy}\cos sx$$

$$\frac{e^{isx+sy} - e^{-isx+sy}}{2i} = e^{sy}\,\frac{e^{isx} - e^{-isx}}{2i} = e^{sy}\sin sx$$

are solutions of the differential equation, and are real when s is real. Hence

$$(1.5.5) \qquad U = e^{sy} \cos sx \,, \quad V = e^{sy} \sin sx \,, \quad -\infty < s < \infty \,,$$

are two families of real-valued solutions.

As in the previous example, we can construct other solutions by using the principle of superposition.

A closely related but different technique for finding particular solutions of *homogeneous* linear partial differential equations is called the **method of separation of variables.** The exponential solutions,

$$e^{rx+sy} = e^{rx}e^{sy} \,,$$

are products of functions of the separate variables. In the method of separation of variables, we begin by seeking solutions which are products of functions of the separate variables.

For example, consider the equation (1.5.4). We seek a solution of the form

$$(1.5.6) \qquad u(x,y) = \varphi(x)\psi(y) \,.$$

Substitution of (1.5.6) into (1.5.4) gives

$$\varphi''(x)\psi(y) + \varphi(x)\psi''(y) = 0 \,.$$

Dividing this equation by $\varphi(x)\psi(y)$ and transposing, we have

$$(1.5.7) \qquad -\frac{\varphi''(x)}{\varphi(x)} = \frac{\psi''(y)}{\psi(y)} \,.$$

Now the left member of (1.5.7) is independent of y so that the right member must also be independent of y. Similarly, the right member is independent of x so that the left member is also independent of x. The two members of Equation (1.5.7) thus depend neither on x nor on y. Their common value is therefore a constant λ, and we have $\varphi''/\varphi = -\lambda$, $\psi''/\psi = \lambda$ or

$$(1.5.8) \qquad \varphi'' + \lambda\varphi = 0 \,,$$

$$(1.5.9) \qquad \psi'' - \lambda\psi = 0 \,.$$

Solutions of (1.5.8) are $\cos x\sqrt{\lambda}$ and $\sin x\sqrt{\lambda}$. Combining these with the solutions $e^{\pm y\sqrt{\lambda}}$ of (1.5.9) we obtain

$$(1.5.10) \qquad e^{\pm y\sqrt{\lambda}} \cos x\sqrt{\lambda} \,, \quad e^{\pm y\sqrt{\lambda}} \sin x\sqrt{\lambda}$$

as families of solutions of (1.5.4). If we set $\pm\sqrt{\lambda} = s$, then the solutions (1.5.10) coincide with the previously obtained solutions (1.5.5).

The method of separating the variables can be used to find solutions of some linear homogeneous equations with nonconstant coefficients. For an equation of the form

$$(1.5.11) \qquad Au_{xx} + Bu_{xy} + Cu_{yy} + Du_x + Eu_y + Fu = 0\,,$$

where A, B, C, D, E, F are functions of x and y, we substitute $u(x,y) = \varphi(x)\psi(y)$ and then try to write the equation in a form where y and ψ do not appear in the left member and x and φ do not appear in the right member; for example, in the form

$$(1.5.12) \qquad \frac{a(x)\varphi'' + b(x)\varphi' + c(x)\varphi}{\varphi} = \frac{\alpha(y)\psi'' + \beta(y)\psi' + \gamma(y)\psi}{\psi}\,.$$

(Other forms will be seen in Exercises 1c.) If this procedure is possible, we say that the equation (1.5.11) is **separable** with respect to the variables x, y. Using the same argument as in the above example, each member of (1.5.12) must equal a constant λ and to find the factors $\varphi(x)$, $\psi(y)$ we have only to solve the *ordinary* linear differential equations

$$a\varphi'' + b\varphi' + c\varphi - \lambda\varphi = 0\,,$$
$$\alpha\psi'' + \beta\psi' + \gamma\psi - \lambda\psi = 0\,.$$

As an example, consider the equation

$$(1.5.13) \qquad x^2u_{xx} + xu_x - u_y = 0\,.$$

Substituting $u = \varphi(x)\psi(y)$ and rearranging gives

$$\frac{x^2\varphi'' + x\varphi'}{\varphi} = \frac{\psi'}{\psi}\,.$$

Thus (1.5.13) is separable and the ordinary differential equations for this case are

$$x^2\varphi'' + x\varphi' - \lambda\varphi = 0\,,$$
$$\psi' - \lambda\psi = 0\,.$$

The first equation is of the Euler type and has solutions $\varphi_1(x) = x^{\sqrt{\lambda}}$, $\varphi_2(x) = x^{-\sqrt{\lambda}}$. The second equation has the solution $\psi = e^{\lambda y}$ and so we obtain two families of solutions,

$$u_\lambda(x,y) = e^{\lambda y}x^{\sqrt{\lambda}}\,, \quad v_\lambda(x,y) = e^{\lambda y}x^{-\sqrt{\lambda}}$$

of (1.5.13). Naturally, the method of seeking exponential solutions does not work in this example. On the other hand, there are equations with constant coefficients that are not separable, in spite of the fact that such an equation always has solutions of the form $u(x,y) = e^{rx}e^{sy} = \varphi(x)\psi(y)$. The reader should try separating the variables in the equation $u_{xx} + u_{xy} + u_{yy} = 0$.

The methods of exponential solution and of separation of variables are applicable to higher-order equations and to equations in any number of independent variables. As an example of the method of separation of variables in the case of three independent variables, we will find solutions of the equation

$$u_x - u_y + 2u_z = 0 . \tag{1.5.14}$$

We substitute $u(x,y,z) = X(x)Y(y)Z(z)$ in (1.5.14) and then divide by XYZ to obtain

$$\frac{X'(x)}{X(x)} - \frac{Y'(y)}{Y(y)} + \frac{2Z'(z)}{Z(z)} = 0 .$$

Transposing the first term, we have

$$\frac{X'(x)}{X(x)} = \frac{Y'(y)}{Y(y)} - \frac{2Z'(z)}{Z(z)} .$$

Since the left member is a function of x alone and the right member is a function of y and z, each must be constant, so that

$$\frac{X'(x)}{X(x)} = \lambda \tag{1.5.15}$$

$$\frac{Y'(y)}{Y(y)} - \frac{2Z'(z)}{Z(z)} = \lambda . \tag{1.5.16}$$

Equation (1.5.15) is an ordinary differential equation for X, whose solution is $X(x) = e^{\lambda x}$. We rewrite Equation (1.5.16) as

$$\frac{Y'(y)}{Y(y)} = \lambda + \frac{2Z'(z)}{Z(z)} .$$

Again, each side of this equation must be constant, and we obtain for Y and Z the ordinary differential equations

$$\frac{Y'(y)}{Y(y)} = \mu \tag{1.5.17}$$

$$\lambda + \frac{2Z'(z)}{Z(z)} = \mu . \tag{1.5.18}$$

Solving these equations, we obtain a family of solutions of (1.5.14) depending on two parameters:

$$u(x,y,z;\lambda,\mu) = e^{\lambda x}e^{\mu y}e^{(\mu-\lambda)z/2} . \tag{1.5.19}$$

EXERCISES 1c

1. Find particular solutions of the following equations by (i) exponential substitution, (ii) separation of variables:

(a) $u_x + 2u_y - u = 0$
(b) $u_{xx} - u_{yy} + u_y = 0$
(c) $u_{xxyy} + u_{xx} + u_{yy} = 0$
(d) $u_{xx} + u_{yy} + u_{zz} - u = 0$
(e) $u_{xy} + u_{yz} + u_{zx} - u = 0.$

2. Find particular solutions of the equation

$$x^2u_x - u_y + u = 0\,.$$

3. Find particular solutions of the equation

$$u_x - u_y + u = 1\,.$$

4. Find solutions by separating the variables:

(a) $u_x + u_y + u_z + u = 0$
(b) $u_{xxxx} + yu_{xxy} = 0$
(c) $u_{xy} + u_x + u_{yy} = 0.$

5. Find solutions by separating the variables:

(a) $x^2u_{xx} + y^2u_{yy} + 5xu_x - 5yu_y + 4u = 0$
(b) $u_{xx} - (1 + y^2)u_{xy} = 0$
(c) $u_{xyz} - xyzu = 0.$

6. Let A be a linear operator such that if u is any real-valued function then Au is real-valued. Show that if the equation $Au = 1$ has a complex-valued solution, then it also has a real-valued solution.

2

Linear equations with constant coefficients in two independent variables

2.1. *AUXILIARY CONDITIONS*

Any equation, other than the differential equation itself, which a solution of a differential equation is required to satisfy is called an **auxiliary condition** for the differential equation. Auxiliary conditions may involve undetermined constants or functions. We shall say that a set of auxiliary conditions involving undetermined elements is **appropriate** for a differential equation if for each specification of these elements there is *one and only one* function which satisfies both the differential equation and the auxiliary conditions.

A fundamental theorem in the theory of ordinary differential equations asserts: *If* $u_0, u_1, \ldots, u_{n-1}$ *are any constants, then the differential equation*

$$(2.1.1) \qquad a_0 u^{(n)} + a_1 u^{(n-1)} + \cdots + a_{n-1} u' + a_n u = f(x) ,$$

where $a_0, \ldots, a_n$ *are given constants,* $a_0 \neq 0$, *and* $f(x)$ *is a given continuous function, has one and only one solution* $u(x)$ *such that*

$$(2.1.2) \qquad u(0) = u_0,\ u'(0) = u_1, \ldots, u^{(n-1)}(0) = u_{n-1} .$$

Each of the equations (2.1.2) is an auxiliary condition for the differential equation (2.1.1), and the content of the stated theorem is that the set of auxiliary conditions (2.1.2) is appropriate for (2.1.1). Thus in the case of ordinary differential equations with constant coefficients there is one set of auxiliary conditions appropriate for all differential equations of the same order.

One might guess that analogous statements hold for partial

differential equations. For first-order equations this is essentially so. The general linear first-order equation with constant coefficients is

$$au_x + bu_y + cu = f(x,y)\,, \tag{2.1.3}$$

where a, b, c are constants, and we assume $f(x,y)$ to be continuous everywhere. In the next section we will see that: *If* $b \neq 0$ *and* $u_0(x)$ *is any everywhere differentiable function, then there is one and only one solution* $u(x,y)$ *of* (2.1.3) *such that, for all* x,

$$u(x,0) = u_0(x)\,. \tag{2.1.4}$$

We will, indeed, be able to find the general solution of (2.1.3), and then the particular solution which satisfies (2.1.4).

For differential equations of second order, no corresponding statement is possible. The general linear second-order equation with constant coefficients is

$$au_{xx} + 2bu_{xy} + cu_{yy} + hu_x + ku_y + eu = f(x,y) \tag{2.1.5}$$

where a, b, c, h, k, e are constants and $f(x,y)$ is assumed continuous. *There is no one set of auxiliary conditions appropriate for all equations of the form* (2.1.5). This necessitates more or less special study of individual second-order equations. In the last section of this chapter, however, we will see that there are essentially three classes of closely related second-order equations, and that every equation of the form (2.1.5) can be transformed into an equation in one of these classes. In succeeding chapters we will study the special classes separately.

2.2. *FIRST-ORDER EQUATIONS*

The possibility of finding the general solution of the first-order differential equation (2.1.3) is based on the fact that *by introducing new coordinates* (ξ,η) *corresponding to a suitable change of axes, every such equation can be transformed into an equation of the form*

$$\omega_\xi + k\omega = \varphi(\xi,\eta)\,, \tag{2.2.1}$$

k constant. The general solution of this last equation (2.2.1) can be obtained by solving it as an ordinary differential equation, treating η as a parameter, so that the arbitrary constants in the solution are replaced by arbitrary functions of η. Thus, returning to the original coordinate system, we obtain the general solution of (2.1.3).

As an illustration, we consider the differential equation

$$3u_x + 4u_y - 2u = 1\,. \tag{2.2.2}$$

If the (ξ,η)-axes are obtained from the (x,y)-axes by rotation through an angle α, then (ξ,η) and (x,y) are related by either of the pair of equations

$$
\xi = x\cos\alpha + y\sin\alpha\,, \qquad x = \xi\cos\alpha - \eta\sin\alpha\,,
$$
$$
\eta = -x\sin\alpha + y\cos\alpha\,, \quad y = \xi\sin\alpha + \eta\cos\alpha\,. \tag{2.2.3}
$$

If $u(x,y)$ is a solution of (2.2.2) and we substitute the second pair of Equations (2.2.3), then u becomes a function $\omega(\xi,\eta)$ of ξ and η. We have

$$
u(x,y) = u(\xi\cos\alpha - \eta\sin\alpha\,,\ \xi\sin\alpha + \eta\cos\alpha) = \omega(\xi,\eta) \tag{2.2.4}
$$

$$
\frac{\partial}{\partial x} = \frac{\partial\xi}{\partial x}\frac{\partial}{\partial\xi} + \frac{\partial\eta}{\partial x}\frac{\partial}{\partial\eta} = \cos\alpha\frac{\partial}{\partial\xi} - \sin\alpha\frac{\partial}{\partial\eta}
$$
$$
\frac{\partial}{\partial y} = \frac{\partial\xi}{\partial y}\frac{\partial}{\partial\xi} + \frac{\partial\eta}{\partial y}\frac{\partial}{\partial\eta} = \sin\alpha\frac{\partial}{\partial\xi} + \cos\alpha\frac{\partial}{\partial\eta}\,. \tag{2.2.5}
$$

Substituting in Equation (2.2.2) we obtain

$$
3(\omega_\xi\cos\alpha - \omega_\eta\sin\alpha) + 4(\omega_\xi\sin\alpha + \omega_\eta\cos\alpha) - 2\omega = 1
$$

or

$$
(3\cos\alpha + 4\sin\alpha)\omega_\xi + (4\cos\alpha - 3\sin\alpha)\omega_\eta - 2\omega = 1\,. \tag{2.2.6}
$$

The coefficient of ω_η in (2.2.6) will vanish if

$$
\tan\alpha = 4/3\,.
$$

To satisfy this we choose α so that

$$
\cos\alpha = 3/5\,, \quad \sin\alpha = 4/5\,,
$$

and Equation (2.2.6) becomes

$$
\omega_\xi - \tfrac{2}{5}\omega = \tfrac{1}{5}\,, \tag{2.2.7}
$$

which is of the form (2.2.1).

We can accomplish a transformation to the same form, in a way which is both more convenient and also extendable to equations in more than two independent variables, by using a general linear change of independent variables (that is, one not necessarily representing a rotation of axes). Let

$$
x = A\xi + B\eta\,,
$$
$$
y = C\xi + D\eta\,, \tag{2.2.8}
$$

where A, B, C, D are constants to be determined, and set

$$
u(x,y) = \omega(\xi,\eta)\,. \tag{2.2.9}
$$

Then, from

$$\frac{\partial \omega}{\partial \xi} = u_x \frac{\partial x}{\partial \xi} + u_y \frac{\partial y}{\partial \xi} \tag{2.2.10}$$

and (2.2.8), (2.2.9), we see that with the choice $A = 3$, $C = 4$, Equation (2.2.2) becomes

$$\omega_\xi - 2\omega = 1\,, \tag{2.2.11}$$

which is again of the form (2.2.1).

The choice of B and D is arbitrary, except for the obvious restriction

$$\begin{vmatrix} A & B \\ C & D \end{vmatrix} \neq 0\,. \tag{2.2.12}$$

Looking ahead to the effort to satisfy the auxiliary condition (2.1.4), we shall choose B and D so that the line $\xi = 0$ is the line on which the auxiliary data is prescribed, namely $y = 0$. This requires $D = 0$, and, since B is arbitrary we make the convenient choice $B = 1$. Then Equations (2.2.8) become explicitly

$$\begin{aligned} x &= 3\xi + \eta\,, & \xi &= \tfrac{1}{4}y \\ y &= 4\xi\,, & \eta &= x - \tfrac{3}{4}y\,. \end{aligned} \tag{2.2.13}$$

Now, to obtain the general solution of Equation (2.2.11) we multiply the equation by the integrating factor

$$\exp\left(\int -2d\xi\right) = e^{-2\xi}\,.$$

The equation may then be written

$$\frac{\partial}{\partial \xi}(e^{-2\xi}\omega) = e^{-2\xi}\,,$$

from which follows

$$e^{-2\xi}\omega = -\tfrac{1}{2}e^{-2\xi} + g(\eta)$$

or

$$\omega(\xi,\eta) = -\tfrac{1}{2} + g(\eta)e^{2\xi}\,, \tag{2.2.14}$$

where g is arbitrary. If we substitute the second pair of Equations (2.2.13) in (2.2.14), we get from (2.2.9)

$$u(x,y) = -\tfrac{1}{2} + g(x - \tfrac{3}{4}y)e^{y/2}\,. \tag{2.2.15}$$

Conversely, it can be verified by differentiation and substitution (see Exercises 1a, Problem 8) that if g is an arbitrary differentiable function of one variable, then (2.2.15) is a solution of (2.2.2). Hence (2.2.15) furnishes the general solution of (2.2.2).

To find the particular solution of (2.2.2) which satisfies the auxiliary condition (2.1.4) we simply substitute (2.2.15) in (2.1.4). We get

$$-\tfrac{1}{2} + g(x) = u_0(x) .$$

Thus we have

$$u(x,y) = -\tfrac{1}{2} + e^{y/2}[u_0(x - \tfrac{3}{4}y) + \tfrac{1}{2}] \tag{2.2.16}$$

as the one and only solution of (2.2.2) which satisfies (2.1.4), so that the auxiliary condition (2.1.4) is appropriate for the differential equation (2.2.2).

The procedure we have followed will furnish the general solution of every equation of the form (2.1.3). The condition that the coefficient of u_y in (2.1.3) be different from zero is sufficient and also necessary for the auxiliary condition (2.1.4) to be appropriate for (2.1.3). If this condition is not satisfied, then the differential equation will have solutions not for arbitrary $u_0(x)$, but only for functions having a special form, and for such functions there will be not one solution of the differential equation satisfying the auxiliary condition, but infinitely many (see Exercises 2a, Problems 4 and 5).

EXERCISES 2a

1. Find the solution $u(x,y)$ of

$$u_x - u_y + u = 1$$

which satisfies the auxiliary condition

$$u(x,0) = \sin x .$$

2. Find the solution of

$$u_x - u_y + u = e^{x+2y}$$
$$u(x,0) = 0 .$$

3. Find the solution of

$$u_x + u_y - u = 0$$
$$u(x,0) = f(x) .$$

4. (i) Find the general solution of

$$au_x + bu_y + cu = 0$$

where a, b, c are constants and b is not zero.

(ii) Find the solution of the equation of (i) which satisfies the auxiliary condition

$$u(x,0) = u_0(x) ,$$

where $u_0(x)$ is differentiable.

5. Find all functions $u_0(x)$ for which the problem

$$u_x - 2u = 0$$
$$u(x,0) = u_0(x)$$

has a solution, and for such functions find all solutions of the problem.

6. Find the solution of $u(x,y)$ of

$$u_x - 3u_y = 0$$

which satisfies the auxiliary condition

$$u(x,x) = x^2 .$$

7. If $P(s,t) = \sum_0^\infty \sum_0^\infty a_{m,n} s^m t^n$ where the coefficients $a_{m,n}$ satisfy the recurrence relation

$$(m + 1)a_{m+1,n} + 5(n + 1)a_{m,n+1} = 0 \qquad m,n \geq 0 ,$$

show that P satisfies the equation

$$P_s + 5P_t = 0 .$$

8. The coefficients of a certain power series

$$P(s,t) = \sum_{m=0}^\infty \sum_{n=0}^\infty a_{m,n} s^m t^n$$

satisfy

$$3(m + 1)a_{m+1,n} - (n + 1)a_{m,n+1} + a_{m,n} = 0$$

and it is known that $P(t,t) = e^{2t}$. Find $P(s,t)$.

2.3. *SECOND-ORDER EQUATIONS*

The situation for second-order equations is far more complicated than that for first-order equations. *By the introduction of new variables ξ, η and ω, every second-order equation*

$$au_{xx} + 2bu_{xy} + cu_{yy} + hu_x + ku_y + eu = f(x,y) , \tag{2.3.1}$$

where a, b, c, h, k, and e are constants, can be transformed into one and only one of the following standard forms:

$$\omega_{\xi\xi} + \omega_{\eta\eta} + \gamma\omega = \varphi(\xi,\eta) , \tag{2.3.2}$$
$$\omega_{\xi\xi} - \omega_{\eta\eta} + \gamma\omega = \varphi(\xi,\eta) , \tag{2.3.3}$$
$$\omega_{\xi\xi} - \omega_{\eta} = \varphi(\xi,\eta) , \tag{2.3.4}$$
$$\omega_{\xi\xi} + \gamma\omega = \varphi(\xi,\eta) , \tag{2.3.5}$$

where γ is a constant with one of the values -1, 0, *or* 1.

Equation (2.3.1) is called **elliptic** if it can be reduced to (2.3.2), and this case occurs if $ac - b^2 > 0$. Equation (2.3.1) is called **hyper-**

bolic if it can be reduced to (2.3.3), and this case occurs if $ac - b^2 < 0$. If $ac - b^2 = 0$, then either (2.3.1) can be reduced to (2.3.4) and it is called **parabolic,** or else it can be reduced to (2.3.5) and is called **degenerate.** The degenerate case is of no further interest, since its solution reduces to that of an ordinary differential equation, with ξ as the independent variable and η a parameter.

In the above list of standard forms there are three elliptic equations, corresponding to the three choices $\gamma = -1, 0, 1$ in (2.3.2). In (2.3.3) the equation with $\gamma = -1$ can be changed into an equation with $\gamma = +1$ by the substitution $\xi' = \eta$, $\eta' = \xi$. Thus there are essentially only two hyperbolic equations, corresponding to $\gamma = 0, 1$. There is one parabolic equation, in standard form.

The determination of the general solutions of the equations (2.3.2), (2.3.3), and (2.3.4) is not possible except in special cases, and even when possible is not widely useful. The importance of the result we have stated on the transformation of second-order equations into standard form is that it indicates that, except for simple changes of variable, there are only six different nondegenerate second-order equations, collected into three classes, so that we may limit our study to these equations.

Further study of the equations reveals that there are essential similarities among equations of the same class, but profound differences between the equations of one class and those of another class. One of the aims of the later chapters in this book is to discover some of these similarities and differences.

To illustrate the procedure for transformation of a second-order equation into standard form, we will transform the equation

$$4u_{xx} - 24u_{xy} + 11u_{yy} - 12u_x - 9u_y - 5u = 0 \tag{2.3.6}$$

into its standard form. In this equation $a = 4$, $2b = -24$, $c = 11$, so that $ac - b^2 = -100$ and the equation is hyperbolic. The transformation is made in three steps.

The first step is the introduction of new coordinates (ξ,η) by rotation of axes so that in the transformed equation the mixed second partial derivative, $\omega_{\xi\eta}$, does not appear. We proceed as in the foregoing discussion of first-order equations. From Equation (2.2.5) we have

$$\frac{\partial^2}{\partial x^2} = \frac{\partial}{\partial x}\frac{\partial}{\partial x} = \left(\cos\alpha\frac{\partial}{\partial\xi} - \sin\alpha\frac{\partial}{\partial\eta}\right)\left(\cos\alpha\frac{\partial}{\partial\xi} - \sin\alpha\frac{\partial}{\partial\eta}\right)$$

so that

$$\frac{\partial^2}{\partial x^2} = \cos^2 \alpha \frac{\partial^2}{\partial \xi^2} - 2 \sin \alpha \cos \alpha \frac{\partial^2}{\partial \xi \, \partial \eta} + \sin^2 \alpha \frac{\partial^2}{\partial \eta^2} .$$

Similarly,

$$\frac{\partial^2}{\partial x \, \partial y} = \sin \alpha \cos \alpha \frac{\partial^2}{\partial \xi^2} + (\cos^2 \alpha - \sin^2 \alpha) \frac{\partial^2}{\partial \xi \, \partial \eta} - \sin \alpha \cos \alpha \frac{\partial^2}{\partial \eta^2},$$

$$\frac{\partial^2}{\partial y^2} = \sin^2 \alpha \frac{\partial^2}{\partial \xi^2} + 2 \sin \alpha \cos \alpha \frac{\partial^2}{\partial \xi \, \partial \eta} + \cos^2 \alpha \frac{\partial^2}{\partial \eta^2} .$$

Substituting these derivative formulas in (2.3.6) we find that the equation satisfied by ω is

(2.3.7) $$\begin{aligned} &[4c^2 - 24sc + 11s^2]\omega_{\xi\xi} + [14sc - 24(c^2 - s^2)]\omega_{\xi\eta} \\ &+ [4s^2 + 24sc + 11c^2]\omega_{\eta\eta} + [-12c - 9s]\omega_{\xi} \\ &+ [12s - 9c]\omega_{\eta} - 5\omega = 0 , \end{aligned}$$

where we have used the abbreviations $s = \sin \alpha$, $c = \cos \alpha$. The coefficient of $\omega_{\xi\eta}$ in (2.3.7) will vanish if α is chosen so that

$$14 \sin \alpha \cos \alpha - 24(\cos^2 \alpha - \sin^2 \alpha) = 0 ,$$

that is,

(2.3.8) $$\tan 2\alpha = 24/7 .$$

We find, using a reference triangle and the half-angle formulas, that (2.3.8) will be satisfied if

(2.3.9) $$\sin \alpha = 3/5 , \quad \cos \alpha = 4/5 .$$

After substitution of (2.3.9) in (2.3.7) the equation satisfied by ω becomes

(2.3.10) $$\omega_{\xi\xi} - 4\omega_{\eta\eta} + 3\omega_{\xi} + \omega = 0 .$$

In analytic geometry, rotation of axes is employed in the transformation of equations of conic sections. If the reader reviews the transformation to standard form of the equation of the conic section

$$ax^2 + 2bxy + cy^2 = 1 ,$$

he will observe the identity of the coefficients of the transformed equation of the conic section with those of the first three terms of (2.3.7) for the special values of a, b, c with which we are dealing, and the identity of the succeeding discussion. It is this correspondence, which evidently holds generally, that justifies the two statements that: (i) every partial differential equation (2.3.1) can be transformed into one in which the mixed derivative is absent, and (ii) the nature of the coefficients of the remaining second-order derivatives depends on the value of $ac - b^2$. The use of the terms *elliptic*, *hyperbolic* and *parabolic* in the classification of (2.3.1) is also based on this formal correspondence. For economy of notation in the remaining discussion, let us assume that the given equation is already in the form (2.3.10) and replace ξ, η and ω by x, y, u, to get

(2.3.11) $$u_{xx} - 4u_{yy} + 3u_x + u = 0 .$$

The second step is a change of *dependent* variable

(2.3.12) $$u = e^{\beta x}\omega ,$$

where β is chosen so that in the transformed equation the coefficient of ω_x vanishes. Differentiating (2.3.12) and substituting in (2.3.11), we obtain for ω the equation

$$\omega_{xx} - 4\omega_{yy} + (2\beta + 3)\omega_x + (\beta^2 + 3\beta + 1)\omega = 0$$

so that, choosing $\beta = -3/2$, we have

(2.3.13) $$\omega_{xx} - 4\omega_{yy} - \tfrac{5}{4}\omega = 0 .$$

Notice that this transformation to an equation lacking the first derivative with respect to x is generally possible when the coefficient of the second derivative with respect to x is not zero, and is otherwise impossible. The same statements hold for derivatives with respect to y.

The final step is a "change of scale,"

(2.3.14) $$x = \mu\xi , \quad y = \nu\eta ,$$

where μ and ν are chosen so that in the transformed equation the coefficients of $\omega_{\xi\xi}$, $\omega_{\eta\eta}$, and ω are equal in absolute value. We have

$$\frac{\partial^2}{\partial x^2} = \frac{1}{\mu^2}\frac{\partial^2}{\partial \xi^2}, \quad \frac{\partial^2}{\partial y^2} = \frac{1}{\nu^2}\frac{\partial^2}{\partial \eta^2},$$

and Equation (2.3.13) becomes

$$\frac{1}{\mu^2}\omega_{\xi\xi} - \frac{4}{\nu^2}\omega_{\eta\eta} - \frac{5}{4}\omega = 0 .$$

The condition

$$\frac{1}{\mu^2} = \frac{4}{\nu^2} = \frac{5}{4}$$

will be satisfied if $\mu = 2/\sqrt{5}$, $\nu = 4/\sqrt{5}$. Thus we obtain as the standard form of the equation (2.3.6) the equation

(2.3.15) $$\omega_{\xi\xi} - \omega_{\eta\eta} - \omega = 0 .$$

We stated above that there is no one set of auxiliary conditions appropriate for all second-order equations with constant coefficients. A fuller statement of the situation is this: For each of the classes, elliptic, hyperbolic, and parabolic, there are several sets of auxiliary conditions which are appropriate for essentially all equations in the class and are not appropriate for the equations in the other two classes. The determination of auxiliary conditions which are plausibly appropriate for one class and not for others is difficult with purely mathematical considerations. We need a guide, and such a guide is furnished by physical problems in whose mathematical descriptions the differential equations appear. Thus we shall reverse the usual procedure followed in the discussion of ordinary differential equations and begin the main part of our study of partial differential equations with the consideration of some of their physical sources and applications. After we have formulated appropriate problems, we shall develop general mathematical procedures for their solution.

The homogeneous equations

$$u_{xx} + u_{yy} = 0 \tag{2.3.16}$$

$$u_{xx} - u_{yy} = 0 \tag{2.3.17}$$

$$u_{xx} - u_y = 0\,, \tag{2.3.18}$$

which are special cases of equations (2.3.2), (2.3.3), and (2.3.4), are particularly important and are essentially typical of their classes. Equation (2.3.16) is called the two-dimensional potential equation or **Laplace's equation**; (2.3.17) is called the one-dimensional **wave equation** or d'Alembert's equation; (2.3.18) is called the one-dimensional **heat equation** or diffusion equation. We shall first examine the one-dimensional heat equation.

EXERCISES 2b

1. Transform the following equations into standard form:

(a) $3u_{xx} + 4u_{yy} - u = 0$

(b) $4u_{xx} + u_{xy} + 4u_{yy} + u = 0$

(c) $u_{xx} + u_{yy} + 3u_x - 4u_y + 25u = 0$

(d) $u_{xx} - 3u_{yy} + 2u_x - u_y + u = 0$

(e) $u_{xx} - 2u_{xy} + u_{yy} + 3u = 0\,.$

2. Show that the equation

$$u_{xx} - u_y + \gamma u = f(x,y)\,,$$

where γ is any constant, can be transformed into

$$\omega_{xx} - \omega_y = \varphi(x,y)\,.$$

3. Show that by rotation of the axes through $\pm 45°$ the equations

$$u_{xx} - u_{yy} = 0$$
$$u_{xy} = 0$$

can be transformed into one another. Find the general solution of both equations.

4. Show that the only second-order linear homogeneous equation with constant coefficients in x and y whose form is unchanged by all rotations of axes is

$$u_{xx} + u_{yy} + ku = 0 .$$

5. Show that the only second-order linear homogeneous equations in x and y whose form is unchanged by all translations of axes are those with constant coefficients.

3

The heat equation and related equations

3.1. *ONE-DIMENSIONAL HEAT FLOW*

Consider a rod composed of a uniform heat-conducting material, with length L and with uniform cross-sectional area A. It will be assumed that the lateral surface of the rod is insulated, and that at any time the temperature in the rod is the same throughout any given cross section but may vary from one cross section to another. Because of these variations in temperature, heat energy will be transported lengthwise along the rod from the hotter parts to the colder parts. Such a unidirectional transfer of heat energy is called a **one-dimensional heat flow.**

Let x be the distance measured along the rod from one end, and let $u(x,t)$ be the temperature at time t in the cross section with coordinate x. Let ρ be the *density* (mass per unit volume) of the substance composing the rod. The heat energy which must be supplied to unit mass of the substance in order to raise it through unit temperature range is a constant, c, called the *specific heat* of the substance. The mass of the rod between the sections x and $x + dx$ is $\rho \cdot A \cdot dx$ where A is the cross-sectional area, and the heat energy which must be supplied to this part of the rod to change its temperature from zero to $u(x,t)$ is therefore $u(x,t) \cdot c\rho A\, dx$. Hence, the heat energy contained in the part of the rod between $x = a$ and $x = b$ at time t is

$$Q(t) = \int_a^b u(x,t) c\rho A\, dx\ . \tag{3.1.1}$$

In writing this equation, the convention has been made that the heat energy is zero when the temperature is everywhere zero.

The quantity $Q(t)$ may increase, either by virtue of heat flowing into the region $a < x < b$ across the end faces $x = a$ and $x = b$, or by virtue of heat being created within the region as a result of, for example, a chemical reaction. The law of conservation of energy thus gives

$$\frac{dQ}{dt} = \text{flux term} + \text{source term} \tag{3.1.2}$$

where "flux term" represents the contribution due to heat flowing in across the boundaries and "source term" represents the contribution of internal heat sources.

The *flux* term is conveniently described by the heat flux function $F(x,t)$ defined as the quantity of heat energy passing unit area of the x cross section per unit time in the positive direction of the x-axis. The rate of flow of heat through the cross section $x = a$ into the region $x > a$ is then $AF(a,t)$, while the rate of flow of heat through the cross section $x = b$ into the region $x < b$ is $-AF(b,t)$ (see Fig. 3.1). Therefore, the flux term is

$$\text{flux term} = -A[F(b,t) - F(a,t)] .$$

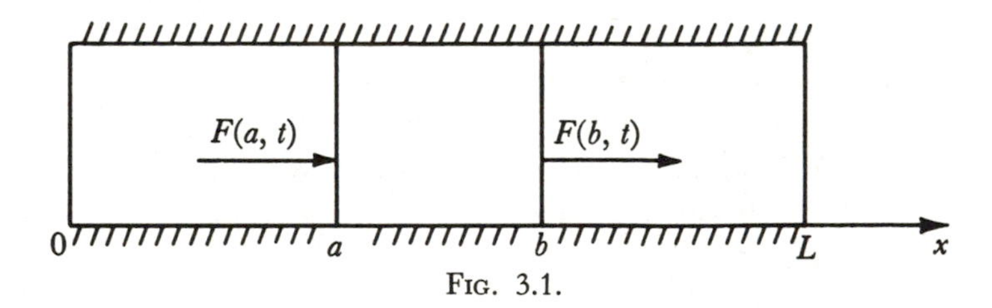

FIG. 3.1.

This can now be related to the temperature function by using an empirical physical law which states that the heat flux at any point is proportional to the temperature gradient at that point. In symbols

$$F(x,t) = -\kappa \frac{\partial u}{\partial x}(x,t) , \qquad \kappa > 0 , \tag{3.1.3}$$

where the proportionality constant κ is characteristic of the material of the rod and is called the *heat conductivity*. The minus sign in (3.1.3) expresses the familiar fact that heat flows in the direction of decreasing temperature. The flux term can now be written as

$$\begin{aligned}\text{flux term} &= A\left[\kappa \frac{\partial u}{\partial x}(x,t)\bigg|_{x=b} - \kappa \frac{\partial u}{\partial x}(x,t)\bigg|_{x=a}\right] \\ &= \int_a^b \frac{\partial}{\partial x}\left(\kappa \frac{\partial u}{\partial x}(x,t)\right) A\, dx .\end{aligned}$$

Using this formula and computing dQ/dt from (3.1.1) the conservation law (3.1.2) becomes

$$\int_a^b \left[c\rho \frac{\partial u}{\partial t}(x,t) - \frac{\partial}{\partial x}\left(\kappa \frac{\partial u}{\partial x}(x,t) \right) \right] A\, dx = \text{source term.} \tag{3.1.4}$$

If there are no heat sources present in the rod, then the source term is zero and the integral in Equation (3.1.4) is zero. In this discussion the interval $[a,b]$ is any subinterval of $[0,L]$. We assume that the integrand in (3.1.4) is a continuous function of x. It then follows that the integrand is zero throughout $[0,L]$. For if the integrand is not zero at some point x_0, we can deduce a contradiction. Suppose the integrand is positive at x_0. Then, by continuity, it will be positive everywhere in a small interval $[a,b]$ containing x_0, and the integral in (3.1.4) could not be zero for this choice of $[a,b]$. This contradiction shows that the integrand cannot be positive at any point; similarly it cannot be negative. Thus

$$c\rho \frac{\partial u}{\partial t}(x,t) - \frac{\partial}{\partial x}\left(\kappa \frac{\partial u}{\partial x}(x,t) \right) = 0, \tag{3.1.5}$$

or

$$\frac{\partial u}{\partial t} = k \frac{\partial^2 u}{\partial x^2}, \tag{3.1.6}$$

where $k = \kappa/c\rho$ is a positive constant called the *diffusivity* of the heat-conducting material. Equation (3.1.6), called the **heat equation,** is linear and homogeneous. It is a parabolic equation.

3.2. *SOURCE TERMS*

If internal sources of heat are present, the flow will still be one-dimensional provided that the rate of production of heat is the same everywhere on any given cross section. It will be assumed that the rate of production of heat energy per unit volume per unit time is a function $q_1(x,t,u)$ of the cross section, the time, and the local temperature. The *source* term for the region $a < x < b$ is the rate of production of heat in the region and is equal to

$$\int_a^b q_1[x,t,u(x,t)] A\, dx. \tag{3.2.1}$$

If this is incorporated in (3.1.4), the resulting equation may be written

$$\int_a^b \left\{ c\rho \frac{\partial u}{\partial t} - \frac{\partial}{\partial x}\left(\kappa \frac{\partial u}{\partial x} \right) - q_1(x,t,u) \right\} A\, dx = 0\,. \tag{3.2.2}$$

If the integrand is continuous, this implies

$$\frac{\partial u}{\partial t} = k\frac{\partial^2 u}{\partial x^2} + q(x,t,u)\,, \tag{3.2.3}$$

where $q(x,t,u) = q_1(x,t,u)/c\rho$. In speaking of sources we have assumed that $q(x,t,u) \geq 0$. The equation is also significant with $q(x,t,u) \leq 0$, however, and describes the situation in which heat is being absorbed internally at the rate $-c\rho q(x,t,u)$ per unit volume, per unit time. In this case we speak of negative sources, or *sinks*. In the general case, the sign of q will vary with (x,t,u).

In general (3.2.3) is a nonlinear equation and its nonlinearity presents difficulties too formidable to be discussed here. However, there are two interesting cases in which the equation is linear. First, the source function may be independent of the local temperature, as would be the case, for example, if the heating were caused by an electric current flowing in the rod. The differential equation (3.2.3) is then of the form

$$\frac{\partial u}{\partial t} = k\frac{\partial^2 u}{\partial x^2} + q(x,t)\,, \tag{3.2.4}$$

which is an inhomogeneous linear equation. Alternatively, the source term may be proportional to the local temperature, $q(x,t,u) = r(x,t)u(x,t)$. For example, the heating might be due to a chemical reaction that takes place at a rate proportional to the local temperature. In this case (3.2.3) becomes

$$\frac{\partial u}{\partial t} = k\frac{\partial^2 u}{\partial x^2} + r(x,t)u\,, \tag{3.2.5}$$

a homogeneous linear equation. More generally these two kinds of source heating may occur in combination, so that the differential equation is of the form

$$\frac{\partial u}{\partial t} = k\frac{\partial^2 u}{\partial x^2} + r(x,t)u + q(x,t)\,. \tag{3.2.6}$$

We see that Equation (2.3.4) of Chapter 2 describes heat conduction in a uniform rod with a given distribution of sources and sinks independent of the temperature.

3.3. *NONUNIFORM ROD*

A still more complicated situation than that which leads to Equation (3.2.3) arises if the material of the rod is uniform over any given cross section but may vary from one cross section to another. The density, specific heat, and conductivity are now functions of x and in place of (3.2.3) we have

$$c(x)\rho(x)\frac{\partial u}{\partial t} - \frac{\partial}{\partial x}\left[\kappa(x)\frac{\partial u}{\partial x}\right] - q_1(x,t,u) = 0\,. \tag{3.3.1}$$

Introducing the specific heat per unit volume $\sigma(x) = c(x)\rho(x)$ and the source function $q(x,t,u) = q_1(x,t,u)/\sigma(x)$, this equation may be written as

$$\frac{\partial u}{\partial t} = \frac{1}{\sigma(x)}\frac{\partial}{\partial x}\left[\kappa(x)\frac{\partial u}{\partial x}\right] + q(x,t,u)\,. \tag{3.3.2}$$

Simplified forms of the source function may occur as before, leading to linear, possibly inhomogeneous, differential equations with variable coefficients.

3.4. *INITIAL AND BOUNDARY CONDITIONS*

It has been shown that, for a rod of given composition with a given distribution of internal sources, the temperature function $u(x,t)$ is a solution of a certain partial differential equation. In any particular problem there will be additional data which will single out that special solution of the differential equation which is the temperature distribution sought. The additional data usually take the form of a pair of boundary conditions and an initial condition. The two **boundary conditions** describe how the two ends of the rod exchange heat energy with the surrounding medium. The **initial condition** prescribes the temperature throughout the rod at some initial time, say at time $t = 0$.

For example, consider a uniform rod with no heat sources. Suppose that the temperature in the rod at time $t = 0$ is $u(x,0) = f(x)$, and the two ends of the rod are suddenly placed in contact with heat reservoirs at temperature zero. The temperature at each end will then be zero at all later times. From a physical point of view, it seems plausible that the temperature function throughout the rod at all times $t > 0$ is completely determined. This temperature function $u(x,t)$ will be a solution of the following problem, called an **initial-boundary value problem** (see Fig. 3.2).

Problem: Find a function $u(x,t)$ defined for $0 \leq x \leq L$, $t \geq 0$, such that

$$\text{D.E.} \qquad \frac{\partial u}{\partial t} = k\frac{\partial^2 u}{\partial x^2}, \qquad 0 < x < L, \quad t > 0,$$

$$\text{B.C.} \qquad \left.\begin{aligned} u(0,t) &= 0 \\ u(L,t) &= 0 \end{aligned}\right\}, \qquad t > 0,$$

$$\text{I.C.} \qquad u(x,0) = f(x), \qquad 0 < x < L.$$

In this problem it is understood that the unknown function u, together with its derivatives $\partial u/\partial t$, $\partial u/\partial x$, $\partial^2 u/\partial x^2$ are continuous functions of x and t for $0 < x < L$ and $t > 0$. The differential equation (D.E.), the boundary conditions (B.C.), and the initial condition (I.C.) must be satisfied only for the indicated values of x and t. These requirements will be discussed in greater detail in Chapter 7.

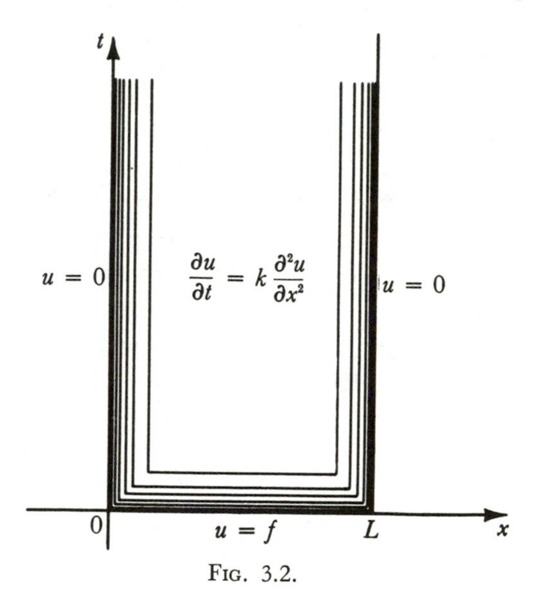

FIG. 3.2.

Other forms of boundary conditions may occur. For example, one or both end faces of the rod may be insulated. If the end at $x = 0$ is insulated, then no heat energy crosses that face; that is, the flux at $x = 0$ is zero, and the corresponding boundary condition is $\kappa u_x(0,t) = 0$, instead of $u(0,t) = 0$. Similarly, if the end $x = L$ is insulated, the boundary condition $u(L,t) = 0$ is to be replaced by $\kappa u_x(L,t) = 0$. Another possibility is that one end of the rod, say the end $x = 0$, is placed in contact with a heat reservoir at a temperature α. The boundary condition at $x = 0$ for this case is $u(0,t) = \alpha$. If the temperature of the heat reservoir is varied in a prescribed manner, then α will be a given function of the time $\alpha = \alpha(t)$, and the boundary condition is said to be *time-dependent*.

All these boundary conditions are *linear*. We will consider still other linear boundary conditions in Section 6. Linear boundary conditions, such as those above, may be classified as *homogeneous* or *inhomogeneous*, and if inhomogeneous, as *time-independent* or *time-dependent*. The boundary conditions

$$u(0,t) = 0\,, \quad \kappa u_x(L,t) = 0$$

are homogeneous, while

$$u(0,t) = 3\,, \quad \kappa u_x(0,t) = e^{-t}$$

are examples of inhomogeneous boundary conditions, the first being time-independent and the second time-dependent.

Let us return to the consideration of the initial-boundary value problem posed above. The B.C. and I.C. there are auxiliary conditions for the D.E. It is physically plausible that the initial temperature function $f(x)$ be any reasonable function; that is, that the problem has a solution for arbitrary $f(x)$. Further, the statement that the temperature function $u(x,t)$ is determined for all $t > 0$ when $f(x)$ is given amounts to the assertion that the problem has only one solution. These propositions are evidently equally valid if the D.E. is replaced by any of the other differential equations of heat conduction discussed in the preceding sections. Thus it appears that the B.C. and I.C. of the initial-boundary value problem constitute an appropriate set of auxiliary conditions for these differential equations. With other boundary conditions we get additional appropriate sets of auxiliary conditions.

EXERCISES 3a

1. A rod of length L, cross section A, whose lateral surface is insulated, is made of a material of thermal constants c, ρ, κ. Heat is produced electrically at a rate β per unit volume. The ends are kept at temperature T and initially the rod is at temperature zero. Formulate the initial-boundary value problem for the temperature in the rod.

2. A rod of length L with insulated lateral surface has thermal constants c, ρ, κ. One end of the rod is kept at temperature zero. The other end is connected to a black box which manages to maintain that end at the instantaneous average temperature of the entire rod. Initially, the rod is at temperature U. Formulate the initial-boundary value problem for the temperature in the rod. Is it a linear problem?

3. (*Radial heat flow in a cylinder*) A case of heat flow which is virtually one-dimensional arises when the conducting medium is a circular cylinder

and the temperature function u depends only on the time t and the distance r from the axis of the cylinder, $u = u(r,t)$. For example, imagine a cylindrical pipe filled with a hot fluid and suppose that one wishes to study the loss of heat through the sides of the pipe.

Let c, ρ, κ, k denote the thermal constants of the cylinder. By considering the heat energy contained in a section of pipe of length H and lying between the radii $r = a$, $r = b$ show that

$$2\pi H \int_a^b \left[c\rho r \frac{\partial u}{\partial t} - \frac{\partial}{\partial r}\left(\kappa r \frac{\partial u}{\partial r}\right)\right] dr = \text{source term}.$$

Hence, obtain the equation for source-free radial heat flow in a cylinder:

$$\begin{aligned} \frac{\partial u}{\partial t} &= \frac{k}{r}\frac{\partial}{\partial r}\left(r\frac{\partial u}{\partial r}\right) \\ &= k\left[u_{rr} + \frac{1}{r}u_r\right]. \end{aligned}$$

4. In a hollow cylinder with thermal constants c, ρ, κ, k and inner radius a, outer radius b, the inner face is kept at temperature U, the outer face at temperature zero and the initial temperature at $t = 0$ is $f(r)$. Formulate the initial-boundary value problem for the temperature $u(r,t)$ at all later times.

5. Transform the equation

$$u_t = k(u_{xx} + u_{yy})$$

to polar coordinates $r = (x^2 + y^2)^{1/2}$, $\theta = \arctan(y/x)$ and specialize the resulting equation to the case when the function u does not depend on the angular variable θ.

6. Derive the differential equation satisfied by the temperature $u(r,t)$ in the case of radial heat flow in a sphere.

3.5. *JUMP CONDITIONS*

In deriving the equation of heat conduction in a rod we assumed that the thermal properties of the rod varied continuously, that is, that $c(x)$, $\rho(x)$, and $\kappa(x)$ were continuous. Problems in which this is not so arise quite naturally. Suppose, for example, that we have two uniform rods of cross section A, one having length L_1, and thermal constants c_1, ρ_1, κ_1, k_1, the second having length L_2 and thermal constants c_2, ρ_2, κ_2, k_2. The first rod has uniform temperature zero and its left end is maintained at that temperature. The second rod has uniform temperature 100 and its right end is maintained at that temperature. At time $t = 0$, the right end of the first rod is placed in contact with the left end of the second. We seek the temperature distribution in the rods at time $t > 0$.

The two joined rods may be considered as a single rod of length $L_1 + L_2$. The temperature distribution $u(x,t)$ in this rod will be a solution of the following problem:

$$\text{D.E.}\qquad u_t = k(x)u_{xx}\,, \qquad 0 < x < L_1\,,\ t > 0\,, \quad \text{and} \quad L_1 < x < L_1 + L_2\,,\ t > 0\,,$$

$$\text{B.C.}\qquad \left.\begin{aligned} u(0,t) &= 0 \\ u(L_1 + L_2,t) &= 100 \end{aligned}\right\} \qquad t > 0\,,$$

$$\text{I.C.}\qquad u(x,0) = f(x)\,, \qquad 0 < x < L_1 + L_2\,,$$

where

$$k(x) = \begin{cases} k_1 & 0 < x < L_1 \\ k_2 & L_1 < x < L_1 + L_2 \end{cases} \tag{3.5.1}$$

and

$$f(x) = \begin{cases} 0 & 0 < x < L_1 \\ 100 & L_1 < x < L_1 + L_2\,. \end{cases} \tag{3.5.2}$$

In each of the intervals $0 < x < L_1$, $L_1 < x < L_1 + L_2$, $u(x,t)$ will be continuous and will have continuous derivatives. However, u or its derivatives may be discontinuous at L_1, and to complete the statement of the problem we must specify the nature of the discontinuities.

Let us write

$$u(x,t) = \begin{cases} u_1(x,t) & 0 < x < L_1 \\ u_2(x,t) & L_1 < x < L_1 + L_2\,. \end{cases} \tag{3.5.3}$$

The rods are said to be in *perfect thermal contact* at $x = L_1$ if both the temperature and the flux are continuous at $x = L_1$, that is, if

$$\begin{aligned} \lim_{\substack{\epsilon \to 0 \\ \epsilon > 0}} u_1(L_1 - \epsilon,t) &= u_1(L_1,t) = u_2(L_1,t) \\ &= \lim_{\substack{\epsilon \to 0 \\ \epsilon > 0}} u_2(L_1 + \epsilon,t)\,, \quad t > 0\,, \end{aligned} \tag{3.5.4}$$

and

$$\begin{aligned} \lim_{\substack{\epsilon \to 0 \\ \epsilon > 0}} \kappa_1 \frac{\partial u_1}{\partial x}(L_1 - \epsilon,t) &= \kappa_1 \frac{\partial u_1}{\partial x}(L_1,t) = \kappa_2 \frac{\partial u_2}{\partial x}(L_1,t) \\ &= \lim_{\substack{\epsilon \to 0 \\ \epsilon > 0}} \kappa_2 \frac{\partial u_2}{\partial x}(L_1 + \epsilon,t)\,, \quad t > 0\,. \end{aligned} \tag{3.5.5}$$

Condition (3.5.5) is also a mathematical consequence of the conservation law (3.1.2) and the assumption that u and its derivatives are bounded. From the discussion in Sections 1 and 2 we have

$$\int_a^b c\rho \frac{\partial u}{\partial t}(x,t)A\,dx = A\left[\kappa \frac{\partial u}{\partial x}(x,t)\right]_a^b$$

where $c(x)$, $\rho(x)$, and $\kappa(x)$ are defined in terms of c_1, c_2, ρ_1, ρ_2, κ_1, κ_2 just as $k(x)$ was defined in terms of k_1 and k_2. Choosing $a = L_1 - \epsilon$, $b = L_1 + \epsilon$ we obtain

$$\kappa_2 \frac{\partial u_2}{\partial x}(L_1 + \epsilon, t) - \kappa_1 \frac{\partial u_1}{\partial x}(L_1 - \epsilon, t) = \int_{L_1-\epsilon}^{L_1+\epsilon} \left[c\rho \frac{\partial u}{\partial t} \right] dx .$$

Letting $\epsilon \to 0$, the integral on the right vanishes since the integrand is bounded, and we obtain

$$\lim_{\epsilon \to 0} \left[\kappa_2 \frac{\partial u_2}{\partial x}(L_1 + \epsilon, t) - \kappa_1 \frac{\partial u_1}{\partial x}(L_1 - \epsilon, t) \right]$$
$$= \left[\kappa_2 \frac{\partial u_2}{\partial x}(L_1, t) - \kappa_1 \frac{\partial u_1}{\partial x}(L_1, t) \right] = 0 .$$

Thus,

$$\kappa_2 \frac{\partial u_2}{\partial x}(L_1, t) = \kappa_1 \frac{\partial u_1}{\partial x}(L_1, t) \qquad t > 0 ,$$

that is, the flux of heat is continuous at L_1.

The two conditions (3.5.4) and (3.5.5) are called **jump conditions.** To complete the formulation of the initial-boundary value problem it must be specified that these two conditions hold, or some alternative conditions hold, at each place where the properties of the heat conducting medium have a discontinuity (see Fig. 3.3.).

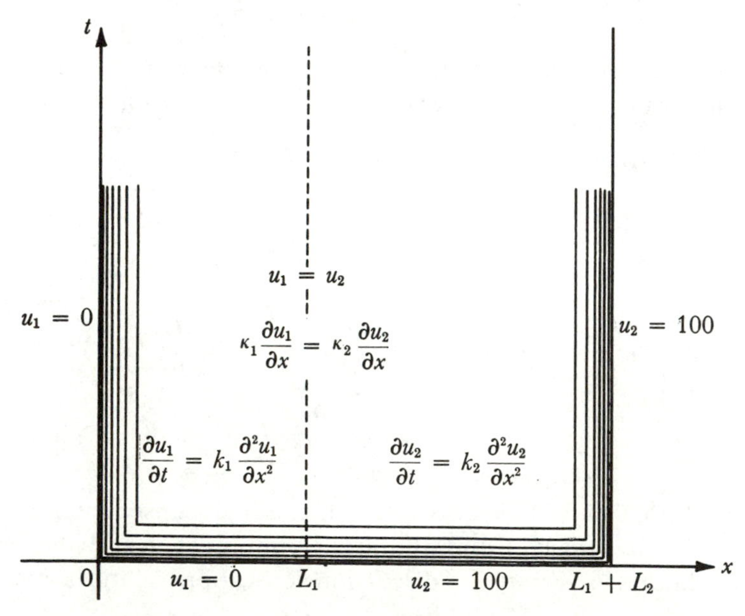

FIG. 3.3.

It is interesting to consider cases when alternative forms of one or both of the jump conditions might be more appropriate. Suppose that instead of perfect thermal contact there is a thin film of foreign material between the two rods. To simplify the problem, we still suppose that the rod consists of two parts and use the heat equation with $k = k_1$ for $x < L_1$, $k = k_2$ for $x > L_1$. However, if the film has either (i) a very low conductivity or (ii) a very high specific heat (or both), then the jump conditions need to be re-examined. In case (i) the film can support high temperature gradients at low flux and condition (3.5.4) is unsatisfactory. In case (ii) the film can store significant quantities of heat without undergoing large temperature changes and condition (3.5.5) is unrealistic. We will not concern ourselves with the exact form of the jump conditions in these cases, but see the next section for further discussion of thin films.

3.6. *BOUNDARY CONDITIONS IN GENERAL*

The boundary conditions discussed in Section 4

$$u(x_0,t) = \alpha(t) \tag{3.6.1}$$

$$\frac{\partial u}{\partial x}(x_0,t) = \beta(t) \tag{3.6.2}$$

are called **boundary conditions of the first** and **second kind** or the **Dirichlet** and **Neumann conditions,** respectively. Both are *linear* boundary conditions. There is another linear boundary condition which occurs often in practice; namely,

$$u(x_0,t) + h\frac{\partial u}{\partial x}(x_0,t) = \gamma(t) \qquad h \neq 0 \tag{3.6.3}$$

which is called the **boundary condition of the third kind** or **Robin's condition.**

We give a simple physical interpretation of condition (3.6.3). Suppose that we have a uniform rod of conductivity κ extending from $x = 0$ to $x = L$, whose left end is in contact with a heat reservoir at temperature $\gamma(t)$, and let $u(x,t)$ be the temperature in the rod. If there is a thin film (for example, of grease or oxide) on the end of the rod, then the statement

$$u(0,t) = \gamma(t)$$

is not strictly correct. Rather we must treat the problem as one of conduction in a composite rod, as in the preceding section. Let κ_0

be the conductivity of the film, L_0 its thickness, and let the temperature in the film be $u_0(x,t)$. Assuming perfect thermal contact between the end of the rod and the film, the jump conditions at $x = 0$ are

$$u(0,t) = u_0(0,t) , \tag{3.6.4}$$

$$\kappa \frac{\partial u}{\partial x}(0,t) = \kappa_0 \frac{\partial u_0}{\partial x}(0,t) . \tag{3.6.5}$$

Further, since the temperature at the end of the film is that of the heat reservoir, we have

$$u_0(-L_0,t) = \gamma(t) .$$

Since L_0 is very small we may approximate the derivative $\partial u_0/\partial x$ in (3.6.5) by the difference quotient

$$\frac{u_0(-L_0,t) - u_0(0,t)}{-L_0} = \frac{\gamma(t) - u(0,t)}{-L_0} .$$

Making this approximation in (3.6.5) we obtain for $u(x,t)$ at $x = 0$ the boundary condition

$$\kappa \frac{\partial u}{\partial x}(0,t) = -\frac{\kappa_0}{L_0}[\gamma(t) - u(0,t)]$$

or

$$u(0,t) - \frac{L_0\kappa}{\kappa_0}\frac{\partial u}{\partial x}(0,t) = \gamma(t) , \tag{3.6.6}$$

which is a boundary condition of the third kind with $h = -L_0\kappa/\kappa_0$.

A boundary condition such as $u(0,t) - u_x(L,t) = 0$, which involves values of u and its derivatives at both ends of the rod, is often called a **mixed boundary condition,** while one such as $u_x(L,t) + hu(L,t) = 0$, which involves the values of u and its derivatives at only one end, is said to be **unmixed** (see Exercises 3b, Problem 6).

For a condition such as

$$u(0,t) - \int_0^L u(x,t)g(x)\,dx = 0 \tag{3.6.7}$$

which relates the boundary value $u(0,t)$ to values $u(x,t)$, $0 < x < L$ in the interior of the rod, the term boundary condition has been deemed inappropriate, and the condition is called a **lateral condition.** (See Exercises 3a, Problem 2.)

In the statement of some boundary conditions the notion of a *heat reservoir* occurs. If the end of a rod is in contact with a body of fluid which is well stirred so that the fluid is at the same temperature v throughout, we say the rod is in contact with a reservoir at temperature v. If the fluid is of mass M and specific heat c, the quantity $C = Mc$, called the *heat capacity* of the reservoir, is the heat energy required to increase the reservoir temperature by one unit. The presence of such a reservoir may lead to a linear boundary condition of the form

$$u(0,t) + h_1 u_x(0,t) + h_2 u_t(0,t) = 0 \,. \tag{3.6.8}$$

In addition, there is a great variety of possible *nonlinear* boundary conditions. Problems involving nonlinear boundary conditions are extremely difficult, and are usually solved approximately by approximating the nonlinear boundary condition by a linear boundary condition. Such approximations lead most often to linear boundary conditions of the third kind.

3.7. *EQUILIBRIUM PROBLEMS; LAPLACE'S EQUATION*

When the temperature at each point in a body is independent of the time, the temperature in the body is said to be in **equilibrium** and the flow of heat in the body is called a **steady-state flow.** In this case $\partial u/\partial t = 0$, so that, in the most general case of heat conduction in a nonuniform rod with internal sources, the differential equation (3.3.2) satisfied by the temperature function becomes

$$\frac{1}{\sigma(x)} \frac{d}{dx}\left[\kappa(x) \frac{du}{dx}\right] + q(x,u) = 0 \,. \tag{3.7.1}$$

The temperature distribution in a state of equilibrium is completely specified by conditions on the boundary. The problem of determining a solution of (3.7.1) which satisfies given boundary conditions is called a **boundary value problem.** For example, if the temperature in a uniform rod of length L with no internal sources is in equilibrium when the temperatures at the left and right ends are a and b, respectively, then the temperature distribution $u(x)$ in the rod is the solution of the boundary value problem

$$\text{D.E.} \qquad \frac{d^2u}{dx^2} = 0 \,, \qquad 0 < x < L$$

$$\text{B.C.} \qquad u(0) = a \,, \qquad u(L) = b \,.$$

Equilibrium problems for a rod are boundary value problems for ordinary differential equations. We present next a problem which leads to a boundary value problem for a partial differential equation, although we will not consider such problems in detail until Chapter 11.

We will say that a heat-conducting substance is **uniform** if it has the same thermal properties at each point, and that it is **isotropic** if at each point it has the same properties in all directions. Now consider a flat plate of a uniform isotropic substance which has no thermal sources and which is in temperature equilibrium. We take the plate to lie on the horizontal (x,y)-plane in (x,y,z)-space and assume that the temperature u is uniform on every vertical line, so that u is a function $u(x,y)$ of x and y. If we make the plausible hypothesis that the temperature function satisfies a linear, homogeneous, second-order, partial differential equation in x and y, we can show that the differential equation is *Laplace's equation*,

$$u_{xx} + u_{yy} = 0 . \tag{3.7.2}$$

For, since the plate is uniform, the differential equation cannot depend on the choice of the origin of the coordinate system; that is, it must be unchanged by translations, and hence must have constant coefficients (see Exercises 2b, Problem 5). Further, the form of the equation cannot depend on the orientation of the axes since the substance is isotropic; that is, the form must be unchanged by rotation of axes, and hence the equation must be

$$u_{xx} + u_{yy} + ku = 0 , \tag{3.7.3}$$

with k some constant (see Exercises 2b, Problem 4). Finally, since a constant temperature is a possible equilibrium temperature, we must have $k = 0$.

Now suppose that the plate is square, of side length L, and take the x- and y-axes along edges of the plate. If the left and right edges of the plate are maintained at temperature 0, and the bottom and top edges at temperatures $f(x)$ and $g(x)$, respectively, then $u(x,y)$ is the solution of the **boundary value problem** (see Fig. 3.4)

D.E. $\quad u_{xx} + u_{yy} = 0 , \qquad 0 < x < L , \quad 0 < y < L$

B.C. $\quad u(0,y) = 0 , \quad u(L,y) = 0 , \qquad 0 < y < L$
$\qquad\;\; u(x,0) = f(x) , \quad u(x,L) = g(x) , \qquad 0 < x < L .$

The physical interpretation of the preceding problem indicates that the auxiliary conditions stated there are appropriate for the differential equation $u_{xx} + u_{yy} = 0$, that is, that for arbitrary $f(x)$ and $g(x)$ there will be one and only one solution of the differential equation which satisfies the auxiliary conditions. It is interesting to

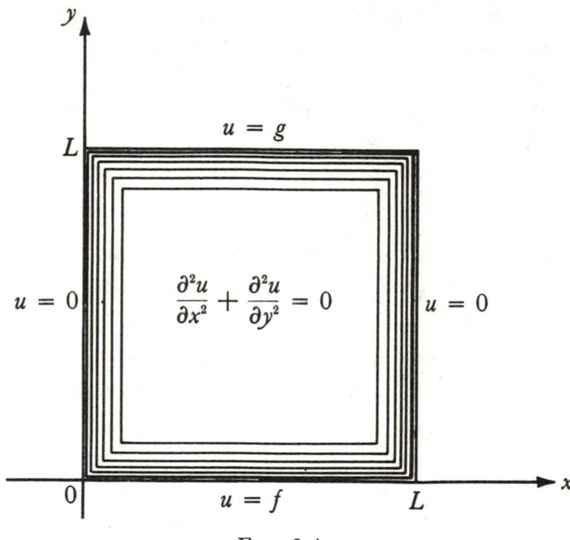

FIG. 3.4.

observe that these conditions are not appropriate for the equation $u_{xx} - u_y = 0$, that is, the boundary value problem

$$\text{D.E.} \quad u_{xx} - u_y = 0\,, \qquad 0 < x < L\,, \quad 0 < y < L$$

$$\text{B.C.} \quad \begin{array}{ll} u(0,y) = 0\,, \quad u(L,y) = 0 & 0 < y < L \\ u(x,0) = f(x)\,, \quad u(x,L) = g(x) & 0 < y < L \end{array}$$

will *not* in general have a solution. This is not mathematically evident, but it is so physically, if we interpret y as the time and $u(x,y)$ as the temperature distribution in a rod. If the problem had a solution in general, then we could find a temperature distribution in a rod whose ends were maintained at temperature 0, for which the temperature had arbitrary prescribed values at time 0 and also at a later time L. We know, however, that when the temperature is prescribed at time 0, it is completely determined at all later times (compare Figs. 3.2 and 3.4).

EXERCISES 3b

1. Find the equilibrium temperature distribution in a uniform rod when
(i) the end $x = 0$ is maintained at temperature 0 and the end $x = L$ is maintained at temperature T;
(ii) the end $x = 0$ is insulated and the end $x = L$ is maintained at temperature T.

2. The end $x = 0$ of a uniform rod is maintained at a constant temperature and the end $x = L$ is maintained at temperature zero. After the steady state has been reached, it is found that the flux out of the end of the rod at $x = L$ has the value F. What is the temperature at $x = 0$?

3. Consider the problem of finding the equilibrium temperature distribution in a uniform rod when the flux at $x = 0$ is F and the flux at $x = L$ is G. Explain physically why this problem does not always have a solution. Also explain why additional information is needed to specify the solution when one exists, and state what information would be suitable.

4. The two ends of a uniform rod are insulated and the temperature is initially $u(x) = f(x)$. Find the equilibrium temperature which the rod assumes after a long time.

5. The right end of a rod of length L_1 and thermal constants c_1, ρ_1, κ_1 is joined to the left end of a rod of length L_2 and thermal constants c_2, ρ_2, κ_2. The left end of the composite rod is maintained at temperature 0 and the right end at temperature U. Find the equilibrium temperature distribution in the composite rod if:

(i) the joined ends are in perfect thermal contact;

(ii) there is a thin film of width L_0 and conductivity κ_0 between the joined ends.

6. A cylindrical wire of length L whose lateral surface is insulated is bent into a circle with the two ends in contact. State an initial-boundary value problem for the determination of the temperature $u(x,t)$ in the wire if the ends are in perfect thermal contact.

7. A heat conducting system consists of three rods, R_1, R_2, R_3, connected by two reservoirs W_1, W_2. Each rod is of length L and has thermal constants ρ, c, κ, k, but the cross sections are A_1, A_2, A_3, respectively. The left end of rod 1 is kept at temperature U and the right end of rod 3 is kept at temperature zero. Find the temperatures of the reservoirs when the equilibrium state is attained.

8. Find the equilibrium temperature for the cylinder in Exercises 3a, Problem 4.

9. A thin wire runs along the axis of a cylinder of radius b. The thermal constants of the cylinder are ρ, c, κ, k and the heat capacity and radius of the wire are negligible. Heat is generated electrically in the wire at a constant rate Q per unit length and the outer face of the cylinder is kept at temperature zero. Find the equilibrium temperature distribution in the cylinder. (See Exercises 3a, Problem 3.)

10. Find the steady-state temperature in a hollow sphere of inner radius a and outer radius b, if the inner surface is kept at temperature U, the outer surface at temperature zero. (See Exercises 3a, Problem 6.)

11. A homogeneous rod of length $L = 5$, with thermal constants c, ρ, κ, k and with lateral surface insulated, extends from $x = 0$ to $x = 5$. Both ends are maintained at temperature zero and in the part between $x = 3$ and $x = 4$ heat is produced at a constant rate Q per unit volume. Find the steady-state temperature $u(x)$.

12. A rod with cross section A, thermal constants c, ρ, κ, extends from $x = 0$ to $x = L$. The end $x = 0$ is in contact with an otherwise insulated reservoir of heat capacity C. Find the boundary condition satisfied by the temperature distribution $u(x,t)$ in the rod at $x = 0$.

13. Mercury flows with velocity V in an insulated cylindrical pipe from a reservoir at temperature 0 at $x = 0$ to a reservoir at temperature T at $x = L$. Initially the mercury in the pipe is at temperature zero. State an initial-boundary value problem for the determination of the temperature distribution in the mercury, assuming the temperature to be constant on each cross section perpendicular to the axis of the pipe. (Note that for an observer traveling to the right with velocity V, the column of mercury appears as a fixed rod.)

4

The method of eigenfunction expansions

4.1. OUTLINE OF THE METHOD

When the differential equation and the boundary conditions of an initial-boundary value problem are homogeneous, we call the problem a **homogeneous initial-boundary value problem.** The simplest homogeneous problem in heat conduction, for example, is

$$\text{(4.1.1)} \quad \text{D.E.} \quad \frac{\partial u}{\partial t} = k\frac{\partial^2 u}{\partial x^2}, \qquad t > 0, \quad 0 < x < L,$$

$$\begin{array}{ll} \text{(4.1.2)} \\ \text{(4.1.3)} \end{array} \quad \text{B.C.} \quad \left.\begin{array}{l} u(0,t) = 0 \\ u(L,t) = 0 \end{array}\right\}, \qquad t > 0,$$

$$\text{(4.1.4)} \quad \text{I.C.} \quad u(x,0) = f(x), \qquad 0 < x < L.$$

In this chapter we shall present and apply, to this and other homogeneous problems of one-dimensional heat flow, a method of solution that was discovered by the French mathematical physicist J. Fourier in 1812. The method is quite general and can be employed for the solution of homogeneous initial-boundary value problems for many other linear partial differential equations with constant coefficients and for some equations with variable coefficients. It can also be employed in the solution of boundary value problems. Finally, its results furnish the basis for the solution of inhomogeneous problems.

In some respects, the procedure is simple. Particular functions which satisfy the differential equation can be found by separating the variables. We shall go one step further and require that the separated solutions satisfy not only the differential equation but also the boundary conditions. For example, for the problem formulated above; we will seek functions of the special form $u(x,t) = T(t)\varphi(x)$ which satisfy (4.1.1), (4.1.2) and (4.1.3). All three of these equations are linear and homogeneous, and hence if a number of such functions can be found, then any linear combination of them will also satisfy the three equations. The idea of the method is to find sufficiently many functions of this special kind that a linear combination of them can be formed which will also satisfy the initial condition (4.1.4). There is little *a priori* basis, mathematical or physical, for the belief that this should be possible for an arbitrary function $f(x)$. Fourier and later workers made the remarkable discovery that it is, in fact, generally possible.

Our discussion in this chapter will be entirely formal; that is, we will ignore questions concerning continuity and convergence. We will also assume without proof the most important part of Fourier's series theorem. These matters will be discussed at length in Chapters 6 and 7.

4.2. *REMARKS ON ORDINARY DIFFERENTIAL EQUATIONS*

In the course of our work we shall need to solve initial and boundary value problems for second order ordinary linear differential equations of the type

$$c_0(x)\,\frac{d^2\varphi}{dx^2} + c_1(x)\,\frac{d\varphi}{dx} + c_2(x)\varphi + \lambda\varphi = 0\,. \tag{4.2.1}$$

We shall be concerned mainly with the special cases of (4.2.1) which can be solved explicitly: for example, $\varphi'' + \lambda\varphi = 0$ or $x^2\varphi'' + x\varphi' + \lambda\varphi = 0$. Nevertheless, it is useful to describe some general aspects of the solution of (4.2.1), and to formulate a single method of procedure that can be used in all cases.

The main general proposition about (4.2.1) is that the equation has solutions. We will state this proposition below, but without proof since, as we have already said, we shall be mainly concerned with the special cases in which the solutions can be found explicitly.

We consider Equation (4.2.1) on an interval $a \leq x \leq b$. The number λ, which may be real or complex, is a constant in the sense that it does not depend on x. Frequently, however, the value of λ will not be determined until a much later stage of the problem, so that λ has to be treated as an undetermined constant or *parameter*.

A solution $\varphi(x)$ of equation (4.2.1) can be determined by prescribing the value of $\varphi(x)$ and of its derivative $\varphi'(x)$ at one point of the interval $[a,b]$. The solution $\varphi(x)$ is primarily a function of x, but of course it also depends on the parameter λ. When the dependence on λ is important, instead of writing $\varphi(x)$, $\varphi'(x)$, etc., we shall write $\varphi(x,\lambda)$, $\varphi'(x,\lambda)$, etc.

The theorem in the theory of ordinary differential equations which explicitly states the proposition above is: *if the coefficients $c_0(x)$, $c_1(x)$, $c_2(x)$ are continuous on the interval $[a,b]$ and $c_0(x)$ is different from zero throughout the interval, then for any fixed point x_0 in the interval $[a,b]$ and for any constants α, β Equation* (4.2.1) *has one and only one solution $\varphi(x) = \varphi(x,\lambda)$ satisfying at x_0 the initial conditions*

$$\varphi(x_0) = \alpha\,, \qquad \varphi'(x_0) = \beta\,. \tag{4.2.2}$$

Moreover $\varphi(x,\lambda)$ and $\varphi'(x,\lambda)$ are continuous and continuously differentiable functions of both x and λ for $a \leq x \leq b$ and all λ.

It is convenient to single out the two special solutions $\varphi_1(x) = \varphi_1(x,\lambda)$, $\varphi_2(x) = \varphi_2(x,\lambda)$ of (4.2.1) which satisfy

$$\begin{aligned} \varphi_1(x_0) &= 1\,, \qquad & \varphi_2(x_0) &= 0\,, \\ \varphi_1'(x_0) &= 0\,, \qquad & \varphi_2'(x_0) &= 1\,. \end{aligned} \tag{4.2.3}$$

We call $\varphi_1(x)$ and $\varphi_2(x)$ the *basic solutions of* (4.2.1) *at* x_0. If these basic solutions are known, it is easy to construct the general solution $\varphi(x)$ which satisfies (4.2.2). In fact the solution

$$\hat{\varphi}(x) = \alpha\varphi_1(x) + \beta\varphi_2(x)$$

satisfies (4.2.2) since

$$\begin{aligned} \hat{\varphi}(x_0) &= \alpha\cdot 1 + \beta\cdot 0 = \alpha\,, \\ \hat{\varphi}'(x_0) &= \alpha\cdot 0 + \beta\cdot 1 = \beta\,. \end{aligned}$$

But the equation has only one solution that satisfies (4.2.2) so that $\varphi(x) = \hat{\varphi}(x)$ or

$$\varphi(x) = \alpha\varphi_1(x) + \beta\varphi_2(x)\,. \tag{4.2.4}$$

Thus the general solution of (4.2.1) is represented in terms of the basic solutions at x_0 by formula (4.2.4).

Let us consider the most important special case of (4.2.1), namely the equation

$$\varphi'' + \lambda\varphi = 0\,. \tag{4.2.5}$$

We begin by finding the basic solutions at $x_0 = 0$. They satisfy

$$\begin{aligned} \varphi_1(0) &= 1\,, \qquad & \varphi_2(0) &= 0\,, \\ \varphi_1'(0) &= 0\,, \qquad & \varphi_2'(0) &= 1\,, \end{aligned} \tag{4.2.6}$$

and, as is easily found, they are given by the formulas

$$\varphi_1(x) = \varphi_1(x,\lambda) = \cos x\sqrt{\lambda}\,, \tag{4.2.7}$$
$$\varphi_2(x) = \varphi_2(x,\lambda) = \frac{\sin x\sqrt{\lambda}}{\sqrt{\lambda}}\,.$$

These formulas are perfectly straightforward if λ is positive. If λ is negative, or complex, we use the Taylor series definitions for the trigonometric functions,

$$\cos z = 1 - \frac{z^2}{2!} + \frac{z^4}{4!} - \frac{z^6}{6!} + \cdots = \sum_{n=0}^{\infty} (-1)^n \frac{z^{2n}}{(2n)!}\,,$$
$$\sin z = z - \frac{z^3}{3!} + \frac{z^5}{5!} - \frac{z^7}{7!} + \cdots = \sum_{n=0}^{\infty} (-1)^n \frac{z^{2n+1}}{(2n+1)!}\,.$$

Thus (4.2.7) is equivalent to

$$\varphi_1(x,\lambda) = 1 - \lambda\frac{x^2}{2!} + \lambda^2\frac{x^4}{4!} - \lambda^3\frac{x^6}{6!} + \cdots\,, \tag{4.2.8}$$
$$\varphi_2(x,\lambda) = x - \lambda\frac{x^3}{3!} + \lambda^2\frac{x^5}{5!} - \lambda^3\frac{x^7}{7!} + \cdots\,,$$

and these series converge for all real or complex values of λ.

When λ is negative, say $\lambda = -s^2$, $s > 0$, it is often useful to express the basic solutions in terms of hyperbolic functions. The differential equation (4.2.5) is

$$\varphi'' - s^2\varphi = 0$$

and the solutions which satisfy (4.26) are

$$\varphi_1(x) = \cosh xs\,, \qquad \varphi_2(x) = \frac{\sinh xs}{s}$$

or, since $s = \sqrt{-\lambda}$,

$$\varphi_1(x) = \varphi_1(x,\lambda) = \cosh x\sqrt{-\lambda}\,, \tag{4.2.9}$$
$$\varphi_2(x) = \varphi_2(x,\lambda) = \frac{\sinh x\sqrt{-\lambda}}{\sqrt{-\lambda}}\,.$$

Let us verify that formulas (4.2.7) and (4.2.9), which look different, are really the same. If we use the formulas

$$\cos\theta = \frac{e^{i\theta} + e^{-i\theta}}{2}\,, \qquad \cosh t = \frac{e^t + e^{-t}}{2}\,,$$
$$\sin\theta = \frac{e^{i\theta} - e^{-i\theta}}{2i}\,, \qquad \sinh t = \frac{e^t - e^{-t}}{2}\,,$$

and set $t = i\theta$, we get

$$\cosh i\theta = \cos\theta\,,$$
$$\sinh i\theta = i\sin\theta\,.$$

Therefore,

$$\cosh x\sqrt{-\lambda} = \cosh ix\sqrt{\lambda} = \cos x\sqrt{\lambda}\,,$$

$$\frac{\sinh x\sqrt{-\lambda}}{\sqrt{-\lambda}} = \frac{\sinh ix\sqrt{\lambda}}{i\sqrt{\lambda}} = \frac{\sin x\sqrt{\lambda}}{\sqrt{\lambda}}\,,$$

which shows that (4.2.7) and (4.2.9) are indeed equivalent.

When $\lambda = 0$, the differential equation (4.2.5) is simply

$$\varphi'' = 0$$

and the solutions of this which satisfy (4.2.6) are

$$\begin{aligned}\varphi_1(x) &= \varphi_1(x,0) = 1\,,\\ \varphi_2(x) &= \varphi_2(x,0) = x\,.\end{aligned} \tag{4.2.10}$$

This agrees with (4.2.8); that is, if we set $\lambda = 0$ in (4.2.8) we obtain (4.2.10). Moreover if we let $\lambda \to 0$ in (4.2.7), we have

$$\lim_{\lambda\to 0}\cos x\sqrt{\lambda} = 1\,, \qquad \lim_{\lambda\to 0}\frac{\sin x\sqrt{\lambda}}{\sqrt{\lambda}} = x\,.$$

Therefore it is correct to interpret the second basic solution $\varphi_2(x,\lambda)$ when $\lambda = 0$ as

$$\varphi_2(x,0) = \lim_{\lambda\to 0}\varphi_2(x,\lambda) = \lim_{\lambda\to 0}\frac{\sin x\sqrt{\lambda}}{\sqrt{\lambda}}\,.$$

In summary, the basic solutions can be regarded as defined by (4.2.7) for all values of λ if we define the trigonometric functions by their Taylor expansions (4.2.8). If λ is positive, the trigonometric functions in (4.2.7) can even be interpreted in the usual elementary sense. If λ is negative, the equivalent formula (4.2.9) may be more convenient than (4.2.7). If $\lambda = 0$, the expression $(\sin x\sqrt{\lambda})/\sqrt{\lambda}$ can be correctly interpreted as

$$\lim_{\lambda\to 0}\frac{\sin x\sqrt{\lambda}}{\sqrt{\lambda}} = x\,.$$

In accordance with (4.2.4), the general solution of the differential equation (4.2.5) is the family of functions

$$\varphi = A\cos x\sqrt{\lambda} + B\frac{\sin x\sqrt{\lambda}}{\sqrt{\lambda}}\,, \tag{4.2.11}$$

where A and B are any constants. It is often incorrectly stated that the general solution of (4.2.5) is

$$\varphi = A \cos x\sqrt{\lambda} + B \sin x\sqrt{\lambda} .$$

But, when $\lambda = 0$, this reduces to

$$\varphi = A\cdot 1 + B\cdot 0 = A ,$$

which is not the general solution of $\varphi'' = 0$, while (4.2.11) reduces to

$$\varphi = A + Bx$$

which is the general solution of $\varphi'' = 0$.

EXERCISES 4a

1. Find the basic solutions at $x_0 = 0$ for

$$\varphi'' + 4\varphi' + \lambda\varphi = 0 .$$

Also find the solution which satisfies $\varphi(0) = 3$, $\varphi'(0) = -1$.

2. Find the basic solutions at $x_0 = 1$ for
(a) $x^2\varphi'' + x\varphi' + \lambda\varphi = 0 ,$
(b) $x^2\varphi'' + 3x\varphi' + \lambda\varphi = 0 .$

3. Comment on the statement: The general solution of $\varphi'' + 6\varphi' + \lambda\varphi = 0$ is

$$\varphi = C_1 e^{(-3+i\sqrt{\lambda-9})x} + C_2 e^{(-3-i\sqrt{\lambda-9})x} .$$

4. Let φ_1, φ_2, φ_3 be the solutions of

$$\varphi''' - \lambda\varphi = 0$$

such that

$$\begin{array}{lll} \varphi_1(0) = 1 , & \varphi_2(0) = 0 , & \varphi_3(0) = 0 , \\ \varphi_1'(0) = 0 , & \varphi_2'(0) = 1 , & \varphi_3'(0) = 0 , \\ \varphi_1''(0) = 0 , & \varphi_2''(0) = 0 , & \varphi_3''(0) = 1 . \end{array}$$

Find, in terms of φ_1, φ_2, φ_3 the solution φ of the equation which satisfies $\varphi(0) = 5$, $\varphi'(0) = -3$, $\varphi''(0) = 9$.

5. Find the solutions φ_1, φ_2, φ_3, φ_4 of

$$\frac{d^4\varphi}{dx^4} - \lambda\varphi = 0$$

which satisfy

$$\begin{array}{llll} \varphi_1(0) = 1 , & \varphi_2(0) = 0 , & \varphi_3(0) = 0 , & \varphi_4(0) = 0 , \\ \varphi_1'(0) = 0 , & \varphi_2'(0) = 1 , & \varphi_3'(0) = 0 , & \varphi_4'(0) = 0 , \\ \varphi_1''(0) = 0 , & \varphi_2''(0) = 0 , & \varphi_3''(0) = 1 , & \varphi_4''(0) = 0 , \\ \varphi_1'''(0) = 0 , & \varphi_2'''(0) = 0 , & \varphi_3'''(0) = 0 , & \varphi_4'''(0) = 1 . \end{array}$$

4.3. *SOLUTION OF THE SIMPLEST PROBLEM; EIGENVALUE PROBLEMS*

In this and the next section we will determine the solution of the homogeneous initial-boundary value problem (4.1.1) to (4.1.4), the problem for the heat equation in which both boundary conditions are of the first kind. We begin by seeking solutions with separated variables of the differential equation (4.1.1), which also satisfy the boundary conditions (4.1.2), (4.1.3).

For a separated solution $u(x,t) = T(t)\varphi(x)$ the differential equation may be written in the form

$$\frac{1}{kT(t)}\frac{dT}{dt} = \frac{1}{\varphi(x)}\frac{d^2\varphi}{dx^2}\,. \tag{4.3.1}$$

By the usual argument, the two members of this equation depend neither on x nor on t, and their common value is a constant $-\lambda$. (The choice of $-\lambda$ instead of λ to represent the constant is arbitrary, but certainly permissible and, as we will see later, convenient.) The factors therefore satisfy the ordinary differential equations

$$\frac{dT}{dt} + \lambda kT = 0\,, \tag{4.3.2}$$

$$\frac{d^2\varphi}{dx^2} + \lambda\varphi = 0\,, \tag{4.3.3}$$

each of which contains the *same* constant λ. For the separated solution, the boundary conditions are

$$T(t)\varphi(0) = 0\,, \qquad t > 0\,, \tag{4.3.4}$$

$$T(t)\varphi(L) = 0\,, \qquad t > 0\,. \tag{4.3.5}$$

The first of these boundary conditions will be satisfied either if $\varphi(0) = 0$ or, alternatively, if $T(t) = 0$ for all $t > 0$. The second alternative leads to the solution $u \equiv 0$ of (4.1.1), (4.1.2), (4.1.3). Since this solution is of no use in building up a function that will also satisfy (4.1.4), we require that $\varphi(0) = 0$. Similarly, the second boundary condition leads us to require that $\varphi(L) = 0$. Thus, φ must satisfy the equations

$$\begin{aligned} &\frac{d^2\varphi}{dx^2} + \lambda\varphi = 0\,, \qquad 0 < x < L\,,\\ &\varphi(0) = 0\,,\\ &\varphi(L) = 0. \end{aligned} \tag{4.3.6}$$

These equations always have a solution, regardless of the value of λ, namely the **trivial solution** $\varphi \equiv 0$. This, however, furnishes the solution $u \equiv 0$ of the partial differential equation, and is of no interest. We are thus led to the following problem, called an **eigenvalue problem:** For what values of λ do the equations (4.3.6) have a **nontrivial** solution φ? A value of λ for which there is a nontrivial solution is called an **eigenvalue** and the nontrivial solution φ is called an **eigenfunction** belonging to that eigenvalue.

The general solution of the differential equation in (4.3.6) is

$$\varphi(x) = A \cos x\sqrt{\lambda} + B \frac{\sin x\sqrt{\lambda}}{\sqrt{\lambda}} \tag{4.3.7}$$

where A and B are independent of x. The boundary condition $\varphi(0) = 0$ requires that $A = 0$ and hence any eigenfunction is of the form $\varphi(x) = B(\sin x\sqrt{\lambda})/\sqrt{\lambda}$. The second boundary condition

$$0 = \varphi(L) = B \frac{\sin L\sqrt{\lambda}}{\sqrt{\lambda}} \tag{4.3.8}$$

could be satisfied by taking $B = 0$ but this leads only to the trivial solution $\varphi \equiv 0$, not to an eigenfunction. For a nontrivial solution, λ must be chosen so that

$$\frac{\sin L\sqrt{\lambda}}{\sqrt{\lambda}} = 0 \,. \tag{4.3.9}$$

Conversely if λ satisfies (4.3.9), then the function $\varphi(x) = B(\sin x\sqrt{\lambda})/\sqrt{\lambda}$ satisfies both boundary conditions even if $B \neq 0$, and hence λ is an eigenvalue. Thus every eigenvalue is a root of (4.3.9) and every root of (4.3.9) is an eigenvalue. In other words, (4.3.9) is an equation whose roots are the eigenvalues.

The equation has infinitely many roots

$$\lambda_n = \left(\frac{n\pi}{L}\right)^2, \qquad n = 1, 2, 3, \ldots \tag{4.3.10}$$

and these are the eigenvalues. (It is quite evident that each of these values of λ is a root of (4.3.9), but it is perhaps less clear that there are no complex roots. This question is discussed in more detail at the end of the section.)

The eigenfunctions belonging to the eigenvalues λ_n are $B(\sin x\sqrt{\lambda_n})/\sqrt{\lambda_n}$ which are constant multiples of

$$\varphi_n(x) = \sin x\sqrt{\lambda_n} = \sin \frac{n\pi x}{L}, \qquad n = 1, 2, \ldots . \tag{4.3.11}$$

The time factor $T(t)$ which must be used with $\varphi_n(x)$ is a solution of (4.3.2) with $\lambda = \lambda_n$, that is,

$$\frac{dT}{dt} + \lambda_n k T = 0 . \tag{4.3.12}$$

Thus this factor is a constant multiple of

$$T_n(t) = e^{-\lambda_n kt} = \exp\left(-\frac{n^2\pi^2 kt}{L^2}\right) \qquad n = 1, 2, \ldots . \tag{4.3.13}$$

Finally, the solutions of the heat equation of the special form desired are found to be

$$\begin{aligned} u_n(x,t) &= T_n(t)\varphi_n(x) \\ &= e^{-\lambda_n kt} \sin x\sqrt{\lambda_n} , \qquad n = 1, 2, \ldots \end{aligned} \tag{4.3.14}$$

where $\lambda_n = (n^2\pi^2)/L^2$, or to be constant multiples of these functions.

We return now to the problem of finding all the eigenvalues, that is, all the solutions of Equation (4.3.9). If $\lambda = 0$ the left member of (4.3.9) is to be interpreted as

$$\lim_{\lambda \to 0} \frac{\sin L\sqrt{\lambda}}{\sqrt{\lambda}} = L \neq 0$$

so that $\lambda = 0$ is not an eigenvalue. A *nonzero* value of λ will be an eigenvalue if and only if

$$\sin L\sqrt{\lambda} = 0 . \tag{4.3.15}$$

We wish to find not only the real eigenvalues, but also the complex eigenvalues if there are any. Let $L\sqrt{\lambda} = \alpha + i\beta$ where α, β are real. It is required to find all real values of α and β such that

$$\begin{aligned} 0 &= \sin(\alpha + i\beta) \\ &= \sin\alpha \cos i\beta + \cos\alpha \sin i\beta \\ &= \sin\alpha \cosh\beta + i\cos\alpha \sinh\beta . \end{aligned}$$

Both the real and the imaginary part of the right member of this equation must be zero; that is, we must have

$$\sin\alpha \cosh\beta = 0 \tag{4.3.16}$$

$$\cos\alpha \sinh\beta = 0 . \tag{4.3.17}$$

Since $\cosh\beta \geq 1$ for real β, Equation (4.3.16) requires that $\sin\alpha = 0$, $\alpha = n\pi$, $n = 0, \pm 1, \pm 2, \ldots$. But if $\alpha = n\pi$, then $\cos\alpha = \pm 1$, so that (4.3.17) becomes $\pm\sinh\beta = 0$. Therefore $\beta = 0$ and

$$L\sqrt{\lambda} = \alpha + i\beta = n\pi , \qquad n = 0, \pm 1, \pm 2, \ldots . \tag{4.3.18}$$

The value $n = 0$ must be excluded because it corresponds to the value $\lambda = 0$, which as we have seen, is not an eigenvalue. The eigenvalues are therefore $\lambda_n = (n\pi/L)^2$, $n = 1, 2, \ldots$ as stated earlier. Note that $L\sqrt{\lambda} = n\pi$ and $L\sqrt{\lambda} = -n\pi$ both lead to the same eigenvalue $\lambda_n = (n\pi/L)^2$. In particular, it has turned out that the eigenvalues are all real.

4.4. *SOLUTION CONTINUED; ORTHOGONALITY*

We have found infinitely many functions $u_n(x,t)$, $n = 1, 2, \ldots$ each of which satisfies Equations (4.1.1), (4.1.2), (4.1.3). Since each of these equations is linear and homogeneous, any finite linear combination of these functions, say

$$u(x,t) = \sum_{n=1}^{N} b_n u_n(x,t) , \tag{4.4.1}$$

will also satisfy (4.1.1), (4.1.2), (4.1.3). The initial value of $u_n(x,t)$ is

$$u_n(x,0) = \sin \frac{n\pi x}{L}$$

and so the initial value of (4.4.1) is

$$u(x,0) = \sum_{n=1}^{N} b_n \sin \frac{n\pi x}{L} .$$

We see that the constants b_n cannot be chosen so that (4.4.1) satisfies (4.1.4) except in the very special circumstance that the function f appearing in (4.1.4) is a finite linear combination of the functions $\sin (n\pi x/L)$, $n = 1, 2, \ldots$. In an effort to satisfy (4.1.4) with more general functions f we consider infinite series solutions

$$\begin{aligned} u(x,t) &= \sum_{n=1}^{\infty} b_n u_n(x,t) \\ &= \sum_{n=1}^{\infty} b_n e^{-\lambda_n k t} \sin \frac{n\pi x}{L} . \end{aligned} \tag{4.4.2}$$

The condition that (4.4.2) satisfies (4.1.4) is formally

$$f(x) = \sum_{n=1}^{\infty} b_n \sin \frac{n\pi x}{L} . \tag{4.4.3}$$

Assuming that $f(x)$ is a function for which (4.4.3) holds we are led to the following question. How can the constants b_n be found, thereby completing the formal solution of the initial-boundary value problem?

The system of functions sin $(n\pi x/L)$ has a remarkable property, called orthogonality, which makes it easy to determine the b_n. We define this property generally as follows.

Two real-valued functions $\varphi(x), \psi(x)$ *defined on an interval* $a \leq x \leq b$ *are said to be* **orthogonal** *on the interval whenever*

$$\int_a^b \varphi(x)\psi(x)\, dx = 0 .$$

A finite or infinite sequence $\varphi_1(x),\ \varphi_2(x),\ \ldots,\ \varphi_n(x),\ \ldots$ *of real-valued functions defined on* $a \leq x \leq b$ *is called an* **orthogonal system on the interval** $[a,b]$ *if and only if*

$$\begin{aligned} &\int_a^b \varphi_n(x)\varphi_m(x)\, dx = 0 \qquad \text{when } n \neq m , \\ &\int_a^b \varphi_n^2(x)\, dx > 0 \qquad \text{for } n = 1, 2, \ldots . \end{aligned} \tag{4.4.4}$$

The fact that the system

$$\varphi_n(x) = \sin \frac{n\pi x}{L} \qquad n = 1, 2, \ldots \tag{4.4.5}$$

is an orthogonal system on the interval $[0,L]$ is easily verified. Indeed, integrals of the form

$$\int \sin \alpha x \sin \beta x\, dx$$

can be evaluated by methods of elementary calculus and we can show that, if m and n are integers,

$$\int_0^L \sin \frac{m\pi x}{L} \cdot \sin \frac{n\pi x}{L}\, dx = 0 \qquad \text{if } m \neq n , \tag{4.4.6}$$

$$\int_0^L \left(\sin \frac{n\pi x}{L}\right)^2 dx = \frac{L}{2}, \qquad n = 1, 2, 3, \ldots . \tag{4.4.7}$$

Using this orthogonality property, the constants b_n in (4.4.3) can be computed. We assume that the series (4.4.3) can be multiplied by sin $(m\pi x/L)$ and integrated term-by-term. This gives

$$\int_0^L f(x) \sin \frac{m\pi x}{L}\, dx = \sum_{n=1}^{\infty} b_n \int_0^L \sin \frac{n\pi x}{L} \sin \frac{m\pi x}{L}\, dx . \tag{4.4.8}$$

In the right-hand member of this equation, every term is zero except the one with $n = m$. Thus the infinite series reduces to just one term,

$$\begin{aligned} \int_0^L f(x) \sin \frac{m\pi x}{L}\, dx &= b_m \int_0^L \left(\sin \frac{m\pi x}{L}\right)^2 dx \\ &= b_m \frac{L}{2} . \end{aligned}$$

Hence

$$b_n = \frac{2}{L} \int_0^L f(x) \sin \frac{n\pi x}{L}\, dx\, . \tag{4.4.9}$$

The series (4.4.2) with λ_n given by (4.3.10) and the constants b_n as in (4.4.9) is the solution of our initial-boundary value problem. The series is called the **eigenfunction expansion** of the solution.

The series (4.4.3) with constants b_n given by (4.4.9) is called the *eigenfunction expansion* of $f(x)$. In Chapter 6 it is shown that if $f(x)$ is one of a large class of functions which includes, in particular, all functions with continuous derivatives on $[0,L]$, then the series (4.4.3) converges to $f(x)$ for $0 < x < L$. In Chapter 7 we will prove that when f is a member of this class of functions the series (4.4.2) converges, and its sum is a solution of the initial-boundary value problem (4.1.1)–(4.1.4). We will also prove that it is the only solution of the problem. In this chapter and the next we will study the *technique* of finding eigenfunction expansions of solutions of a variety of other initial-boundary value problems. The further discussion of these solutions will also be deferred to Chapter 7.

If the function $f(x)$ is sufficiently simple, the constants b_n defined by (4.4.9) can be evaluated explicitly. For example if $f(x) = 1$, $0 \leq x \leq L$, then

$$\begin{aligned} b_n &= \frac{2}{L} \int_0^L \sin \frac{n\pi x}{L}\, dx \\ &= \frac{2}{n\pi}(1 - \cos n\pi) \\ &= \frac{2}{n\pi}[1 - (-1)^n] \\ &= \begin{cases} 0 & \text{if } n \text{ is even} \\ \dfrac{4}{n\pi} & \text{if } n \text{ is odd}\, . \end{cases} \end{aligned}$$

Thus $b_{2m} = 0$, $b_{2m+1} = 4/(2m+1)\pi$ and the series for $f(x)$ is

$$f(x) = 1 = \frac{4}{\pi} \sum_{m=0}^{\infty} \frac{1}{2m+1} \sin \frac{(2m+1)\pi x}{L}\, , \qquad 0 < x < L\, .$$

When $x = 0$ and when $x = L$ every term of the series is zero so the series does not converge to 1. But, and this will be proved in Chapter 6, the series does converge to 1 if $0 < x < L$.

The solution of problem (4.1.1)–(4.1.4) when $f(x) = 1$ is thus

$$u(x,t) = \frac{4}{\pi}\sum_{m=0}^{\infty}\frac{e^{-\lambda_{2m+1}kt}}{2m+1}\sin\frac{(2m+1)\pi x}{L}, \qquad (4.4.10)$$
$$\lambda_{2m+1} = \left[\frac{(2m+1)\pi}{L}\right]^2 .$$

4.5. *BOUNDARY CONDITIONS OF THE SECOND KIND*

When both ends of a rod are insulated, the temperature function is determined by solving the following problem:

(4.5.1) D.E. $u_t = ku_{xx}$, $t > 0$, $0 < x < L$,

(4.5.2) (4.5.3) B.C. $\begin{cases} u_x(0,t) = 0 \\ u_x(L,t) = 0 \end{cases}$ $t > 0$,

(4.5.4) I.C. $u(x,0) = f(x)$, $0 < x < L$.

As in Section 3 we seek nontrivial functions $u(x,t) = T(t)\varphi(x)$ which satisfy the differential equation and the boundary conditions. We find that $T(t)$ must be a solution of

$$\frac{dT}{dt} + \lambda kT = 0, \qquad t > 0, \qquad (4.5.5)$$

and for $\varphi(x)$ we find the following eigenvalue problem: For what values of λ does the system

$$\begin{aligned} \varphi'' + \lambda\varphi &= 0, \qquad 0 < x < L, \\ \varphi'(0) &= 0 \\ \varphi'(L) &= 0 \end{aligned} \qquad (4.5.6)$$

have a nontrivial solution? As before the constant λ in (4.5.5) is the same as that in (4.5.6).

The general solution of the differential equation in (4.5.6) is

$$\varphi(x) = A\cos x\sqrt{\lambda} + B\frac{\sin x\sqrt{\lambda}}{\sqrt{\lambda}}.$$

The boundary condition $\varphi'(0) = 0$ requires that $B = 0$ and hence any eigenfunction is of the form

$$\varphi(x) = A\cos x\sqrt{\lambda}.$$

The second boundary condition of (4.5.6) becomes

$$0 = \varphi'(L) = -A\sqrt{\lambda}\sin L\sqrt{\lambda}.$$

This must be satisfied by choosing λ so that

$$\sqrt{\lambda}\sin L\sqrt{\lambda} = 0\,. \tag{4.5.7}$$

The roots of (4.5.7) are the eigenvalues of the problem (4.5.6). The left member of (4.5.7) vanishes if either λ or $\sin L\sqrt{\lambda}$ is zero and we see that the eigenvalues are

$$\lambda_n = \left(\frac{n\pi}{L}\right)^2, \qquad n = 0, 1, 2, \ldots. \tag{4.5.8}$$

The eigenfunction corresponding to $\lambda = \lambda_n$ is $A \cos x\sqrt{\lambda_n}$ which is a multiple of

$$\varphi_n(x) = \cos\frac{n\pi x}{L}, \qquad n = 0, 1, 2, \ldots. \tag{4.5.9}$$

Integrals of the form $\int \cos \alpha x \cos \beta x\, dx$ may be evaluated by elementary methods. The reader should verify the following orthogonality relation:

$$\int_0^L \varphi_m(x)\varphi_n(x)\, dx = 0 \qquad \text{if } m \neq n\,, \tag{4.5.10}$$

and also

$$\int_0^L \varphi_0^2(x)\, dx = L\,, \tag{4.5.11}$$

$$\int_0^L \varphi_n^2(x)\, dx = \frac{L}{2}, \qquad n = 1, 2, \ldots. \tag{4.5.12}$$

As in Section 4, we can use the equations (4.5.10) to (4.5.12) to determine the coefficients A_n of an *eigenfunction expansion*

$$f(x) = \sum_{n=0}^{\infty} A_n\varphi_n(x) = A_0 + \sum_{n=1}^{\infty} A_n \cos\frac{n\pi x}{L} \tag{4.5.13}$$

of a function $f(x)$. We find that

$$A_0 = \frac{1}{L}\int_0^L f(x)\varphi_0(x)\, dx = \frac{1}{L}\int_0^L f(x)\, dx$$

and

$$A_n = \frac{2}{L}\int_0^L f(x)\varphi_n(x)\, dx = \frac{2}{L}\int_0^L f(x)\cos\frac{n\pi x}{L}\, dx\,, \qquad n \neq 0\,.$$

The difference in the formulas for A_0 and A_n, $n \neq 0$, is a consequence of the difference between (4.5.11) and (4.5.12). As a convenience, we will write instead of (4.5.13)

$$f(x) = \frac{a_0}{2} + \sum_{n=1}^{\infty} a_n \cos\frac{n\pi x}{L}, \tag{4.5.14}$$

so that for the coefficients a_n we have the single formula

$$(4.5.15) \qquad a_n = \frac{2}{L}\int_0^L f(x)\cos\frac{n\pi x}{L}\,dx \qquad n = 0, 1, 2, \ldots .$$

Solutions of the partial differential equation (4.5.1) that satisfy the boundary conditions (4.5.2) and (4.5.3) can now be obtained by multiplying $\varphi_n(x)$ by a solution of (4.5.5) with $\lambda = \lambda_n$. Thus we get the sequence of functions

$$(4.5.16) \qquad u_n(x,t) = e^{-\lambda_n kt}\cos\frac{n\pi x}{L}, \qquad n = 0, 1, 2, \ldots .$$

Noting that $u_0(x,t) = 1$, we form the infinite series

$$(4.5.17) \qquad u(x,t) = \frac{1}{2}a_0 + \sum_{n=1}^{\infty} a_n e^{-\lambda_n kt}\cos\frac{n\pi x}{L}.$$

Formally this series satisfies the initial condition (4.5.4) if

$$(4.5.18) \qquad f(x) = \frac{1}{2}a_0 + \sum_{n=1}^{\infty} a_n\cos\frac{n\pi x}{L}.$$

We therefore choose the constants a_n as in (4.5.15). Finally, then, (4.5.17) and (4.5.15) provide the solution to the problem formulated in Equations (4.5.1) to (4.5.4).

EXERCISES 4b

1. The following problems concern the temperature distribution $u(x,t)$ in a rod of length L, cross section A, conductivity κ, and diffusivity k, whose ends are maintained at temperature 0, and which is initially at temperature 1. For this case, $u(x,t)$ is given by Equation (4.4.10). Assume that this series may be integrated and differentiated term by term. The numerical equation

$$\sum_0^{\infty}\frac{1}{(2m+1)^2} = \frac{\pi^2}{8}$$

(which we will establish later) may be used.

(a) Find the temperature at time t at the center of the rod, $x = L/2$. Use a theorem on alternating series to obtain an approximation to this temperature, together with an estimate of the error in the approximation. Show that the relative error approaches 0 as $t \to \infty$. Show, in particular, that the relative error in the approximation is less than 0.01 per cent for $t \geq 10L^2/8\pi^2k$.

(b) Find the average temperature $U(t)$ in the rod at time t. Show that

$$\frac{8}{\pi^2}e^{-(\pi^2/L^2)kt} \leq U(t) \leq e^{-(\pi^2/L^2)kt}.$$

(c) The initial heat content of the rod is

$$Q = \frac{\kappa}{k} \int_0^L f(x) A \, dx = \frac{\kappa L}{k} A \, ,$$

since $f(x) = 1$. In the course of time, all of this heat is lost through the two ends of the rod. The rate at which heat is *lost* through the end $x = 0$ at time t is $A\kappa u_x(0,t)$, so that the eventual net loss through this end is

$$Q_1 = \int_0^\infty A\kappa u_x(0,t) \, dt \, .$$

Calculate Q_1 and compare this value with Q. Account for the relation between Q_1 and Q.

2. Evaluate the constants b_n in (4.4.2) for the special case $f(x) = \sin(3\pi x/L)$. On a single graph plot the solution $u(x,t)$ for $t = 0$, $t = 1$, $t = 2$, $t = 3$, assuming $L = \pi$, $k = 1$.

3. Evaluate the constants b_n in (4.4.2) when

$$f(x) = \begin{cases} 0 \, , & 0 \le x < L/3 \, , \\ 1 \, , & L/3 \le x < L/2 \, , \\ 0 \, , & L/2 \le x \le L \, . \end{cases}$$

4. Show that if $f(x)$ is bounded, say $|f(x)| \le M$ for all x in $[0,L]$, then the series solution of the problem of Sections 3 and 4 is convergent when $t > 0$ and $0 \le x \le L$.

5. (a) Let $u(x,t)$ be the solution of the problem of Sections 3 and 4. What is

$$\lim_{t \to \infty} u(x,t) \, ?$$

(b) Let $u(x,t)$ be the solution of the problem of Section 5. What is

$$\lim_{t \to \infty} u(x,t) \, ?$$

(c) Give a physical explanation of the results of (a) and (b).

6. For the eigenvalue problem

$$\begin{aligned} \varphi'' + \lambda\varphi &= 0 \, , \qquad 0 < x < L \\ \varphi'(0) &= 0 \, , \\ \varphi(L) &= 0 \, , \end{aligned}$$

do the following.

(a) Find an equation whose roots are the eigenvalues.

(b) Find the eigenvalues λ_n.

(c) Use the method of Section 3 to show that the eigenvalues are all real.

(d) Find the eigenfunctions φ_n.

(e) Show that the eigenfunctions are orthogonal.

(f) Evaluate the integrals

$$\int_0^L \varphi_n^2(x) \, dx \, .$$

7. Use your results in Problem 6 to solve the problem

$$\begin{array}{lll} \text{D.E.} & u_t = ku_{xx}\,, & t > 0\,,\quad 0 < x < L \\ \text{B.C.} & \left.\begin{array}{l} u_x(0,t) = 0 \\ u(L,t) = 0 \end{array}\right\} & t > 0\,, \\ \text{I.C.} & u(x,0) = L - x & 0 < x < L\,. \end{array}$$

(Evaluate all constants.)

8. For each of the following eigenvalue problems, find an equation whose roots are the eigenvalues.

(a) $\varphi'' + 2\varphi' + \lambda\varphi = 0,\ 0 < x < L,$
$\varphi(0) = 0,$
$\varphi'(L) = 0.$

(b) $x^2\varphi'' + x\varphi' + \lambda\varphi = 0,\ 1 < x < a,$
$\varphi'(1) = 0,$
$\varphi'(a) = 0.$

(c) $\varphi'''' - \lambda\varphi = 0,\ 0 < x < L,$
$\varphi(0) = \varphi'(0) = \varphi(L) = \varphi'(L) = 0.$

9. For the initial-boundary value problem

$$\begin{array}{lll} \text{D.E.} & u_t = ku_{xx}\,, & 0 < x < L\,,\quad t > 0 \\ \text{B.C.} & \left.\begin{array}{l} u(0,t) = 0 \\ u_x(L,t) = 0 \end{array}\right\} & t > 0\,, \\ \text{I.C.} & u(x,0) = f(x) & 0 < x < L \end{array}$$

carry out the following steps.

(a) Formulate the eigenvalue problem obtained by separation of the variables.
(b) Find an equation whose roots are the eigenvalues.
(c) Find the eigenvalues and eigenfunctions and show that the eigenfunctions belonging to different eigenvalues are orthogonal.
(d) Write out the eigenfunction expansion of the solution $u(x,t)$.
(e) Evaluate the constants in part (d) for the special case $f(x) = 1$.

10. What heat flow problem corresponds to the initial-boundary value problem in Problem 9? With $f(x) = 1$ calculate the heat lost through the end $x = 0$ in two ways.

11. In a rod of length L with insulated lateral surface and thermal constants c, ρ, κ, k heat is generated at a rate uniformly proportional to the temperature, that is, at the rate $ru(x,t)$ per unit volume per unit time, where r is a constant and $u(x,t)$ is the temperature function of the rod. The ends of the rod are maintained at temperature 0 and initially the rod has uniform temperature 1.

(a) Formulate an initial-boundary value problem for the determination of $u(x,t)$.

(b) Solve the problem formulated in (a), following the outline in Problem 9.

(c) Find a simple approximation for the temperature function valid for large values of t.

12. Let $\varphi_1(x), \ldots, \varphi_n(x)$ be an orthogonal system on the interval $[a,b]$. Show that the n functions $\varphi_1(x), \ldots, \varphi_n(x)$ are linearly independent on $[a,b]$.

4.6. *GREEN'S FORMULA AND SOME APPLICATIONS TO EIGENVALUE PROBLEMS*

In the eigenvalue problems considered so far, we have been able to solve completely and explicitly the equation that determined the eigenvalues. However, this will not be possible in many of the problems we shall study later, and in general it is necessary to employ approximation methods, graphical or numerical procedures, to determine the eigenvalues. Thus, it is important to establish in advance of the determination of the eigenvalues that, for example, all eigenvalues are real.

In this section we will study a formula which has wide application to eigenvalue problems and eigenfunction expansion problems. The formula was first stated and used by Lagrange, but later generalized by Green, and is called Green's formula. With its aid we can establish several important properties of eigenvalues and eigenfunctions in advance of the determination of these eigenvalues and eigenfunctions. In particular, we will rediscover properties of the solutions of the eigenvalue problems we have already considered, but without making use of our explicit knowledge of the solutions.

Let f and g be functions with continuous first and second derivatives on an interval $a \leq x \leq b$. Then

$$f''g - fg'' = \frac{d}{dx}(f'g - fg') \tag{4.6.1}$$

and hence

$$\int_a^b [f''(x)g(x) - f(x)g''(x)]\, dx = [f'(x)g(x) - f(x)g'(x)]\Big|_{x=a}^{x=b}. \tag{4.6.2}$$

We define the linear operator A by $Af = -f''$ or symbolically

$$A = -\frac{d^2}{dx^2}. \tag{4.6.3}$$

Then (4.6.2) may be written as

$$\int_a^b [(Af)g - f(Ag)]\, dx = -[f'g - fg']_a^b. \tag{4.6.4}$$

Equation (4.6.4) is *Green's formula for the operator A on the interval* $[a,b]$, or briefly, **Green's formula.**

In the sequel we will need another property of the operator A. Let f be a complex-valued function of the real variable x, say

$$f(x) = f_1(x) + if_2(x)\ ,$$

where f_1 and f_2 are real. Then the *conjugate* function $\bar{f}(x)$ is defined by

$$\bar{f}(x) = f_1(x) - if_2(x)\ .$$

If f_1 and f_2 have continuous first and second derivatives, then so do both f and $\bar{f}$ and

$$Af = -f'' = -f_1'' - if_2''\ ,$$
$$A\bar{f} = -\bar{f}'' = -(f_1'' - if_2'') = -f_1'' + if_2''\ .$$

Consequently,

$$A\bar{f} = \overline{Af}\ . \tag{4.6.5}$$

Now consider the eigenvalue problem (4.3.6) which can be written

$$\begin{aligned} A\varphi &= \lambda\varphi \\ \varphi(0) &= 0 \\ \varphi(L) &= 0\ . \end{aligned} \tag{4.6.6}$$

We will employ Green's formula in the study of this problem, and later see how our results can be extended to problems with other boundary conditions. Our first proposition based on Green's formula follows.

The eigenvalues of the problem (4.6.6) *are all real.*

Proof: We observe that if λ is a complex eigenvalue of the problem with eigenfunction φ, then the conjugate $\bar{\lambda}$ is also an eigenvalue with eigenfunction $\bar{\varphi}$. For we have

$$A\bar{\varphi} = \overline{A\varphi} = \overline{\lambda\varphi} = \bar{\lambda}\bar{\varphi}$$

so that $\bar{\varphi}$ satisfies the differential equation with λ replaced by $\bar{\lambda}$. Moreover

$$\bar{\varphi}(0) = \overline{\varphi(0)} = 0$$
$$\bar{\varphi}(L) = \overline{\varphi(L)} = 0$$

so that $\bar{\varphi}$ satisfies the boundary conditions. Finally, φ is not identically zero (it is an eigenfunction) so that $\bar{\varphi}$ is also not identically zero. Hence $\bar{\lambda}$ is an eigenvalue with eigenfunction $\bar{\varphi}$.

Now let λ be any eigenvalue and φ a corresponding eigenfunction. We compute the integral

$$I = \int_0^L [(A\varphi)\bar{\varphi} - \varphi(A\bar{\varphi})]\, dx \tag{4.6.7}$$

in two different ways. Using Green's formula, with $f = \varphi$, $g = \bar{\varphi}$

$$\begin{aligned} I &= -[\varphi'(x)\bar{\varphi}(x) - \varphi(x)\bar{\varphi}'(x)]_0^L \\ &= -[\varphi'(L)\bar{\varphi}(L) - \varphi(L)\bar{\varphi}'(L)] \\ &\quad + [\varphi'(0)\bar{\varphi}(0) - \varphi(0)\bar{\varphi}'(0)] \end{aligned}$$

and this is zero because $\varphi(0) = \varphi(L) = \bar{\varphi}(0) = \bar{\varphi}(L) = 0$, that is, because of the boundary conditions. Hence $I = 0$. On the other hand, substituting $A\varphi = \lambda\varphi$, $A\bar{\varphi} = \bar{\lambda}\bar{\varphi}$ in (4.6.7) gives

$$\begin{aligned} I &= \int_0^L [(\lambda\varphi)\bar{\varphi} - \varphi(\bar{\lambda}\bar{\varphi})]\, dx \\ &= (\lambda - \bar{\lambda}) \int_0^L |\varphi(x)|^2\, dx\,, \end{aligned}$$

since $\varphi(x)\bar{\varphi}(x) = |\varphi(x)|^2$. Since $I = 0$ we have

$$(\lambda - \bar{\lambda}) \int_0^L |\varphi(x)|^2\, dx = 0\,. \tag{4.6.8}$$

Now φ is a continuous function not zero everywhere on $0 \leq x \leq L$, so that

$$\int_0^L |\varphi(x)|^2\, dx \neq 0\,.$$

It therefore follows from (4.6.8) that $\lambda - \bar{\lambda} = 0$, $\lambda = \bar{\lambda}$ and so λ is real. This completes the proof.

The collection of all eigenvalues of an eigenvalue problem is sometimes called the *spectrum* of the problem, and the eigenfunction expansion of a function is called the *spectral representation* of the function. For example, when stated in this terminology, the above proposition asserts that the spectrum of problem (4.6.6) lies on the real axis.

An argument similar to that employed above establishes the next proposition.

For the problem (4.6.6) *eigenfunctions belonging to different eigenvalues are orthogonal.*

Proof: Let φ_j and φ_k be eigenfunctions belonging to distinct eigenvalues λ_j and λ_k, respectively. Then

$$\int_0^L [(A\varphi_j)\varphi_k - \varphi_j(A\varphi_k)]\, dx = (\lambda_j - \lambda_k) \int_0^L \varphi_j(x)\varphi_k(x)\, dx\,,$$

and using Green's formula we get

$$\int_0^L [(A\varphi_j)\varphi_k - \varphi_j(A\varphi_k)]\, dx = -[\varphi_j'(x)\varphi_k(x) - \varphi_j(x)\varphi_k'(x)]_0^L\,. \tag{4.6.9}$$

The expression on the right side of (4.6.9) is zero because φ_j and φ_k both vanish at $x = 0$ and $x = L$. Hence

$$(\lambda_j - \lambda_k) \int_0^L \varphi_j(x)\varphi_k(x)\, dx = 0 ,$$

and since λ_k and λ_j are *different* eigenvalues, $\lambda_j - \lambda_k \neq 0$, we must have

$$\int_0^L \varphi_j(x)\varphi_k(x)\, dx = 0 . \tag{4.6.10}$$

This proves the proposition.

In the proof of the last proposition the expression

$$[\varphi_j'(x)\varphi_k(x) - \varphi_j(x)\varphi_k'(x)]_0^L \tag{4.6.11}$$

was found to vanish because both φ_j and φ_k satisfied the boundary conditions. The expression consists of four terms and each term separately was zero. Another case in which each of the four terms is zero arises if instead of $\varphi(0) = 0$, $\varphi(L) = 0$ we have the boundary conditions $\varphi'(0) = 0$, $\varphi'(L) = 0$. Also, if we are dealing with the eigenvalue problem (h is a real constant)

$$\begin{aligned} A\varphi &= \lambda\varphi \\ \varphi'(0) &= 0 \\ \varphi'(L) + h\varphi(L) &= 0 , \end{aligned} \tag{4.6.12}$$

then the expression (4.6.11) is still zero provided φ_j and φ_k are eigenfunctions. To see this we write

$$\varphi_j'(L)\varphi_k(L) - \varphi_j(L)\varphi_k'(L) = -[h\varphi_j(L)]\varphi_k(L) + \varphi_j(L)[h\varphi_k(L)] = 0 .$$

Thus we see that the last proposition is valid not only for the eigenvalue problems (4.3.6) and (4.5.6) which we discussed earlier, but also for other problems including the problem (4.6.12). It is not difficult to show that the first proposition of this section also extends to these other problems. We shall learn in Section 11 that both of the propositions carry over to a wide class of eigenvalue problems which includes the above-mentioned problems as special cases.

4.7. *FURTHER APPLICATION OF GREEN'S FORMULA; NORMALIZING CONSTANTS*

If $\{\varphi_n(x)\}$ is an orthogonal system of functions on the interval $[0,L]$, then the values of the integrals

$$\int_0^L [\varphi_n(x)]^2\, dx \tag{4.7.1}$$

are sometimes called the **normalizing constants** for the orthogonal system. They occur in the calculation of the coefficients b_n in an expansion

$$f(x) = \sum_{n=1}^{\infty} b_n \varphi_n(x) .$$

In fact we have

$$b_n = \int_0^L f(x)\varphi_n(x)\, dx \Big/ \int_0^L \varphi_n^2(x)\, dx .$$

If $\varphi_n(x)$ is an eigenfunction of (4.3.6) belonging to an eigenvalue λ_n, and C is any nonzero constant, then $\varphi_n^*(x) = C\varphi_n(x)$ is also an eigenfunction belonging to the same eigenvalue. We could use either $\varphi_n(x)$ or $\varphi_n^*(x)$ as the representative eigenfunction, but naturally we do not use both since they are not linearly independent. Of course, if $\varphi_n^*(x)$ rather than $\varphi_n(x)$ were chosen as the representative eigenfunction, then the normalizing constant would be different. Thus, the normalizing constants depend on the choices of the eigenfunctions.

To calculate the integrals (4.7.1) in the special case $\varphi_n(x) = \sin(n\pi x/L)$ we used our special knowledge of trigonometric functions. An alternate method, which can be adapted to many cases when the $\varphi_n(x)$ are eigenfunctions, employs Green's formula. We will illustrate the method in connection with the eigenvalue problem (4.3.6).

In solving the eigenvalue problem we found that any eigenfunction is a multiple of the function

$$\varphi(x,\lambda) = \frac{\sin x\sqrt{\lambda}}{\sqrt{\lambda}} \tag{4.7.2}$$

which is defined for all values of λ, not just eigenvalues. It is the solution of

$$\begin{aligned} \varphi'' + \lambda\varphi &= 0 , \\ \varphi(0) &= 0 , \\ \varphi'(0) &= 1 . \end{aligned} \tag{4.7.3}$$

From Green's formula we get

$$\begin{aligned} (\mu - \lambda) \int_0^L \varphi(x,\lambda)\varphi(x,\mu)\, dx &= \int_0^L [\varphi''(x,\lambda)\varphi(x,\mu) - \varphi(x,\lambda)\varphi''(x,\mu)]\, dx \\ &= [\varphi'(x,\lambda)\varphi(x,\mu) - \varphi(x,\lambda)\varphi'(x,\mu)]_0^L \end{aligned}$$

which gives, since $\varphi(0,\lambda) = \varphi(0,\mu) = 0$,

$$\int_0^L \varphi(x,\lambda)\varphi(x,\mu)\, dx = \frac{\varphi'(L,\lambda)\varphi(L,\mu) - \varphi(L,\lambda)\varphi'(L,\mu)}{\mu - \lambda} , \tag{4.7.4}$$

valid when $\mu \neq \lambda$. We emphasize that here μ and λ are any two values for which $\mu \neq \lambda$, and are not necessarily eigenvalues. From Equation (4.7.2) it follows that

$$\lim_{\mu \to \lambda} \varphi(x,\mu) = \varphi(x,\lambda) ,$$

and the convergence is uniform for $0 \leq x \leq L$, so that

$$\lim_{\mu \to \lambda} \int_0^L \varphi(x,\lambda)\varphi(x,\mu)\, dx = \int_0^L [\varphi(x,\lambda)]^2\, dx . \tag{4.7.5}$$

Now combining (4.7.4) and (4.7.5) we obtain

$$\int_0^L [\varphi(x,\lambda)]^2\, dx = \lim_{\mu \to \lambda} \frac{\varphi'(L,\lambda)\varphi(L,\mu) - \varphi(L,\lambda)\varphi'(L,\mu)}{\mu - \lambda} .$$

Using l'Hospital's rule to evaluate this limit we have

$$\int_0^L [\varphi(x,\lambda)]^2\, dx = \varphi'(L,\lambda)\, \frac{\partial \varphi(L,\lambda)}{\partial \lambda} - \varphi(L,\lambda)\, \frac{\partial \varphi'(L,\lambda)}{\partial \lambda} , \tag{4.7.6}$$

which holds for any value λ. Now if λ is an eigenvalue of problem (4.3.6), $\lambda = \lambda_n$, then $\varphi(x,\lambda)$ also satisfies the second boundary condition, that is, $\varphi(L,\lambda_n) = 0$, since $\varphi(x,\lambda_n)$ is an eigenfunction belonging to λ_n. Then from (4.7.6) we obtain

$$\int_0^L [\varphi(x,\lambda_n)]^2\, dx = \varphi'(L,\lambda_n) \left[\frac{\partial \varphi(L,\lambda)}{\partial \lambda}\right]_{\lambda=\lambda_n} . \tag{4.7.7}$$

From (4.7.2) we get

$$\varphi'(L,\lambda_n) = \cos L\sqrt{\lambda_n} ,$$

$$\frac{\partial}{\partial \lambda}\varphi(L,\lambda) = \frac{\partial}{\partial \lambda}\frac{\sin L\sqrt{\lambda}}{\sqrt{\lambda}}$$

$$= -\frac{1}{2}\lambda^{-3/2}\sin L\sqrt{\lambda} + \frac{L}{2\lambda}\cos L\sqrt{\lambda}$$

and since $\sin L\sqrt{\lambda_n} = 0$,

$$\left[\frac{\partial}{\partial \lambda}\varphi(L,\lambda)\right]_{\lambda=\lambda_n} = \frac{L}{2\lambda_n}\cos L\sqrt{\lambda_n} .$$

Hence (4.7.7) becomes

$$\int_0^L \left[\frac{\sin x\sqrt{\lambda_n}}{\sqrt{\lambda_n}}\right]^2 dx = \frac{L}{2\lambda_n}\cos^2 L\sqrt{\lambda_n} , \tag{4.7.8}$$

which agrees with the result of our more elementary calculation because $\cos^2 L\sqrt{\lambda_n} = 1$.

We briefly indicate another way in which formula (4.7.6) can be obtained. The function (4.7.2) satisfies

$$\varphi'' + \lambda\varphi = 0\,, \tag{4.7.9}$$

and differentiating this equation with respect to λ,

$$\left(\frac{\partial\varphi}{\partial\lambda}\right)'' + \lambda\frac{\partial\varphi}{\partial\lambda} + \varphi = 0\,. \tag{4.7.10}$$

Application of Green's formula to the two functions φ, $\partial\varphi/\partial\lambda$ gives

$$\int_0^L \left[\varphi''\frac{\partial\varphi}{\partial\lambda} - \varphi\left(\frac{\partial\varphi}{\partial\lambda}\right)''\right]dx = \left[\varphi'\frac{\partial\varphi}{\partial\lambda} - \varphi\left(\frac{\partial\varphi}{\partial\lambda}\right)'\right]_0^L. \tag{4.7.11}$$

Substitution of (4.7.9) and (4.7.10) in the left member of (4.7.11) leads to

$$\int_0^L \left[\varphi''\frac{\partial\varphi}{\partial\lambda} - \varphi\left(\frac{\partial\varphi}{\partial\lambda}\right)''\right]dx = \int_0^L \varphi^2\,dx\,.$$

In the right member of (4.7.11) the terms at $x = 0$ vanish because $\varphi(0,\lambda) = 0$ and

$$\left(\frac{\partial\varphi}{\partial\lambda}\right)_{x=0} = \frac{\partial}{\partial\lambda}\,\varphi(0,\lambda) = 0\,.$$

Hence (4.7.11) is equivalent to

$$\int_0^L \varphi^2\,dx = \left[\varphi'\frac{\partial\varphi}{\partial\lambda} - \varphi\left(\frac{\partial\varphi}{\partial\lambda}\right)'\right]_{x=L}$$

which is the same as (4.7.6).

Observe that the function $\varphi(x,\lambda)$ involved in formula (4.7.6) is defined for all values of λ. It satisfies the boundary condition $\varphi(0) = 0$, but it does not satisfy the boundary condition $\varphi(L) = 0$ except if λ is an eigenvalue of problem (4.3.6). Consequently formula (4.7.6) can be applied to other eigenvalue problems in which the boundary condition at $x = 0$ is $\varphi(0) = 0$ but the boundary condition at $x = L$ may be different from $\varphi(L) = 0$. On the other hand in passing from (4.7.6) to (4.7.7) we use the boundary condition $\varphi(L) = 0$ and (4.7.7) is valid only if λ_n is an eigenvalue of the special problem (4.3.6).

EXERCISES 4c

1. Find by three methods a Green's formula for the operator $B = d^4/dx^4$. Use each of the following hints:

(i) Apply integration by parts four times to the integral

$$\int_a^b f''''g\,dx\,.$$

(ii) Notice that $B = A^2$ where $A = -(d^2/dx^2)$.

(iii) Show that $f''''g - fg''''$ is a derivative [analogous to formula (4.6.1)].

2. Consider the eigenvalue problem in which the differential equation is

$$\varphi'''' - \lambda\varphi = 0\,, \qquad 0 < x < L\,,$$

and the boundary conditions are *one* of the following sets:

$$\begin{aligned}
&\text{(a)}\ \varphi(0) = \varphi''(0) = \varphi(L) = \varphi''(L) = 0\,;\\
&\text{(b)}\ \varphi(0) = \varphi'(0) = \varphi(L) = \varphi'(L) = 0\,;\\
&\text{(c)}\ \varphi(0) = \varphi'''(0) = \varphi(L) = \varphi'''(L) = 0\,.
\end{aligned}$$

(i) In which of these cases can you use the Green's formula of Problem 1 to show that the eigenvalues are real?

(ii) In which of the three cases can you show, using the Green's formula, that eigenfunctions belonging to different eigenvalues are orthogonal?

3. Let $\theta(x,\lambda)$ be the solution of

$$\begin{aligned}
\theta'' + \lambda\theta &= 0\,,\\
\theta(0) &= 1\,,\\
\theta'(0) &= 0\,.
\end{aligned}$$

Use Green's formula to evaluate

$$\int_0^L \theta^2(x,\lambda)\, dx$$

and by specializing the result evaluate

$$\int_0^L \cos^2 \frac{n\pi x}{L}\, dx\,, \quad \int_0^L \cos^2 \left(n + \frac{1}{2}\right)\frac{\pi x}{L}\, dx$$

where n is an integer.

4. For each of the following differential operators L show that there is a linear differential operator M such that an identity like (4.6.1) of the form

$$(Lf)g - f(Mg) = \frac{d}{dx}\{\cdots\}$$

is valid:

(i) $L = a(x)\, D$

(ii) $L = a(x)\, D^2$

(iii) $L = a(x)\, D^3$

(iv) $L = a(x)\, D^4$.

The identity is called a *Lagrange identity* and the differential operator M is called the Lagrange *adjoint* of L. From the identity, a Green's formula can be obtained by integration. Find the Lagrange identity and the adjoint operator M when

(v) $L = a_0(x)\, D^4 + a_1(x)\, D^3 + a_2(x)\, D^2 + a_3(x)\, D + a_4(x)$.

4.8. *GREEN'S FIRST FORMULA; POSITIVE AND NEGATIVE EIGENVALUES*

The solutions of the heat flow problems we have encountered so far have been of the form

$$u(x,t) = \sum c_n e^{-\lambda_n kt} \varphi_n(x)$$

and the eigenvalues λ_n have always been nonnegative. Eventually we will encounter problems in which one or more of the eigenvalues may be negative.

If λ_n is negative, then $e^{-\lambda_n kt} \to \infty$ as $t \to \infty$, while if $\lambda_n \geq 0$, then $e^{-\lambda_n kt}$ remains bounded as $t \to \infty$. Thus negative eigenvalues are associated with temperature functions which become indefinitely large after a long time, and we therefore expect negative eigenvalues only in connection with systems that have a source of energy.

In the case of heat flow in a rod, the temperature may increase without limit either (a) because there is internal production of heat in the rod or (b) because one of the ends of the rod is connected to a mechanism which enables the rod to extract heat from its surroundings indefinitely. In an eigenvalue problem, therefore, the presence of negative eigenvalues is to be associated with either (a) a special source term in the differential equation, or (b) the special nature of one or more of the boundary conditions.

If the temperature $u(x,t)$, $0 \leq x \leq L$, $t > 0$, in a rod satisfies the boundary condition

$$(\kappa u_x - hu)\big|_{x=0} = 0 \tag{4.8.1}$$

with *positive* h, then heat flows *out* of the rod at $x = 0$ at a rate proportional to the temperature of the rod. In this case if the rod is hot, it tends to cool by giving off heat to its surroundings. However, if (4.8.1) is satisfied with *negative* h, then heat flows *into* the rod at $x = 0$ at a rate proportional to the temperature of the rod. In this case if the rod is hot, it tends to get hotter by extracting more heat from its surroundings. Such a heat flow problem seems rather unrealistic. Nevertheless, boundary conditions like (4.8.1) with negative h do occur, in particular as the result of transformation of other problems, and in Section 9 we will make a detailed study of an eigenvalue problem with such a boundary condition.

The above discussion suggests the following conclusion. By briefly examining the differential equation and boundary conditions in an eigenvalue problem it may be possible to ascertain that there are no negative eigenvalues. Such information is of great value when one tries to compute the eigenvalues. Our speculation is supported by the following proposition.

Let $\varphi(x)$ be a nontrivial solution of

$$\varphi'' + \lambda\varphi = 0\,, \qquad 0 < x < L\,.$$

We have

(a) *if* $-\bar{\varphi}\varphi'|_0^L$ *is real then* λ *is real,*

(b) *if* $-\bar{\varphi}\varphi'|_0^L \geq 0$ *then* $\lambda \geq 0$,

(c) *if* $-\bar{\varphi}\varphi'|_0^L > 0$ *then* $\lambda > 0$.

Before proving the proposition let us apply it to the eigenvalue problem (4.6.12). If λ is an eigenvalue and φ a corresponding eigenfunction, then φ is a nontrivial solution of $\varphi'' + \lambda\varphi = 0$ on $0 < x < L$, and from $\varphi'(0) = 0$, $\varphi'(L) = -h\varphi(L)$ we find

$$\begin{aligned} -\bar{\varphi}\varphi'|_0^L &= \bar{\varphi}(0)\varphi'(0) - \bar{\varphi}(L)\varphi'(L) \\ &= h|\varphi(L)|^2\,. \end{aligned} \tag{4.8.2}$$

Since the constant h in (4.8.2) is assumed to be real, $-\bar{\varphi}\varphi'|_0^L$ is real and hence, by part (a) of the proposition, λ is real. Thus the eigenvalues of (4.6.12) are all real. (This much we could already prove by the method of Section 6.) If, in addition, it is known that $h \geq 0$ then from (4.8.2) we conclude that $-\bar{\varphi}\varphi'|_0^L \geq 0$, and so by part (b) of the proposition, $\lambda \geq 0$. *Hence if* $h \geq 0$, *then the eigenvalues of* (4.6.12) *are all real and nonnegative.* For an application of part (c) of the proposition, see Exercises 4d, Problem 1.

In proving the above proposition we make use of a formula called *Green's first formula.* The Green's formula employed in the preceding sections, is an immediate consequence of the first formula and is accordingly often called Green's *second* formula. However, since the second formula has much wider application than the first, an unqualified reference to Green's formula in this or other texts usually is a reference to the second formula.

We establish Green's first formula by transforming the integral

$$-\int_0^L (Af)g\,dx = \int_0^L f''g\,dx$$

by integration by parts. We get

$$\begin{aligned} \int_0^L f''g\,dx &= \int_0^L \left[\frac{d}{dx}(f'g) - f'g'\right] dx \\ &= \int_0^L \frac{d}{dx}(f'g)\,dx - \int_0^L f'g'\,dx\,. \end{aligned}$$

Thus, using the definition (4.6.3),

$$\int_0^L (Af)g\,dx = \int_0^L f'g'\,dx - [f'g]_0^L\,, \tag{4.8.3}$$

and this is **Green's first formula.** Similarly we have

$$\int_0^L f(Ag)\,dx = \int_0^L f'g'\,dx - [fg']_0^L\,, \tag{4.8.4}$$

and Green's second formula is obtained when (4.8.4) is subtracted from (4.8.3).

To prove the proposition stated earlier, we use (4.8.3) with $f = \varphi$, $g = \bar{\varphi}$. Thus, utilizing the relation $A\varphi = \lambda\varphi$, we obtain

$$\lambda \int_0^L \varphi\bar{\varphi}\,dx = \int_0^L \varphi'\bar{\varphi}'\,dx - [\varphi'\bar{\varphi}]_0^L\,,$$

or

$$\lambda \int_0^L |\varphi|^2\,dx = -\bar{\varphi}\varphi'|_0^L + \int_0^L |\varphi'|^2\,dx\,. \tag{4.8.5}$$

Since φ is nontrivial by hypothesis,

$$\int_0^L |\varphi|^2\,dx > 0\,,$$

and since $\int_0^L |\varphi'|^2\,dx \geq 0$ we see from (4.8.5) that λ is real if $-\bar{\varphi}\varphi'|_0^L$ is real, $\lambda \geq 0$ if $-\bar{\varphi}\varphi'|_0^L \geq 0$, and $\lambda > 0$ if $-\bar{\varphi}\varphi'|_0^L > 0$.

EXERCISES 4d

1. In the eigenvalue problem (4.6.12) show that
  (i) if φ is an eigenfunction then $\varphi(L) \neq 0$;
  (ii) if $h > 0$ and λ is an eigenvalue then $\lambda > 0$.

2. Find a Green's first formula for the operator $B = d^4/dx^4$ and use the formula to show that the eigenvalues of the problem

$$B\varphi = \lambda\varphi\,, \quad 0 < x < L\,,$$
$$\varphi(0) = \varphi''(0) = \varphi(L) = \varphi''(L) = 0$$

are all real and nonnegative.

3. Use Green's formula to show that for the problem

$$\varphi'' + q(x)\varphi + \lambda\varphi = 0\,, \qquad 0 < x < L\,,$$
$$\varphi'(0) - \varphi(0) = 0$$
$$\varphi'(L) = 0$$

all eigenvalues are real, and eigenfunctions belonging to different eigenvalues are orthogonal. Assume $q(x)$ is a real-valued function.

4. Use the method of Section 8 to show that if $q(x) \leq 0$ for $0 \leq x \leq L$ then the eigenvalues in Problem 3 are all nonnegative. What is the physical significance of the condition $q(x) \leq 0$?

5. Show that if the function $q(x)$ in Problem 3 satisfies $q(x) \leq M$ where M is a real number then any eigenvalue λ satisfies

$$\lambda \geq -M\,.$$

6. Let $\varphi(x,\lambda)$ be the solution of

$$\begin{aligned} \varphi'' + q(x)\varphi + \lambda\varphi &= 0\,, \qquad 0 < x < L\,, \\ \varphi(0,\lambda) &= 1\,, \\ \varphi'(0,\lambda) &= 1\,. \end{aligned}$$

Evaluate $\int_0^L [\varphi(x,\lambda)]^2\,dx$ by the method of Section 7 and discuss the relation between this problem and Problem 3.

7. Show that for the eigenvalue problem

$$\begin{aligned} \varphi'''' - \varphi'' - \lambda\varphi &= 0\,, \qquad 0 < x < L\,, \\ \varphi(0) = \varphi'(0) = \varphi(L) &= \varphi'(L) = 0 \end{aligned}$$

the eigenvalues are real and nonnegative, and eigenfunctions belonging to different eigenvalues are orthogonal on $[0,L]$.

4.9. *PROBLEMS WITH BOUNDARY CONDITIONS OF THE THIRD KIND*

We consider now the eigenvalue problem

$$\begin{aligned} \varphi'' + \lambda\varphi &= 0\,, \qquad 0 < x < L\,, \\ \varphi'(0) &= 0\,, \\ \varphi'(L) + h\varphi(L) &= 0\,, \end{aligned} \tag{4.9.1}$$

where h is a real constant. Using the methods of either Section 6 or Section 8 we can easily show that the eigenvalues are all real. Furthermore it was shown in Section 8 that there are no negative eigenvalues if $h \geq 0$. In the special case $h = 0$, the problem is the same as (4.5.6), which was solved in Section 5. We have still to solve the problem when $h > 0$ or $h < 0$.

The general solution of the differential equation is

$$\varphi = A\cos x\sqrt{\lambda} + B\,\frac{\sin x\sqrt{\lambda}}{\sqrt{\lambda}}\,.$$

If this is to satisfy $\varphi'(0) = 0$ we must have $B = 0$, and so any eigenfunction is of the form

$$\varphi = A\cos x\sqrt{\lambda}\,. \tag{4.9.2}$$

With $A \neq 0$ this can satisfy the second boundary condition if and only if

$$-\sqrt{\lambda}\sin L\sqrt{\lambda} + h\cos L\sqrt{\lambda} = 0\,, \tag{4.9.3}$$

and so (4.9.3) is an equation whose roots are the eigenvalues. If $\cos L\sqrt{\lambda} = 0$ then λ is not a root, because $\sin L\sqrt{\lambda} = \pm 1$, and $\sqrt{\lambda} \neq 0$, so that clearly (4.9.3) is not satisfied. Hence we can divide (4.9.3) by $\cos L\sqrt{\lambda}$ and obtain another equation with the same roots, namely

$$L\sqrt{\lambda}\tan L\sqrt{\lambda} = Lh\,. \tag{4.9.4}$$

Assuming h real we seek the roots of (4.9.4) and we already know these roots are real.

Case (i): $h > 0$. In this case we know the roots of (4.9.4) are all nonnegative. Setting $L\sqrt{\lambda} = s$ or $\lambda = (s/L)^2$ we have to find the real roots of

$$s\tan s = Lh\,. \tag{4.9.5}$$

Since $\tan(-s) = -\tan s$ we see that if s is a root of (4.9.5) then $-s$ is also a root. But these two values of s give the same eigenvalue $\lambda = (s/L)^2 = (-s/L)^2$. Hence it is only necessary to find the nonnegative roots of (4.9.5). The nonnegative roots of (4.9.5) may be found graphically by plotting on a single diagram the graphs of the functions

$$y_1 = \tan s \qquad\qquad 0 \leq s < \infty\,.$$
$$y_2 = \frac{Lh}{s}$$

See Figure 4.1, illustrating the case $Lh = \pi/2$. The roots of (4.9.5) are the abscissas of the intersection points of the two graphs. In

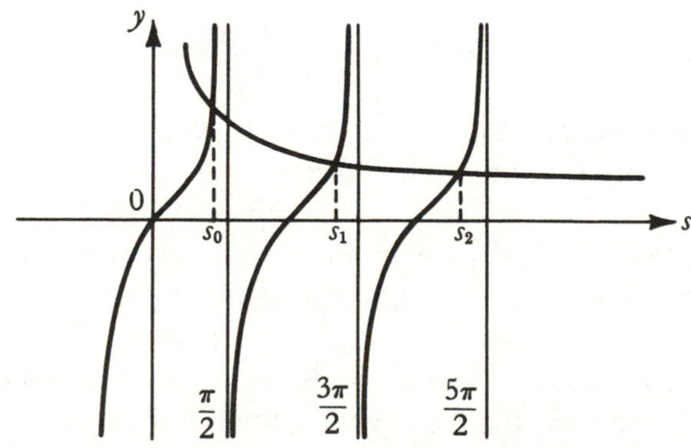

FIG. 4.1. Graphic solution of $s\tan s = Lh$ $(Lh = \pi/2)$.

each interval $n\pi < s < (n + \frac{1}{2})\pi$ the function y_1 increases steadily from 0 to $+\infty$ while the function y_2 is positive and steadily decreasing. Hence there is exactly one intersection point in the interval, corresponding to a value

$$s_n = n\pi + \alpha_n, \qquad n = 0, 1, 2, \ldots \tag{4.9.6}$$

where $0 < \alpha_n < \pi/2$. There are no intersection points in the intervals $(n + \frac{1}{2})\pi \leq s \leq (n + 1)\pi$. Thus there are infinitely many eigenvalues given by

$$\sqrt{\lambda_n} = \frac{s_n}{L} = \frac{n\pi}{L} + \sigma_n, \qquad n = 0, 1, 2, \ldots \tag{4.9.7}$$

where $\sigma_n = \alpha_n/L$. Inspection of Figure 4.1 shows that

$$\frac{\pi}{2L} > \sigma_0 > \sigma_1 \cdots > \sigma_n > \sigma_{n+1} > \cdots \tag{4.9.8}$$

and

$$\lim_{n\to\infty} \sigma_n = 0. \tag{4.9.9}$$

For the eigenfunctions we have the formula

$$\varphi_n(x) = \cos x\sqrt{\lambda_n} = \cos\left(\frac{n\pi}{L} + \sigma_n\right)x. \tag{4.9.10}$$

The normalizing constants

$$\int_0^L [\varphi_n(x)]^2\,dx = \int_0^L \cos^2 x\sqrt{\lambda_n}\,dx$$

can be evaluated either by using the double-angle formulas of trigonometry, or by the method of Section 7. One finds

$$\int_0^L \cos^2 x\sqrt{\lambda_n}\,dx = \frac{L}{2} + \frac{h}{2\lambda_n}[\varphi_n(L)]^2 \tag{4.9.11}$$

or

$$\int_0^L [\varphi_n(x)]^2\,dx = \left\{\frac{L}{2} + \frac{h(1 + Lh)}{2\lambda_n}\right\}[\varphi_n(L)]^2 \tag{4.9.12}$$

and these are equivalent because λ_n is a solution of (4.9.4). It is important to notice that these normalizing constants really vary with n.

Case (ii): $h < 0$. In this case we know that the roots of (4.9.4) are all real, but we do not yet know whether there are any negative eigenvalues. We will see shortly that for any $h < 0$ there is exactly one negative eigenvalue. (In Exercises 4e there are similar problems in which the number of negative eigenvalues depends on h.) To check for negative roots of (4.9.4) we make the substitution $L\sqrt{\lambda} = iu$ where u is real and not zero,

$$L\sqrt{\lambda}\tan L\sqrt{\lambda} = iu\frac{\sin iu}{\cos iu} = -u\tanh u\,.$$

Thus we have to find real roots of

$$u\tanh u = -Lh\,, \tag{4.9.13}$$

and, as with (4.9.5), it suffices to find the positive roots. This is done graphically (Fig. 4.2) by finding the intersection points of the two curves

$$\begin{aligned} y_1 &= \tanh u \\ y_2 &= -Lh/u \end{aligned} \qquad 0 < u < \infty\,.$$

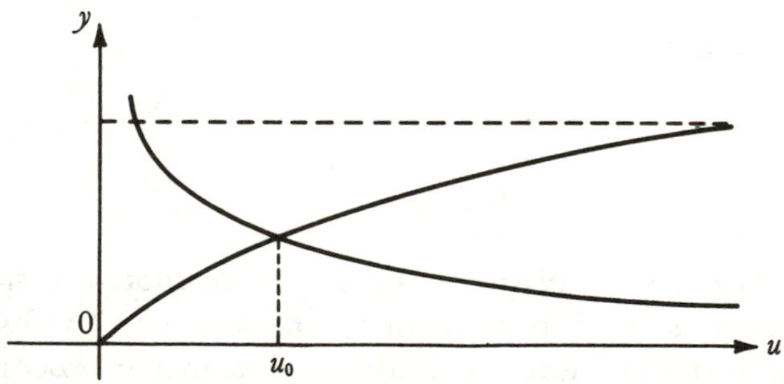

FIG. 4.2. Graphic solution of $u \tanh u = -Lh \quad (Lh = -\pi/2)$.

As u runs from 0 to $+\infty$ the curve y_1 increases steadily from 0 to 1 while the hyperbola y_2 decreases steadily from $+\infty$ to 0 ($h < 0$). Hence there is exactly one intersection, at say $u = u_0$, corresponding to a negative eigenvalue,

$$\lambda_0 = -u_0^2/L^2\,. \tag{4.9.14}$$

The eigenfunction corresponding to the eigenvalue λ_0 is

$$\varphi_0(x) = \cos x\sqrt{\lambda_0} = \cosh(u_0 x/L)\,. \tag{4.9.15}$$

To find the positive eigenvalues we set $L\sqrt{\lambda} = s$ which leads again to equation (4.9.5), but with $h < 0$. The graphic solution is illustrated in Figure 4.3. There is one positive root of (4.9.5) in each interval $(n - \frac{1}{2})\pi < s < n\pi$, $n = 1, 2, 3, \ldots$. Calling this root s_n and the corresponding eigenvalue $\lambda_n = s_n^2/L_n^2$, we have

$$\sqrt{\lambda_n} = \frac{n\pi}{L} + \sigma_n\,, \qquad n = 1, 2, \ldots \tag{4.9.16}$$

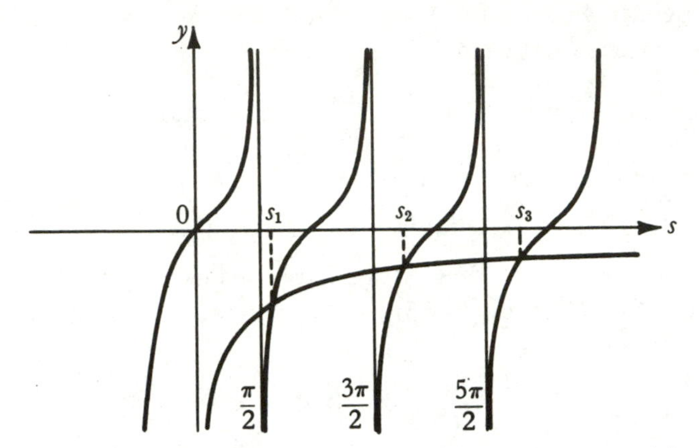

FIG. 4.3. Graphic solution of $s \tan s = Lh$ $(Lh = -\pi/2)$.

where $\sigma_n < 0$ and

$$-\frac{\pi}{2L} < \sigma_1 < \sigma_2 < \cdots < \sigma_n < \sigma_{n+1} < \cdots, \tag{4.9.17}$$

$$\lim_{n\to\infty} \sigma_n = 0. \tag{4.9.18}$$

When one has found the equation whose roots are the eigenvalues of a problem, there is often a temptation to "simplify" the equation immediately. Caution must be used in the process of simplification. For example, the equation

$$\sin^2 \sqrt{\lambda} + \sin \sqrt{\lambda} \cos \sqrt{\lambda} = 0$$

might be simplified by dividing by $\sin \sqrt{\lambda}$ to obtain the new equation

$$\sin \sqrt{\lambda} + \cos \sqrt{\lambda} = 0.$$

But the new equation does not have the same roots as the original equation; in fact, every root of $\sin \sqrt{\lambda} = 0$ is a root of the original equation but is not a root of the new equation. In general, if we divide an equation by a factor $F(\lambda)$ we may lose roots, namely the roots of $F(\lambda)$. Similarly if we multiply an equation by a factor $F(\lambda)$ we may gain roots, namely the roots of $F(\lambda)$. For example, if the equation

$$\frac{\sin \sqrt{\lambda}}{\sqrt{\lambda}} + \cos \sqrt{\lambda} = 0$$

is simplified by multiplying by $\sqrt{\lambda}$ to obtain the new equation

$$\sin \sqrt{\lambda} + \sqrt{\lambda} \cos \sqrt{\lambda} = 0,$$

then $\lambda = 0$ is a root of the new equation but not of the original equation.

EXERCISES 4e

1. For the eigenvalue problem

$$\varphi'' + \lambda\varphi = 0, \qquad 0 < x < L,$$
$$\varphi(0) = 0,$$
$$\varphi'(L) + h\varphi(L) = 0,$$

where h is real, do the following:

(a) Show that the eigenvalues are all nonnegative if $h \geq 0$.

(b) Suppose $\lambda = 0$ is an eigenvalue. By solving $\varphi'' = 0$, calculate the value of h.

(c) It is known that the lowest eigenvalue λ_0 is a continuous function of h for $-\infty < h < \infty$. Use this fact together with the results of (a) and (b) to show that the eigenvalues are all nonnegative if $h \geq -1/L$.

2. For the eigenvalue problem in Problem 1:

(a) Find an equation whose roots are the eigenvalues.

(b) By graphic methods show that there are infinitely many eigenvalues.

(c) State the approximate value of λ_n for large n, when $h > 0$.

3. In the eigenvalue problem

$$\varphi'' + \lambda\varphi = 0, \qquad 0 < x < L,$$
$$\varphi'(0) - h\varphi(0) = 0,$$
$$\varphi'(L) + h\varphi(L) = 0,$$

h is a real constant.

(a) Show that the eigenvalues are real and if $h \geq 0$ they are nonnegative (method of Section 8).

(b) For what values of h is $\lambda = 0$ an eigenvalue of this problem?

(c) Find an equation whose roots are the eigenvalues of the problem, and find a formula for the eigenfunctions.

(d) Assuming that $h > 0$, use the trigonometric identity $\cot 2\theta = (\cot^2\theta - 1)/2\cot\theta$ to show that the eigenvalues are the roots of

$$\left(\cot\frac{L\sqrt{\lambda}}{2} - \frac{\sqrt{\lambda}}{h}\right)\left(\cot\frac{L\sqrt{\lambda}}{2} + \frac{h}{\sqrt{\lambda}}\right) = 0.$$

With the aid of a sketch discuss the solution of this equation, showing there are infinitely many eigenvalues

$$\lambda_0 < \lambda_1 < \cdots < \lambda_n < \cdots.$$

(e) When n is large, what is the approximate value of λ_n?

(f) How many negative eigenvalues are there for various negative values of h?

4. Find Green's formulas (first and second) for the operator $Bf = d^4f/dx^4$ by applying integration by parts twice to the integral

$$\int_a^b \frac{d^4f}{dx^4} g \, dx .$$

5. Use the result of Problem 4 to show that the eigenvalues of

$$\frac{d^4\varphi}{dx^4} - \lambda\varphi = 0 , \qquad 0 < x < L ,$$

$$\varphi'''(0) + h_1\varphi(0) = 0 \quad \varphi'''(L) - h_3\varphi(L) = 0$$
$$\varphi''(0) - h_2\varphi'(0) = 0 \quad \varphi''(L) + h_4\varphi'(L) = 0$$

are (a) all real if h_1, h_2, h_3, h_4 are real, and (b) all nonnegative if h_1, h_2, h_3, h_4 are nonnegative.

6. Use the result of Problem 4 to show that for Problem 5, if the constants h_1, h_2, h_3, h_4 are real, then eigenfunctions belonging to different eigenvalues are orthogonal on the interval $[0,L]$.

7. Let $\varphi_n(x)$ be the eigenfunctions of problem (4.9.1) with $h > 0$. Find the coefficients c_n in the expression

$$f(x) = \sum c_n\varphi_n(x) , \qquad 0 < x < L ,$$

for the special case $f(x) = 1$.

4.10. *MIXED BOUNDARY CONDITIONS; MULTIPLE EIGENVALUES*

We consider a one-dimensional heat flow in a thin insulated wire bent into the shape of a circle. If the wire is of length $2L$ and x is distance measured along the wire, then $x = -L$ and $x = L$ correspond to the same point on the wire. If we assume perfect thermal contact between the faces at $x = -L$ and $x = L$, and use the jump conditions, we are led to the following problem for the temperature function $u(x,t)$ in the wire:

$$\text{D.E.} \quad \frac{\partial u}{\partial t} = k\frac{\partial^2 u}{\partial x^2}, \qquad -L < x < L , \quad t > 0 ,$$

$$(4.10.1) \quad \text{B.C.} \quad \left.\begin{aligned} u(-L,t) &= u(L,t) \\ \kappa u_x(-L,t) &= \kappa u_x(L,t) \end{aligned}\right\} \qquad t > 0 ,$$

$$\text{I.C.} \quad u(x,0) = f(x) , \qquad -L < x < L .$$

This is one of the first problems treated by Fourier in his study of heat conduction.

On separation of the variables (4.10.1) leads to the eigenvalue problem

$$\begin{aligned} \varphi'' + \lambda\varphi &= 0\,, \qquad -L < x < L\,, \\ \varphi(-L) - \varphi(L) &= 0\,, \\ \varphi'(-L) - \varphi'(L) &= 0\,. \end{aligned} \tag{4.10.2}$$

The general solution of the differential equation of (4.10.2) is

$$\varphi = A\cos x\sqrt{\lambda} + B\frac{\sin x\sqrt{\lambda}}{\sqrt{\lambda}}\,, \tag{4.10.3}$$

and λ is an eigenvalue if and only if the constants A and B in (4.10.3) can be chosen (not both zero) so that φ satisfies both boundary conditions. We substitute (4.10.3) into the two boundary conditions, obtaining the two equations

$$\begin{aligned} &\left[A\cos(-L\sqrt{\lambda}) + B\frac{\sin(-L\sqrt{\lambda})}{\sqrt{\lambda}}\right] \\ &\qquad\qquad - \left[A\cos L\sqrt{\lambda} + B\frac{\sin L\sqrt{\lambda}}{\sqrt{\lambda}}\right] = 0 \\ &[-A\sqrt{\lambda}\sin(-L\sqrt{\lambda}) + B\cos(-L\sqrt{\lambda})] \\ &\qquad\qquad - [-A\sqrt{\lambda}\sin L\sqrt{\lambda} + B\cos L\sqrt{\lambda}] = 0\,. \end{aligned} \tag{4.10.4}$$

In these equations A and B are the unknowns, and we know that λ is an eigenvalue of (4.10.2) if and only if there is a pair of numbers A, B not both zero which satisfy the two equations. But this is a system of two homogeneous linear equations in the two unknowns A and B, and so there is a nontrivial solution if and only if the determinant of the system is zero. The two equations may be rewritten in the form

$$\begin{aligned} 0\cdot A - 2\frac{\sin L\sqrt{\lambda}}{\sqrt{\lambda}}\cdot B &= 0 \\ 2\lambda\frac{\sin L\sqrt{\lambda}}{\sqrt{\lambda}}\cdot A + 0\cdot B &= 0\,. \end{aligned} \tag{4.10.5}$$

Equating the determinant to zero we have an equation whose roots are the eigenvalues, namely,

$$\begin{vmatrix} 0 & -2\dfrac{\sin L\sqrt{\lambda}}{\sqrt{\lambda}} \\ 2\lambda\dfrac{\sin L\sqrt{\lambda}}{\sqrt{\lambda}} & 0 \end{vmatrix} = 0$$

or

$$4\lambda\left(\frac{\sin L\sqrt{\lambda}}{\sqrt{\lambda}}\right)^2 = 0 . \tag{4.10.6}$$

The roots of this equation are

$$\lambda_n = \left(\frac{n\pi}{L}\right)^2, \qquad n = 0, 1, 2, \ldots . \tag{4.10.7}$$

To find the eigenfunctions belonging to the eigenvalue λ_n we set $\lambda = \lambda_n$ in the system (4.10.5) and then find the nontrivial solutions A, B of the system. With these values of A and B and $\lambda = \lambda_n$, (4.10.3) is then an eigenfunction.

If $\lambda = \lambda_0 = 0$, then (4.10.5) has the solution $B = 0$, A arbitrary, and so the eigenfunctions are of the form $A \cos x\sqrt{\lambda_0} = A$. Hence there is just one linearly independent eigenfunction, which we take as

$$\varphi_0(x) = 1 . \tag{4.10.8}$$

If $\lambda = \lambda_n = (n\pi/L)^2$, $n \geq 1$, then $(\sin L\sqrt{\lambda})/\sqrt{\lambda} = 0$ and (4.10.5) is satisfied with *both* A and B arbitrary. Hence there are two linearly independent eigenfunctions, which we choose in the form

$$\begin{aligned} \varphi_{n,1}(x) &= \cos\frac{n\pi x}{L} , \\ \varphi_{n,2}(x) &= \sin\frac{n\pi x}{L} . \end{aligned} \tag{4.10.9}$$

Any linear combination $A\varphi_{n,1}(x) + B\varphi_{n,2}(x)$ is, if A, B are not both zero, also an eigenfunction belonging to the same eigenvalue.

If $\varphi_m(x)$ and $\varphi_n(x)$ are two of the eigenfunctions belonging to eigenvalues λ_m, λ_n, respectively, then

$$(\lambda_m - \lambda_n)\int_{-L}^{L} \varphi_m(x)\varphi_n(x)\,dx = -[\varphi_m'\varphi_n - \varphi_m\varphi_n']_{-L}^{L} = 0$$

because, by virtue of the boundary conditions, the terms at $x = L$ cancel against those at $x = -L$. If $\lambda_m \neq \lambda_n$ it follows that

$$\int_{-L}^{L} \varphi_m(x)\varphi_n(x)\,dx = 0 ;$$

that is, eigenfunctions belonging to *different* eigenvalues are orthogonal. However, in the above eigenvalue problem, for each eigenvalue λ_n, $n \geq 1$, we found two linearly independent eigenfunctions $\varphi_{n,1}$, $\varphi_{n,2}$ and our argument does not show that these are orthogonal to one another. We easily verify that they are orthogonal since

(4.10.10)
$$\begin{aligned}\int_{-L}^{L} \varphi_{n,1}(x)\varphi_{n,2}(x)\,dx &= \int_{-L}^{L} \cos\frac{n\pi x}{L}\sin\frac{n\pi x}{L}\,dx \\ &= \frac{1}{2}\int_{-L}^{L} \sin\frac{2n\pi x}{L}\,dx = 0\,.\end{aligned}$$

Thus the system composed of the functions $\varphi_0(x)$ and $\varphi_{n,1}(x)$, $\varphi_{n,2}(x)$ for $n \geq 1$, is an orthogonal system. The normalizing constants are found to be

(4.10.11)
$$\begin{aligned}&\int_{-L}^{L} \varphi_0^2(x)\,dx = 2L\,,\\ &\int_{-L}^{L} \varphi_{n,1}^2(x)\,dx = \int_{-L}^{L} \varphi_{n,2}^2(x)\,dx = L\,, \qquad n \geq 1\,.\end{aligned}$$

We return to the initial-boundary value problem (4.10.1). Its formal solution is

(4.10.12)
$$u(x,t) = \frac{1}{2}a_0 + \sum_{n=1}^{\infty} e^{-\lambda_n kt}\left(a_n \cos\frac{n\pi x}{L} + b_n \sin\frac{n\pi x}{L}\right),$$

where $\lambda_n = (n\pi/L)^2$. The constants a_n, b_n are determined by orthogonality from

(4.10.13)
$$f(x) = \frac{1}{2}a_0 + \sum_{n=1}^{\infty}\left(a_n \cos\frac{n\pi x}{L} + b_n \sin\frac{n\pi x}{L}\right)$$

which gives, for $n = 0, 1, 2, \ldots$,

(4.10.14)
$$\begin{aligned}a_n &= \frac{1}{L}\int_{-L}^{L} f(x)\cos\frac{n\pi x}{L}\,dx\,,\\ b_n &= \frac{1}{L}\int_{-L}^{L} f(x)\sin\frac{n\pi x}{L}\,dx\,.\end{aligned}$$

These last two formulas are frequently written in the more compact form

$$\begin{Bmatrix} a_n \\ b_n \end{Bmatrix} = \frac{1}{L}\int_{-L}^{L} f(x)\begin{Bmatrix} \cos\dfrac{n\pi x}{L} \\ \sin\dfrac{n\pi x}{L}\end{Bmatrix} dx\,.$$

In the eigenvalue problem (4.10.2) there is only one linearly independent eigenfunction belonging to the eigenvalue λ_0 and for this reason λ_0 is called an eigenvalue of *multiplicity one*. An eigenvalue to which there belong k linearly independent eigenfunctions is said to be of **multiplicity** k. In problem (4.10.2) each eigenvalue λ_n, $n \geq 1$ is of *multiplicity two*. In the eigenvalue problems we have solved thus far, the differential equation was a second-order ordinary differential equation with only two linearly independent solutions. The multiplicity of an eigenvalue could therefore not be greater than two. Later we shall meet eigenvalue problems in which the differential equation is a partial differential equation and in which there may be eigenvalues of arbitrarily high multiplicity.

After choosing the eigenfunctions $\varphi_{n,1}$, $\varphi_{n,2}$ belonging to the eigenvalue λ_n, $n \geq 1$, in problem (4.10.2) we could be sure, because of a Green's formula argument, that these two eigenfunctions were orthogonal to all eigenfunctions belonging to eigenvalues λ_m different from λ_n. But the Green's formula method fails to show that $\varphi_{n,1}$ and $\varphi_{n,2}$ are orthogonal to one another, and to verify that this was actually the case we had to resort to explicit formulas for the eigenfunctions. That $\varphi_{n,1}$ and $\varphi_{n,2}$ were orthogonal to one another was fortunate, because to calculate the coefficients a_n, b_n in (4.10.13) we had to use this orthogonality.

If in choosing the eigenfunctions belonging to the eigenvalue λ_n we had chosen some linearly independent pair other than (4.10.9), then this other pair would not necessarily be orthogonal. For example, we might have chosen

$$\begin{aligned} \varphi^*_{n,1}(x) &= \cos\frac{n\pi x}{L} \\ \varphi^*_{n,2}(x) &= 3\cos\frac{n\pi x}{L} + \sin\frac{n\pi x}{L} \end{aligned} \tag{4.10.15}$$

instead of (4.10.9). These are linearly independent eigenfunctions belonging to the eigenvalue λ_n and every eigenfunction belonging to the eigenvalue λ_n is of the form

$$A\varphi^*_{n,1}(x) + B\varphi^*_{n,2}(x) .$$

However $\varphi^*_{n,1}$ and $\varphi^*_{n,2}$ are not orthogonal; in fact,

$$\int_{-L}^{L} \varphi^*_{n,1}\varphi^*_{n,2}\,dx = 3L .$$

Nevertheless, if we have found $\varphi^*_{n,1}$ and $\varphi^*_{n,2}$ then we can recombine them linearly to obtain the orthogonal pair $\varphi_{n,1}$ and $\varphi_{n,2}$ as follows:

$$\begin{aligned} \varphi_{n,1} &= \varphi^*_{n,1} \\ \varphi_{n,2} &= \varphi^*_{n,2} - 3\varphi^*_{n,1} . \end{aligned}$$

Indeed we can show by a general argument that: *If φ^*_1 and φ^*_2 are two linearly independent real eigenfunctions belonging to the same eigenvalue λ,* *then there are linear combinations* φ_1 *and* φ_2 *of* φ_1^* *and* φ_2^* *which are linearly independent eigenfunctions belonging to* λ *and which are orthogonal.*

For the proof, we observe first that since the D.E. and B.C. of the eigenvalue problem are linear and homogeneous, any linear combination of solutions of these equations is also a solution and hence, unless it is identically zero, is an eigenfunction. We set $\varphi_1 = \varphi_1^*$, $\varphi_2 = A\varphi_1^* + B\varphi_2^*$, and seek to choose A and B so that φ_1 and φ_2 are orthogonal; that is,

$$\begin{aligned} 0 &= \int_{-L}^{L} \varphi_1\varphi_2\,dx = \int_{-L}^{L} \varphi_1^*(A\varphi_1^* + B\varphi_2^*)\,dx \\ &= A\int_{-L}^{L} (\varphi_1^*)^2\,dx + B\int_{-L}^{L} \varphi_1^*\varphi_2^*\,dx\,. \end{aligned} \tag{4.10.16}$$

Now (4.10.16) is a linear equation of the form $0 = \alpha A + \beta B$ for the unknowns A, B. Since α is not zero, we can choose B and then solve the equation for A. One convenient choice is $B = \alpha$ and then we get $A = -\beta$, that is,

$$A = -\int_{-L}^{L} \varphi_1^*\varphi_2^*\,dx\,, \quad B = \int_{-L}^{L} (\varphi_1^*)^2\,dx\,. \tag{4.10.17}$$

With this choice of A and B, the functions φ_1 and φ_2 are orthogonal eigenfunctions. To prove that they are linearly independent, we will verify that φ_1 and φ_2 are an orthogonal system on $[-L,L]$, and linear independence will follow from Exercises 4b, Problem 12. First since $\varphi_1 = \varphi_1^*$ is real-valued, continuous, and not identically zero, we have

$$\int_{-L}^{L} \varphi_1^2(x)\,dx > 0\,. \tag{4.10.18}$$

Since φ_1^* and φ_2^* are linearly independent and by (4.10.17) the constant B in $\varphi_2 = A\varphi_1^* + B\varphi_2^*$ is not zero, it follows that φ_2 is not identically zero. Also since A and B are real and φ_1^* and φ_2^* are real-valued and continuous, φ_2 is real-valued and continuous. Therefore,

$$\int_{-L}^{L} \varphi_2^2(x)\,dx > 0\,. \tag{4.10.19}$$

We already know that φ_1 and φ_2 are orthogonal, so (4.10.18) and (4.10.19) show that φ_1, φ_2 is an orthogonal system on $[-L,L]$. The proof is now complete.

Using the values of A and B given by (4.10.17) we have the following formulas for φ_1 and φ_2:

$$\varphi_1(x) = \varphi_1^*(x)\,, \quad \varphi_2(x) = \begin{vmatrix} c_{11} & \varphi_1^*(x) \\ c_{21} & \varphi_2^*(x) \end{vmatrix} \tag{4.10.20}$$

where

$$c_{ij} = \int_{-L}^{L} \varphi_i^* (x) \varphi_j^* (x) \, dx \, .$$

The proof, as well as the final formulas (4.10.20), can be generalized to the case of eigenvalues of higher multiplicity (see Exercises 4f).

EXERCISES 4f

1. For the eigenvalue problem

$$\varphi'' + \lambda\varphi = 0 \, , \qquad 0 < x < L$$
$$\varphi'(0) = \varphi'(L) \, , \qquad \varphi'(0) = \frac{\varphi(L) - \varphi(0)}{L} \, ,$$

(i) Find an equation whose roots are the eigenvalues.
(ii) Show that $\lambda = 0$ is an eigenvalue of multiplicity two and find an orthogonal pair of eigenfunctions belonging to this eigenvalue.
(iii) Show that all eigenvalues are real.
(iv) Show that all eigenvalues are nonnegative.

2. For the eigenvalue problem

$$\varphi'' + \lambda\varphi = 0 \, , \qquad -L < x < L \, ,$$
$$\varphi(L) + \varphi(-L) = 0 \, , \quad \varphi'(L) + \varphi'(-L) = 0$$

find all the eigenvalues and find an orthogonal system of eigenfunctions.

3. For the eigenvalue problem

$$\varphi'' + \lambda\varphi = 0 \, , \qquad 0 < x < L$$
$$\varphi(L) + \varphi(0) = 0 \, ,$$
$$\varphi'(L) - \varphi(0) + \varphi'(0) = 0 \, ,$$

(i) Show that the eigenvalues are real and nonnegative.
(ii) Show that eigenfunctions belonging to different eigenvalues are orthogonal.
(iii) Find an equation whose roots are the eigenvalues.
(iv) Determine whether zero is an eigenvalue.
(v) If the eigenvalues are arranged in increasing order

$$\lambda_0 < \lambda_1 < \lambda_2 < \lambda_3 < \cdots \, ,$$

show that λ_{2n} can be found exactly and λ_{2n+1} can be found approximately. (What is the approximate value of λ_{2n+1} when n is large?)
(vi) For each eigenvalue, determine the multiplicity of the eigenvalue and find formulas for the corresponding eigenfunctions.

4. The polynomials $P_0(x) = a$, $P_1(x) = bx + c$, $P_2(x) = dx^2 + ex + f$, (a, b, c, d, e, f constants) are an orthogonal system on $[0,1]$ and $P_2(0) = 1$. Find $P_2(x)$.

5. Let $f_1(x)$, $f_2(x)$, $f_3(x)$ be linearly independent real-valued continuous functions on an interval $[a,b]$, and let

$$c_{ij} = \int_a^b f_i(x) f_j(x)\, dx\,.$$

Show that the functions

$$\varphi_1(x) = f_1(x)\,,$$

$$\varphi_2(x) = \begin{vmatrix} c_{11} & f_1(x) \\ c_{21} & f_2(x) \end{vmatrix},$$

$$\varphi_3(x) = \begin{vmatrix} c_{11} & c_{12} & f_1(x) \\ c_{21} & c_{22} & f_2(x) \\ c_{31} & c_{32} & f_3(x) \end{vmatrix}$$

form an orthogonal system on $[a,b]$.

Two complex-valued functions $f(x)$, $g(x)$ are said to be **orthogonal on** $[a,b]$ **in the complex sense** if

$$\int_a^b f(x)\overline{g(x)}dx = 0\,,$$

where $\overline{g(x)}$ is the conjugate of $g(x)$. A system $\{\varphi_n(x)\}$ of complex-valued functions is called an orthogonal system on $[a,b]$ in the complex sense if

$$\int_a^b \varphi_m(x)\overline{\varphi_n(x)}dx = 0\,, \qquad \text{when } m \neq n\,,$$

$$\int_a^b \varphi_m(x)\overline{\varphi_m(x)}dx \neq 0\,.$$

If the last integral is not zero it is positive, since $\varphi_m(x)\overline{\varphi_m(x)} = |\varphi_m(x)|^2$ is nonnegative. If the functions $\varphi_n(x)$ are real-valued, then $\overline{\varphi_n(x)} = \varphi_n(x)$, and hence a system of real-valued functions, orthogonal in the ordinary sense, is also orthogonal in the complex sense.

6. (i) Show that the functions

$$\varphi_n(x) = e^{i(n\pi x/L)}\,, \qquad n = \ldots, -2, -1, 0, 1, 2, \ldots$$

are complex-valued eigenfunctions of (4.10.2) and form an orthogonal system on $[-L,L]$ in the complex sense.

(ii) Show by Green's formula that if $\psi_1(x)$ and $\psi_2(x)$ are complex-valued eigenfunctions of (4.10.2) belonging to *different* eigenvalues, then they are orthogonal in the complex sense on $[-L,L]$.

7. If $\varphi_n(x)$, $n = 0, 1, 2, \ldots$ is an orthogonal system on $[a,b]$ in the complex sense, find a formula for the coefficients c_n in an expansion

$$f(x) = \sum_{n=0}^{\infty} c_n \varphi_n(x)\,.$$

8. Find the eigenvalues and eigenfunctions of the problem

$$\text{D.E.} \quad -i\varphi' = \lambda\varphi\,, \qquad -\pi < x < \pi\,,$$
$$\text{B.C.} \quad \varphi(-\pi) = \varphi(\pi)\,,$$

and find the eigenfunction expansion of $f(x) = e^{ix/2}$.

4.11. *NONUNIFORM ROD; STURM-LIOUVILLE PROBLEMS*

The homogeneous linear initial-boundary value problem,

$$\text{D.E.} \quad \frac{\partial u}{\partial t} = \frac{1}{\sigma(x)}\frac{\partial}{\partial x}\left[\kappa(x)\frac{\partial u}{\partial x}\right] + q(x)u\,, \quad a < x < b\,,\ t > 0\,,$$

$$(4.11.1) \quad \text{B.C.} \quad \left.\begin{aligned} u(a,t) &= 0 \\ u(b,t) &= 0 \end{aligned}\right\} \qquad t > 0\,,$$

$$\text{I.C.} \quad u(x,0) = f(x)\,, \qquad a < x < b\,,$$

describes the cooling of a nonuniform rod with a source whose strength is proportional to the local temperature. When the variables are separated the eigenvalue problem obtained is

$$(4.11.2) \quad \begin{aligned} &\text{D.E.} \quad \frac{1}{\sigma(x)}\frac{d}{dx}\left[\kappa(x)\frac{d\varphi}{dx}\right] + q(x)\varphi + \lambda\varphi = 0 \quad a < x < b \\ &\text{B.C.} \quad \begin{aligned} \varphi(a) &= 0\,, \\ \varphi(b) &= 0\,. \end{aligned} \end{aligned}$$

Here the differential equation can be written as $S\varphi = \lambda\varphi$ where

$$(4.11.3) \qquad S\varphi = \frac{-1}{\sigma(x)}\frac{d}{dx}\left[\kappa(x)\frac{d\varphi}{dx}\right] - q(x)\varphi\,.$$

Alternative forms of the boundary conditions in the initial-boundary value problem (4.11.1) lead to eigenvalue problems with the same differential equation $S\varphi = \lambda\varphi$, but with more general boundary conditions. Our purpose in this section is to describe in outline the theory of such eigenvalue problems. The general form of the problem we will consider is

$$(4.11.4) \quad \begin{aligned} &\text{D.E.} \quad \frac{1}{\sigma(x)}\frac{d}{dx}\left[\kappa(x)\frac{d\varphi}{dx}\right] + q(x)\varphi + \lambda\varphi = 0\,, \quad a < x < b\,, \\ &\text{B.C.} \quad \begin{aligned} c_{11}\varphi(a) + c_{12}\varphi'(a) + c_{13}\varphi(b) + c_{14}\varphi'(b) &= 0\,, \\ c_{21}\varphi(a) + c_{22}\varphi'(a) + c_{23}\varphi(b) + c_{24}\varphi'(b) &= 0\,. \end{aligned} \end{aligned}$$

Such a problem is called a **Sturm-Liouville problem** after the two mathematicians who first studied such general eigenvalue problems.

In our discussion we assume that the eight constants c_{11}, c_{12}, . . . , c_{24} are real, that the interval $[a,b]$ is finite, and that

(4.11.5) $\sigma(x)$, $\kappa(x)$, $q(x)$ are real-valued continuous functions on the closed interval $[a,b]$ and $\kappa(x)$ has a continuous derivative on $[a,b]$,

(4.11.6) $\sigma(x) > 0$ and $\kappa(x) > 0$ at every point of the closed interval $[a,b]$.

When these assumptions are valid, the Sturm-Liouville problem is called **regular.**

In this section we confine our attention to regular problems. However, we will later encounter **singular** Sturm-Liouville problems in which either the interval (a,b) is infinite or conditions (4.11.5), (4.11.6) fail to hold. For example, the study of radial heat flow in cylinders leads to the operator

$$S\varphi = -\frac{1}{x}\frac{d}{dx}\left(x\frac{d\varphi}{dx}\right)$$

(x is the radial coordinate usually denoted by r). In this case $\sigma(x) = x$ and $\kappa(x) = x$. An eigenvalue problem for this operator on the interval $[0,b]$ will be singular because $\sigma(x)$ and $\kappa(x)$ vanish at $x = 0$ and (4.11.6) is violated. On the other hand, such a problem on the interval $[a,b]$ where $0 < a < b < \infty$ will be regular. The theory of regular Sturm-Liouville problems, outlined below, does not apply to singular problems, although the main results extend to some singular problems.

In a regular problem the differential equation $S\varphi = \lambda\varphi$ can be written as

$$\frac{\kappa(x)}{\sigma(x)}\varphi'' + \frac{\kappa'(x)}{\sigma(x)}\varphi' + q(x)\varphi + \lambda\varphi = 0\,. \tag{4.11.7}$$

Since $\sigma(x)$ is continuous, $1/\sigma(x)$ is also continuous at each point where $\sigma(x)$ is not zero. But by (4.11.6), $\sigma(x)$ is different from zero at every point of $[a,b]$. It follows that the coefficients κ/σ, κ'/σ, q are continuous functions of x on $a \leq x \leq b$ and κ/σ is everywhere different from zero. Equation (4.11.7) is therefore of the same type as equation (4.2.1) and the discussion following (4.2.1) can be applied to equation (4.11.7), or to the D.E. in (4.11.4). To form the general solution of this D.E. we begin by choosing the basic solutions $\varphi_1(x) = \varphi_1(x,\lambda)$ and $\varphi_2(x) = \varphi_2(x,\lambda)$ at some conveniently chosen fixed point x_0 of $[a,b]$. These are the solutions which satisfy

$$\varphi_1(x_0) = 1\,, \quad \varphi_2(x_0) = 0\,,$$
$$\varphi_1'(x_0) = 0\,, \quad \varphi_2'(x_0) = 1\,.$$

The solution $\varphi(x)$ such that $\varphi(x_0) = A$, $\varphi'(x_0) = B$ is then given by

$$\varphi(x) = A\varphi_1(x) + B\varphi_2(x) . \tag{4.11.8}$$

Formula (4.11.8) with A and B arbitrary constants represents the general solution of $S\varphi = \lambda\varphi$.

To obtain a *Green's formula* for the operator S, we take a pair of functions f, g and compute $(Sf)g - f(Sg)$, obtaining

$$\begin{aligned}(Sf)g - f(Sg) &= -\left\{\frac{1}{\sigma}(\kappa f')' + qf\right\}g + f\left\{\frac{1}{\sigma}(\kappa g')' + qg\right\} \\ &= -\frac{1}{\sigma}\{(\kappa f')'g - f(\kappa g')'\} \\ &= -\frac{1}{\sigma}\frac{d}{dx}\{(\kappa f')g - f(\kappa g')\}\end{aligned}$$

and hence the identity

$$(Sf)g - f(Sg) = -\frac{1}{\sigma}\frac{d}{dx}\{\kappa(f'g - fg')\} . \tag{4.11.9}$$

Assuming that f and g are continuous together with their first and second derivatives on $[a,b]$, we can multiply both sides of (4.11.9) by $\sigma(x)$ and then integrate. The result is the **Green's formula**

$$\int_a^b [(Sf)g - f(Sg)]\sigma(x)\,dx = -\{\kappa(x)(f'g - fg')\}\big|_a^b . \tag{4.11.10}$$

For the eigenvalue problems considered previously, the Green's formula could be used to show that the eigenvalues are real and that eigenfunctions belonging to different eigenvalues are orthogonal. We can therefore hope to use (4.11.10) to achieve the same results for the general Sturm-Liouville problem.

Let $\varphi_m(x)$ and $\varphi_n(x)$ be real eigenfunctions of Problem (4.11.4) belonging to real eigenvalues λ_m and λ_n, respectively. Substituting $f = \varphi_m$, $g = \varphi_n$ in (4.11.10) and using $S\varphi_m = \lambda_m\varphi_m$, $S\varphi_n = \lambda_n\varphi_n$ we obtain

$$\begin{aligned}(\lambda_m - \lambda_n)\int_a^b \varphi_m(x)\varphi_n(x)\sigma(x)\,dx \\ = -\{\kappa(x)(\varphi_m'\varphi_n - \varphi_m\varphi_n')\}\big|_a^b .\end{aligned} \tag{4.11.11}$$

The right member of (4.11.11) in expanded form is

$$-\kappa(b)[\varphi_m'(b)\varphi_n(b) - \varphi_m(b)\varphi_n'(b)] + \kappa(a)[\varphi_m'(a)\varphi_n(a) - \varphi_m(a)\varphi_n'(a)] .$$

In the previous eigenvalue problems we could always say that the right member of (4.11.11) was zero simply because $\varphi_m(x)$ and $\varphi_n(x)$ satisfied the boundary conditions of the eigenvalue problem. As we shall see in Exercises 4g, such a statement does not hold for all eigenvalue problems. By means of the following definition we single out those eigenvalue problems for which the statement is valid. A regular Sturm-Liouville problem (4.11.4) is called **self-adjoint** if

$$[\kappa(x)\{f'(x)g(x) - f(x)g'(x)\}]_a^b = 0 \tag{4.11.12}$$

for every pair of differentiable functions f, g which satisfy the boundary conditions of the problem (f, g are not necessarily eigenfunctions).

If the first boundary condition in (4.11.4) is either of the conditions $\varphi(a) = 0$ or $\varphi'(a) - h_1\varphi(a) = 0$, ($h_1$ real) and if the second boundary condition is either of the conditions $\varphi(b) = 0$ or $\varphi'(b) + h_2\varphi(b) = 0$, ($h_2$ real) then each of the expressions

$$\kappa(b)[f'(b)g(b) - f(b)g'(b)]\,, \quad \kappa(a)[f'(a)g(a) - f(a)g'(a)]$$

which make up (4.11.12) will be zero and hence the problem will be self-adjoint. In addition (see Exercises 4g), there are numerous pairs of mixed boundary conditions which give rise to self-adjoint problems.

When the problem is self-adjoint, the right member of (4.11.11) is zero and we have

$$(\lambda_m - \lambda_n) \int_a^b \varphi_m(x)\varphi_n(x)\sigma(x)\,dx = 0\,.$$

Hence if $\lambda_m \neq \lambda_n$, then

$$\int_a^b \varphi_m(x)\varphi_n(x)\sigma(x)\,dx = 0\,. \tag{4.11.13}$$

This is not an orthogonality relation in the sense we used earlier, but it will serve the same purpose, and accordingly we will interpret (4.11.13) as an orthogonality relation by means of the following definition. Two real-valued functions $\varphi(x)$, $\psi(x)$ are called **orthogonal relative to the weight function** $\sigma(x)$ on the interval $[a,b]$ if

$$\int_a^b \varphi(x)\psi(x)\sigma(x) = 0\,.$$

A finite or infinite sequence of real-valued functions $\varphi_1(x)$, $\varphi_2(x)$, . . . is called an **orthogonal system on** $[a,b]$ **relative to the weight function** $\sigma(x)$ if and only if

$$\begin{aligned} &\int_a^b \varphi_m(x)\varphi_n(x)\sigma(x)\,dx = 0 \qquad && m \neq n\,, \\ &\int_a^b \varphi_n^2(x)\sigma(x)\,dx > 0\,, && n = 1, 2, \ldots. \end{aligned} \tag{4.11.14}$$

The above discussion constitutes a proof of part (ii) of the next theorem. The proof of part (i) is left as an exercise for the reader.

Theorem. *If the regular Sturm-Liouville problem* (4.11.2) *is self-adjoint then*

(i) *All the eigenvalues are real;*

(ii) *Eigenfunctions belonging to different eigenvalues are orthogonal on* $[a,b]$ *relative to the weight function* $\sigma(x)$.

The above theorem asserts that all eigenvalues of a self-adjoint problem are real, but it leaves open the question of whether there are any eigenvalues. Detailed analysis of this question is beyond the scope of our discussion. However the question is answered by the next theorem which is stated without proof. The interested reader may find proofs of the theorem in references listed at the end of the chapter.

Theorem. *A self-adjoint regular Sturm-Liouville problem has infinitely many eigenvalues. The eigenvalues can be arranged in an increasing sequence*

$$\lambda_0 < \lambda_1 < \lambda_2 < \cdots < \lambda_n < \cdots \tag{4.11.15}$$

and

$$\lambda_n \to \infty \quad \text{as} \quad n \to \infty\,. \tag{4.11.16}$$

The phenomenon of multiplicity can be discussed along the same lines as in Section 4.10. If the boundary conditions are unmixed then each eigenvalue is of multiplicity one. In this case to each of the eigenvalues λ_n in (4.11.15) there belongs only one linearly independent eigenfunction $\varphi_n(x)$. Thus corresponding to (4.11.15) there is a sequence of eigenfunctions

$$\varphi_0(x),\ \varphi_1(x),\ \varphi_2(x),\ \ldots,\ \varphi_n(x),\ \ldots \tag{4.11.17}$$

and any two of these are orthogonal on $[a,b]$ relative to the weight function $\sigma(x)$. Each $\varphi_n(x)$ is continuous and not identically zero on $[a,b]$. Since $\sigma(x)$ is continuous and positive at every point, it follows that

$$\int_a^b \varphi_n^2(x)\sigma(x)\,dx > 0 \qquad n = 0, 1, 2, \ldots.$$

Hence (4.11.17) is an orthogonal system on $[a,b]$ relative to the weight function $\sigma(x)$.

If the boundary conditions are mixed, there may or may not be multiple eigenvalues. However since the differential operator S is of the second order, the equation $S\varphi = \lambda\varphi$ has, for any fixed λ, only two linearly independent solutions, and consequently the multiplicity of an eigenvalue cannot be greater than two. If λ is an eigenvalue of multiplicity two and $\varphi_1^*(x)$, $\varphi_2^*(x)$ are any two linearly independent real eigenfunctions belonging to the eigenvalue λ, then the two functions

$$\varphi_1(x) = \varphi_1^*(x)\,, \quad \varphi_2(x) = \begin{vmatrix} c_{11}\,, & \varphi_1^*(x) \\ c_{21}\,, & \varphi_2^*(x) \end{vmatrix}$$

(4.11.18)

$$c_{ij} = \int_a^b \varphi_i^*(x)\varphi_j^*(x)\sigma(x)\,dx\,,$$

are a pair of eigenfunctions belonging to the eigenvalue λ which are orthogonal on $[a,b]$ relative to the weight function $\sigma(x)$. Therefore, when problem (4.11.4) has multiple eigenvalues, the eigenfunctions can be chosen so as to form an orthogonal system on $[a,b]$ relative to the weight function $\sigma(x)$.

Let us return to the initial-boundary value problem (4.11.1). We assume that the interval $[a,b]$ is finite and conditions (4.11.5), (4.11.6) are satisfied, so that the Sturm-Liouville problem (4.11.2) is regular. Since the boundary conditions in (4.11.2) are of the first kind, the problem is self-adjoint, and moreover all eigenvalues are of multiplicity one. Let the eigenvalues be

$$\lambda_0 < \lambda_1 < \cdots < \lambda_n < \cdots$$

with corresponding eigenfunctions

$$\varphi_0(x),\ \varphi_1(x),\ \ldots,\ \varphi_n(x),\ \ldots.$$

For each n the function

$$u_n(x,t) = e^{-\lambda_n t}\varphi_n(x)$$

is a solution of the D.E. and the B.C. Formally the solution of the initial boundary value problem is

$$u(x,t) = \sum_{n=0}^{\infty} c_n e^{-\lambda_n t}\varphi_n(x) \tag{4.11.19}$$

where the constants c_n are to be determined so that the initial condition

$$f(x) = \sum_{n=0}^{\infty} c_n \varphi_n(x) \tag{4.11.20}$$

is satisfied. If (4.11.20) is multiplied by $\varphi_m(x)\sigma(x)$ and integrated over $[a,b]$ term-by-term, there results a formula for c_m,

$$c_m = \frac{\int_a^b f(x)\varphi_m(x)\sigma(x)\,dx}{\int_a^b \varphi_m^2(x)\sigma(x)\,dx}\,. \tag{4.11.21}$$

Formulas (4.11.19) and (4.11.21) provide the solution of the initial-boundary value problem. The convergence of series such as (4.11.19) and (4.11.20) will be discussed in Section 6.9.

In some eigenvalue problems, the differential equation is of the form

$$\alpha(x)\varphi'' + \beta(x)\varphi' + q(x)\varphi + \lambda\varphi = 0\,, \qquad a < x < b\,. \tag{4.11.22}$$

This can usually be written in the form (4.11.7). In fact (4.11.22) is the same as (4.11.7), provided that

$$\frac{\kappa(x)}{\sigma(x)} = \alpha(x)\,, \quad \frac{\kappa'(x)}{\sigma(x)} = \beta(x)\,.$$

From these equations we get

$$\frac{\kappa'(x)}{\kappa(x)} = \frac{\beta(x)}{\alpha(x)}$$

or

$$\begin{aligned} \kappa(x) &= C \exp\left(\int_{x_0}^{x} \frac{\beta(\xi)}{\alpha(\xi)}\, d\xi\right), \\ \sigma(x) &= \frac{C}{\alpha(x)} \exp\left(\int_{x_0}^{x} \frac{\beta(\xi)}{\alpha(\xi)}\, d\xi\right), \end{aligned} \tag{4.11.23}$$

where C is a constant and x_0 is some point, $a \leq x_0 \leq b$. These formulas will be meaningful, and $\kappa(x)$, $\sigma(x)$ will satisfy (4.11.5) and (4.11.6), provided that $\alpha(x) > 0$ on $a \leq x \leq b$, and the constant C is chosen to be positive. For example, the differential equation

$$\varphi'' + \frac{\gamma}{x}\varphi' + \lambda\varphi = 0 \tag{4.11.24}$$

leads to

$$\kappa(x) = Ce^{\gamma(\log x - \log x_0)} = C\frac{x^\gamma}{x_0^\gamma}\,,$$

$$\sigma(x) = C\frac{x^\gamma}{x_0^\gamma}\,.$$

If we take $C = x_0^\gamma$, then we have simply $\kappa(x) = x^\gamma$, $\sigma(x) = x^\gamma$ and the differential equation (4.11.24) can be written as

$$\frac{1}{x^\gamma}\frac{d}{dx}\left(x^\gamma \frac{d\varphi}{dx}\right) + \lambda\varphi = 0\,.$$

In particular, the weight function associated with Equation (4.11.24) is $\sigma(x) = x^\gamma$.

EXERCISES 4g

1. Show that the problem

$$\varphi'' + \lambda\varphi = 0 , \qquad a < x < b ,$$
$$\varphi(b) + \alpha\varphi(a) + \beta\varphi'(a) = 0 ,$$
$$\varphi'(b) + \gamma\varphi(a) + \delta\varphi'(a) = 0 ,$$

where α, β, γ, δ are real constants, is self-adjoint if $\alpha\delta - \beta\gamma = 1$, and is not self-adjoint if $\alpha\delta - \beta\gamma \neq 1$.

2. Find a *Green's first formula* for the operator S defined in (4.11.3). Use your result to obtain information about the eigenvalues of the problem (assumed to be regular)

$$S\varphi = \lambda\varphi , \qquad a < x < b ,$$
$$\varphi(a) = 0 , \quad \varphi'(b) = 0$$

for the case when $q(x) = Q$, a constant.

3. For the eigenvalue problem

$$x^2\varphi'' + x\varphi' + \lambda\varphi = 0 , \qquad 1 < x < b ,$$
$$\varphi(1) = 0 ,$$
$$\varphi(b) = 0 .$$

(i) Find an equation whose roots are the eigenvalues.
(ii) Find the eigenvalues and eigenfunctions.
(iii) Find the constants in the eigenfunction expansion of the function $f(x) = \log x$.

4. Solve

$$\text{D.E.} \quad u_t = u_{xx} + 2\beta u_x , \qquad 0 < x < L , \quad t > 0 ,$$
$$\text{B.C.} \quad u(0,t) = 0 , \quad u(L,t) = 0 , \qquad t > 0 ,$$
$$\text{I.C.} \quad u(x,0) = f(x) , \qquad 0 < x < L ,$$

where β is a constant.

REFERENCES.

Birkhoff, Garrett and Gian-Carlo Rota, *Ordinary Differential Equations*, Ginn, Boston, 1962, pp. 247–264.

Coddington, Earl A. and Norman Levinson, *Theory of Ordinary Differential Equations*, McGraw-Hill, New York, 1955, Chapters 7 and 8.

Ince, E. L., *Ordinary Differential Equations*, Dover, New York, 1956, Chapters 10 and 11.

Miller, Kenneth S., *Linear Differential Equations in the Real Domain*, W. W. Norton, New York, 1963, pp. 103–118.

5

The solution of inhomogeneous problems

5.1. *TRANSFORMATION OF PROBLEMS*

A linear initial-boundary value problem in which either the differential equation or a boundary condition is inhomogeneous, is called an **inhomogeneous initial-boundary value problem.** As an example, consider

$$
(5.1.1)\quad
\begin{array}{lll}
\text{D.E.} & u_t - ku_{xx} = q(x,t) & 0 < x < L, \quad t > 0 \\
\text{B.C.} & \left.\begin{array}{l} u_x(0,t) = A(t) \\ u(L,t) = B(t) \end{array}\right\} & t > 0 \\
\text{I.C.} & u(x,0) = f(x) & 0 < x < L,
\end{array}
$$

in which $q(x,t)$, $A(t)$, $B(t)$, $f(x)$ are given functions (which may be independent of either x or t, or both), such that $q(x,t)$, $A(t)$ and $B(t)$ do not all vanish identically.

The problem obtained from an inhomogeneous problem by replacing each inhomogeneous equation (except the initial condition) by the related homogeneous equation is called the **related homogeneous problem.** Thus, for example, if $q(x,t)$, $A(t)$, $B(t)$, $f(x)$ are any functions, the related homogeneous problem for the problem (5.1.1) is

$$
(5.1.2)\quad
\begin{array}{lll}
\text{D.E.} & v_t - kv_{xx} = 0 & 0 < x < L, \quad t > 0 \\
\text{B.C.} & \left.\begin{array}{l} v_x(0,t) = 0 \\ v(L,t) = 0 \end{array}\right\} & t > 0 \\
\text{I.C.} & v(x,0) = g(x) & 0 < x < L,
\end{array}
$$

where $g(x)$ is an arbitrary function, not necessarily the same as $f(x)$.

Our aim in this chapter is to develop methods for the solution of inhomogeneous problems. We shall see that inhomogeneous problems can always be solved in terms of expansions in the eigenfunctions associated with the related homogeneous problem.

In this section, as a preliminary, we consider the transformation of initial-boundary value problems by subtraction of known functions. Suppose, for example, that $u(x,t)$ is a solution of the following problem,

$$
\text{(5.1.3)}\quad
\begin{array}{lll}
\text{D.E.} & u_t - 5u_{xx} = t^2 & 0 < x < 1\,,\quad t > 0 \\
\text{B.C.} & \left.\begin{array}{l} u_x(0,t) = \sin t \\ u(1,t) = 2 \end{array}\right\} & t > 0 \\
\text{I.C.} & u(x,0) = e^x & 0 < x < 1\,,
\end{array}
$$

and let $K(x,t)$ be any known function. Then the function $v(x,t)$, defined by

$$
\text{(5.1.4)}\qquad v(x,t) = u(x,t) - K(x,t)\,,
$$

is a solution of another initial-boundary value problem which we can write immediately if we recall that all the equations of (5.1.3) are linear, and that if S is a linear operator, then

$$
Su = S(v + K) = Sv + SK\,.
$$

We obtain

$$
\begin{array}{lll}
\text{D.E.} & (v_t - 5v_{xx}) + (K_t - 5K_{xx}) = t^2 & 0 < x < 1\,,\quad t > 0 \\
\text{B.C.} & \left.\begin{array}{l} v_x(0,t) + K_x(0,t) = \sin t \\ v(1,t) + K(1,t) = 2 \end{array}\right\} & t > 0 \\
\text{I.C.} & v(x,0) + K(x,0) = e^x & 0 < x < 1\,.
\end{array}
$$

This can be written in the form

$$
\text{(5.1.5)}\quad
\begin{array}{lll}
\text{D.E.} & v_t - 5v_{xx} = t^2 - (K_t - 5K_{xx}) & 0 < x < 1\,,\quad t > 0 \\
\text{B.C.} & \left.\begin{array}{l} v_x(0,t) = \sin t - K_x(0,t) \\ v(1,t) = 2 - K(1,t) \end{array}\right\} & t > 0 \\
\text{I.C.} & v(x,0) = e^x - K(x,0) & 0 < x < 1\,,
\end{array}
$$

which is, since $K(x,t)$ is a known function, an initial-boundary value problem of the form (5.1.1) for v. Conversely, the same considerations show that if v is a solution of (5.1.5), then u determined from (5.1.4) is a solution of (5.1.3). Thus the problems (5.1.3) and (5.1.5) are *equivalent* in the sense that a solution of one immediately furnishes, via (5.1.4), a solution of the other.

The usefulness of this transformation lies in the fact that by suitable choice of the function $K(x,t)$ we can make one or more of the equations of (5.1.5) homogeneous. If $K(x,t)$ is any function satisfying

$$K(1,t) = 2\,, \tag{5.1.6}$$

then the boundary condition at $x = 1$ of (5.1.5) will be homogeneous. An obvious choice to accomplish this is $K(x,t) = 2$. To make the boundary condition at $x = 0$ homogeneous we must choose K such that

$$K_x(0,t) = \sin t\,, \tag{5.1.7}$$

and this is easily satisfied by choosing $K(x,t) = x \sin t$. We can even choose K so that both (5.1.6) and (5.1.7) are satisfied, namely by taking

$$K(x,t) = x \sin t - \sin t + 2\,.$$

With this choice of K, (5.1.5) becomes

$$\begin{array}{lll} \text{D.E.} & v_t - 5v_{xx} = t^2 + \cos t - x \cos t & 0 < x < 1\,, \quad 0 < t \\ \text{B.C.} & \left.\begin{array}{l} v_x(0,t) = 0 \\ v(1,t) = 0 \end{array}\right\} & 0 < t \\ \text{I.C.} & v(x,0) = e^x - 2 & 0 < x < 1\,, \end{array}$$

and the problem (5.1.3), in which the D.E. and both B.C. are inhomogeneous, is thus transformed into one in which only the D.E. is inhomogeneous.

Alternatively, we can choose $K(x,t)$ so that in the transformed problem the D.E. is homogeneous, although the B.C. are not. For the D.E. to be homogeneous we must have

$$K_t - 5K_{xx} = t^2 \qquad 0 < t\,, \quad 0 < x < 1\,. \tag{5.1.8}$$

Any function which satisfies this equation will suffice, and an obvious choice is a function independent of x,

$$K(x,t) = t^3/3\,.$$

With this choice of K the transformed problem is

$$\begin{array}{lll} \text{D.E.} & v_t - 5v_{xx} = 0 & 0 < x < 1\,, \quad 0 < t \\ \text{B.C.} & \left.\begin{array}{l} v_x(0,t) = \sin t \\ v(1,t) = 2 - t^3/3 \end{array}\right\} & 0 < t \\ \text{I.C.} & v(x,0) = e^x & 0 < x < 1\,. \end{array}$$

At this point it is natural to consider the possibility of finding a function K which satisfies the three equations (5.1.6), (5.1.7) and

(5.1.8). If this were possible, the corresponding transformed problem would be a homogeneous problem which could be solved by the method of eigenfunction expansions: thus, the solution of the original problem could be found. For the present problem this cannot be achieved without considerable difficulty. However, we will consider in the next section a class of problems for which this is a feasible program of solution.

5.2. *USE OF ASYMPTOTIC SOLUTIONS*

Consider an initial-boundary value problem such as (5.1.1) in which the source term and boundary terms *do not depend on* t; that is, in which the problem has the form

$$(5.2.1)\quad \begin{array}{lll} \text{D.E.} & u_t = ku_{xx} + q(x) & 0 < x < L\,, \quad t > 0 \\ \text{B.C.} & \left.\begin{array}{l} u_x(0,t) = A \\ u(L,t) = B \end{array}\right\} & t > 0 \\ \text{I.C.} & u(x,0) = f(x) & 0 < x < L \end{array}$$

where A, B are constants and $q(x)$ is independent of t. A problem of this special kind *usually* has an equilibrium solution $U(x)$ which satisfies the differential equation and the boundary conditions (however, see Exercises 5a, Problem 4).

To find $U(x)$, we solve the problem

$$(5.2.2)\quad \begin{cases} \text{D.E.} & 0 = kU''(x) + q(x)\,, \qquad 0 < x < L\,, \\ \text{B.C.} & U'(0) = A\,, \quad U(L) = B\,. \end{cases}$$

The solution is

$$(5.2.3)\qquad U(x) = A(x - L) + B + \frac{1}{k}\int_x^L d\xi \int_0^\xi q(\eta)\, d\eta\,.$$

Now let

$$(5.2.4)\qquad v(x,t) = u(x,t) - U(x)$$

where $u(x,t)$ is the solution of (5.2.1) and $U(x)$ the solution (5.2.3) of (5.2.2). Following the procedure of Section 1, we get

$$\begin{aligned} v_t - kv_{xx} &= (u_t - ku_{xx}) - (-\,kU'') = q(x) - q(x) = 0 \\ v_x(0,t) &= u_x(0,t) - U'(0) = A - A = 0 \\ v(L,t) &= u(L,t) - U(L) = B - B = 0 \\ v(x,0) &= u(x,0) - U(x) = f(x) - U(x)\,. \end{aligned}$$

Hence $v(x,t)$ is the solution of the problem

$$
\text{(5.2.5)}\quad
\begin{array}{lll}
\text{D.E.} & v_t = kv_{xx} & 0 < x < L\,,\quad 0 < t \\
\text{B.C.} & \left.\begin{array}{l} v_x(0,t) = 0 \\ v(L,t) = 0 \end{array}\right\} & 0 < t \\
\text{I.C.} & v(x,0) = g(x) & 0 < x < L\,,
\end{array}
$$

where $g(x) = u(x,0) - U(x) = f(x) - U(x)$ is a known function. Now problem (5.2.5) can be solved and $v(x,t)$ found by the method of Chapter 4. We obtain

$$v(x,t) = \sum_{n=0}^{\infty} b_n e^{-\lambda_n kt} \cos\left(n + \frac{1}{2}\right)\frac{\pi x}{L}$$

where $\lambda_n = [(n + \frac{1}{2})\pi/L]^2$ and

$$b_n = \frac{2}{L}\int_0^L g(x) \cos\left(n + \frac{1}{2}\right)\frac{\pi x}{L}\,dx\,.$$

Finally, the solution $u(x,t)$ of (5.2.1) is obtained from (5.2.4),

$$u(x,t) = v(x,t) + U(x)\,,$$

where $v(x,t)$, $U(x)$ have been found.

Observe that since $\lambda_0 > 0$, we have

$$\lim_{t\to\infty} v(x,t) = 0$$

so that from (5.2.4),

$$\lim_{t\to\infty} [u(x,t) - U(x)] = 0\,.$$

Thus, for the physical problem described by (5.2.1), the temperature distribution is given approximately for large values of the time by an equilibrium temperature distribution, which is the same for all initial temperature distributions.

In general, given an initial-boundary value problem with solution $u(x,t)$, we shall call a function $U(x,t)$ an **asymptotic solution** of the problem if $U(x,t)$ satisfies the D.E. and B.C. of the problem for all t, and

$$\text{(5.2.6)}\qquad \lim_{t\to\infty} [u(x,t) - U(x,t)] = 0\,.$$

In the problems we have considered so far, an *equilibrium solution* $U(x)$ is an asymptotic solution, and indeed is the simplest possible asymptotic solution. In this case the difference $u(x,t) - U(x)$ is called the *transient part* of the solution. We emphasize, however, that some problems which do not have an equilibrium solution may have an asymptotic solution of particularly simple form. Sometimes, in this case, the term transient part is applied to the difference $u(x,t) - U(x,t)$.

It is clear from the discussion above that if we can find an asymptotic solution of a problem, then we can determine the full solution. Indeed, for this purpose the requirement (5.2.6) is unnecessary. [Although, for the example (5.1.1) which we have been considering it is automatically satisfied.] Since, however, an asymptotic solution furnishes an approximation to the full solution for large values of the time, the determination of such a solution, particularly one of simple form, may be a problem of importance equal to or greater than the solution of the full problem.

EXERCISES 5a

1. (i) If A and B are constants, find a function $K(x)$ that satisfies

$$K(0) = A\,, \quad K(L) = B\,.$$

(ii) If $A(t)$ and $B(t)$ are any differentiable functions of t, find a function $K(x,t)$ that satisfies

$$K(0,t) = A(t)\,, \quad K(L,t) = B(t)\,.$$

(iii) Given the inhomogeneous problem

$$\begin{array}{lll} \text{D.E.} & u_t - ku_{xx} = q(x,t) & 0 < x < L\,, \quad 0 < t \\ \text{B.C.} & \left.\begin{array}{l} u(0,t) = A(t) \\ u(L,t) = B(t) \end{array}\right\} & 0 < t \\ \text{I.C.} & u(x,0) = f(x) & 0 < x < L\,, \end{array}$$

find an equivalent problem in which the B.C. are homogeneous.

2. Transform the problem

$$\begin{array}{lll} \text{D.E.} & u_t - ku_{xx} + ru = q(x,t) & 0 < x < L\,, \quad 0 < t \\ \text{B.C.} & \left.\begin{array}{l} u(0,t) = A(t) \\ u_x(L,t) + hu(L,t) = B(t) \end{array}\right\} & 0 < t \\ \text{I.C.} & u(x,0) = f(x) & 0 < x < L\,, \end{array}$$

(where h, k and r are constants and $A(t)$ and $B(t)$ are differentiable) into an equivalent problem in which the B.C. are homogeneous.

3. Transform the problem

$$\begin{array}{lll} \text{D.E.} & u_t - ku_{xx} = q(x,t) & 0 < x < L\,, \quad 0 < t \\ \text{B.C.} & \left.\begin{array}{l} u_x(0,t) = A(t) \\ u_x(L,t) = B(t) \end{array}\right\} & 0 < t \\ \text{I.C.} & u(x,0) = f(x) & 0 < x < L\,, \end{array}$$

$[A(t), B(t)$ differentiable$]$ into a problem in which the B.C. are homogeneous.

4. Show that the problem

$$\begin{array}{lll} \text{D.E.} & u_t - ku_{xx} = q(x) & 0 < x < L, \quad 0 < t \\ \text{B.C.} & \left.\begin{array}{l} u_x(0,t) = A \\ u_x(L,t) = B \end{array}\right\} & 0 < t \\ \text{I.C.} & u(x,0) = f(x) & 0 < x < L \end{array}$$

$(A, B$ constant) does *not* have an equilibrium solution in general, and find a relation between A, B, L, k and q which must hold for an equilibrium solution to exist.

5. Transform the problem

$$\begin{array}{lll} \text{D.E.} & u_t - ku_{xx} = 2x^3 - 5\sin t & 0 < x < L, \quad 0 < t \\ \text{B.C.} & \left.\begin{array}{l} u(0,t) = 0 \\ u(L,t) = 0 \end{array}\right\} & 0 < t \\ \text{I.C.} & u(x,0) = 7 & 0 < x < L \end{array}$$

into an equivalent problem in which the D.E. is homogeneous.

6. Find the solution of the problem

$$\begin{array}{lll} \text{D.E.} & u_t - ku_{xx} = Q & 0 < x < L, \quad 0 < t \\ \text{B.C.} & \left.\begin{array}{l} u(0,t) = A \\ u(L,t) = B \end{array}\right\} & 0 < t \\ \text{I.C.} & u(x,0) = 0 & 0 < x < L, \end{array}$$

where A, B and Q are constant.

7. (i) Use the fact that any function of t alone satisfies the B.C. of the following problem, to solve the problem

$$\begin{array}{lll} \text{D.E.} & u_t - ku_{xx} = A\cos\omega t & 0 < x < \pi, \quad 0 < t \\ \text{B.C.} & \left.\begin{array}{l} u_x(0,t) = 0 \\ u_x(\pi,t) = 0 \end{array}\right\} & 0 < t \\ \text{I.C.} & u(x,0) = f(x) & 0 < x < \pi, \end{array}$$

where A and ω are constants, and $f(x)$ is an arbitrary function.

(ii) Determine an asymptotic solution of the problem of part (i) (of simplest possible form) and discuss its dependence on A, ω and $f(x)$.

5.3. *THE METHOD OF VARIATION OF PARAMETERS*

We turn now to the solution of general inhomogeneous problems, taking (5.1.1) as our example. First, let us observe that, if $A(t)$ and

$B(t)$ are differentiable for $t > 0$, we can, by subtraction of a known function, transform this problem into one in which both boundary conditions are homogeneous. In Equation (5.1.4) we set

$$K(x,t) = xA(t) - LA(t) + B(t) .$$

Then the initial-boundary value problem satisfied by $v(x,t)$ is

$$\begin{aligned} &\text{D.E.} \quad v_t - kv_{xx} \\ &\qquad = q(x,t) - [xA'(t) - LA'(t) + B'(t)] \qquad 0 < x < L\,,\ 0 < t \\ &\text{B.C.} \quad \left.\begin{aligned} v_x(0,t) &= 0 \\ v(L,t) &= 0 \end{aligned}\right\} \qquad t > 0 \\ &\text{I.C.} \quad v(x,0) = f(x) - [xA(0) - LA(0) + B(0)] \quad 0 < x < L\,. \end{aligned}$$

Thus, if in the original problem the boundary conditions are not homogeneous, we can transform the problem into an equivalent problem in which the boundary conditions are homogeneous. It suffices, therefore, for the solution of the problem (5.1.1) in general, to determine the solutions of problem

$$(5.3.1)\qquad \begin{aligned} &\text{D.E.} \quad u_t - ku_{xx} = q(x,t) \qquad 0 < x < L\,,\ 0 < t \\ &\text{B.C.} \quad \left.\begin{aligned} u_x(0,t) &= 0 \\ u(L,t) &= 0 \end{aligned}\right\} \qquad t > 0 \\ &\text{I.C.} \quad u(x,0) = f(x)\,. \end{aligned}$$

For problem (5.3.1) the related homogeneous problem is problem (5.1.2). The solution of this latter problem by the method of eigenfunction expansions leads to the eigenvalue problem

$$(5.3.2)\qquad \begin{cases} \text{D.E.} & \varphi'' + \lambda\varphi = 0 \qquad 0 < x < L \\ \text{B.C.} & \varphi'(0) = 0\,,\quad \varphi(L) = 0 \end{cases}$$

for which the eigenvalues and eigenfunctions are

$$\lambda_n = \left[\left(n + \frac{1}{2}\right)\frac{\pi}{L}\right]^2, \quad \varphi_n(x) = \cos\left(n + \frac{1}{2}\right)\frac{\pi}{L}x\,, \quad n = 0, 1, 2, \ldots .$$

For these eigenfunctions the normalizing constants are

$$\int_0^L \varphi_n^2(x)\, dx = \frac{L}{2}\cdot$$

The solution of (5.1.2) is of the form

$$(5.3.3)\qquad v(x,t) = \sum_{n=0}^{\infty} a_n e^{-\lambda_n kt}\varphi_n(x)\,,$$

where the parameters a_n are determined by the initial conditions. It is proposed now to find a solution of the inhomogeneous problem

(5.3.1) in the form of a series like (5.3.3), but in which the parameters a_n are replaced by functions of t. The product $a_n e^{-\lambda_n kt}$ will then become a function $T_n(t)$ so that the solution will be a series

$$u(x,t) = \sum_{n=0}^{\infty} T_n(t)\varphi_n(x) \,. \tag{5.3.4}$$

This proposal is similar to that which in the study of ordinary differential equations leads to the method of variation of parameters. Hence the procedure we are going to describe bears the same name. It should be emphasized, however, that this procedure and the method of variation of parameters as usually presented in courses in ordinary differential equations have only this feature in common, and are otherwise unrelated.

It is not hard to see that *if* (5.3.1) *has a solution then that solution can be represented by a series of the form* (5.3.4). In fact, for any fixed $t > 0$ the solution will be a continuously differentiable function of x on the interval $0 \leq x \leq L$. Hence, if any such function of x can be expanded in a series of eigenfunctions $\{\varphi_n(x)\}$, such an expansion of $u(x,t)$ is possible for each fixed value of t. There remains the problem of finding the coefficients $T_n(t)$. Because of the orthogonality of the eigenfunctions $\varphi_n(x)$, the coefficients $T_n(t)$ are related to the (still unknown) solution $u(x,t)$ by the formula

$$T_n(t) = \frac{2}{L}\int_0^L u(x,t)\varphi_n(x)\,dx \,. \tag{5.3.5}$$

We assume that $u_t(x,t)$ is a continuous function in the region $t > 0$, $0 \leq x \leq L$. Under these circumstances, the integral in (5.3.5) has a derivative with respect to t which can be calculated by differentiation under the integral sign. We get, referring to the D.E. of (5.3.1),

$$T_n'(t) = \frac{2}{L}\int_0^L u_t(x,t)\varphi_n(x)\,dx = \frac{2}{L}\int_0^L [ku_{xx}(x,t) + q(x,t)]\varphi_n(x)\,dx \,,$$

that is,

$$T_n'(t) = \frac{2k}{L}\int_0^L u_{xx}(x,t)\varphi_n(x)\,dx + \frac{2}{L}\int_0^L q(x,t)\varphi_n(x)\,dx \,. \tag{5.3.6}$$

The last term of (5.3.6)

$$q_n(t) = \frac{2}{L}\int_0^L q(x,t)\varphi_n(x)\,dx \tag{5.3.7}$$

is a known function of t, since $q(x,t)$ is given. The first term will be transformed using Green's formula. To justify this operation, we assume that for every fixed $t > 0$, u, u_x and u_{xx} are continuous on the interval $0 \leq x \leq L$. We find

$$\frac{2k}{L}\int_0^L u_{xx}\varphi_n\,dx = \frac{2k}{L}\left[u_x(x,t)\varphi_n(x) - u(x,t)\varphi_n'(x)\right]_0^L + \frac{2k}{L}\int_0^L u(x,t)\varphi_n''(x)\,dx\,. \tag{5.3.8}$$

The terms to be evaluated at the boundary can be calculated. In fact, since $\varphi_n'(0) = 0$, $\varphi_n(L) = 0$, and $u_x(0,t) = 0$, $u(L,t) = 0$, each of these terms *vanishes*. Further, because $\varphi_n'' = -\lambda_n\varphi_n$, the integral on the right of (5.3.8) can be written as

$$\frac{2k}{L}\int_0^L u\varphi_n''\,dx = -\lambda_n k\frac{2}{L}\int_0^L u\varphi_n\,dx = -\lambda_n kT_n(t)\,. \tag{5.3.9}$$

Combining (5.3.6)–(5.3.9) we obtain

$$T_n'(t) + \lambda_n kT_n(t) = q_n(t)\,. \tag{5.3.10}$$

Equation (5.3.10) is a linear first-order ordinary differential equation for $T_n(t)$, in which $q_n(t)$ is a known function. Setting $t = 0$ in (5.3.5) we get the initial condition

$$T_n(0) = \frac{2}{L}\int_0^L f(x)\varphi_n(x)\,dx = c_n\,, \tag{5.3.11}$$

where c_n is known, since $f(x)$ is given. The solution of the problem

$$\begin{aligned} T'(t) + \lambda kT(t) &= F(t) \\ T(0) &= c \end{aligned} \tag{5.3.12}$$

is

$$T(t) = ce^{-\lambda kt} + \int_0^t e^{-\lambda k(t-\tau)}F(\tau)\,d\tau\,. \tag{5.3.13}$$

Applying this result to the problem (5.3.10), (5.3.11) we have

$$T_n(t) = c_ne^{-\lambda_n kt} + \int_0^t e^{-\lambda_n k(t-\tau)}q_n(\tau)\,d\tau\,. \tag{5.3.14}$$

The coefficients $T_n(t)$ in (5.3.4) are now completely known and problem (5.3.1) has been solved. We have

$$u(x,t) = \sum_{n=0}^{\infty}\left[c_ne^{-\lambda_n kt} + \int_0^t e^{-\lambda_n k(t-\tau)}q_n(\tau)\,d\tau\right]\varphi_n(x)\,, \tag{5.3.15}$$

where c_n and q_n are given by (5.3.11) and (5.3.7), respectively. The procedure we have described—the method of **variation of parameters** for problems with *inhomogeneous D.E.* and *homogeneous B.C.*—can be employed for any kind of homogeneous B.C. and for a large class of D.E., including those with constant coefficients. The essential feature is that the solution of the problem is expanded in a series of *eigenfunctions associated with the related homogeneous problem.*

If we set

$$v(x,t) = \sum_{n=0}^{\infty} c_n e^{-\lambda_n k t} \varphi_n(x)$$

$$U(x,t) = \sum_{n=0}^{\infty} \left[\int_0^t e^{-\lambda_n k(t-\tau)} q_n(\tau)\, d\tau \right] \varphi_n(x) ,$$

then (5.3.15) can be written

$$u(x,t) = U(x,t) + v(x,t)$$

and clearly $v(x,t)$ is a transient solution, $U(x,t)$ an asymptotic solution, of (5.3.1). For an important class of special functions $q(x,t)$, the asymptotic solution $U(x,t)$ can be determined in closed form: namely, for the class of functions of the form

$$q(x,t) = Q(x) t^{\nu} e^{\alpha t} \begin{Bmatrix} \cos \beta t \\ \sin \beta t \end{Bmatrix} , \tag{5.3.16}$$

where ν is a nonnegative integer, α and β are any constants, and $Q(x)$ any continuous function. We can see most readily that this is so, and why it is so, by returning to the discussion above at the differential equation (5.3.10).

Under the assumption (5.3.16), Equation (5.3.10) becomes, referring to (5.3.7),

$$\frac{dT_n}{dt} + \lambda_n k T_n = Q_n t^{\nu} e^{\alpha t} \begin{Bmatrix} \cos \beta t \\ \sin \beta t \end{Bmatrix} \tag{5.3.17}$$

where the constant Q_n is given by

$$Q_n = \frac{2}{L} \int_0^L Q(x) \varphi_n(x)\, dx .$$

Now, for each n, (5.3.17) is an ordinary linear differential equation with constant coefficients, *whose right side has the form that permits the calculation of the general solution of the equation by "the method of undetermined coefficients."* This is the essential fact.

For simplicity, and because this is the case of greatest importance, we assume that $\nu = 0$ in the remaining discussion. Then (except in the special case $\beta = 0$, $\alpha = -k\lambda_n$) the general solution of (5.3.17) is

$$T_n(t) = C_n e^{-k\lambda_n t} + e^{\alpha t}(A_n \cos \beta t + B_n \sin \beta t) , \tag{5.3.18}$$

where C_n is arbitrary and A_n and B_n are determined by substituting the second term of (5.3.18) in (5.3.17). It is not necessary for us to perform this computation: only its possibility is significant for us.

Substituting (5.3.18) in (5.3.4) and collecting terms we obtain

$$(5.3.19)\qquad u(x,t) = \left[\sum_{n=0}^{\infty} A_n\varphi_n(x)\right] e^{\alpha t}\cos\beta t + \left[\sum_{n=0}^{\infty} B_n\varphi_n(x)\right] e^{\alpha t}\sin\beta t + \sum_{n=0}^{\infty} C_n e^{-k\lambda_n t}\varphi_n(x)\,,$$

where the constants C_n are to be determined so that the initial condition of (5.3.1) is satisfied. We now observe that the two series in square brackets in (5.3.19) are functions of x alone, say $V(x)$ and $W(x)$, respectively, while the last series is a transient solution of our problem. Thus (5.3.1) has an asymptotic solution of the form

$$(5.3.20)\qquad U(x,t) = V(x)e^{\alpha t}\cos\beta t + W(x)e^{\alpha t}\sin\beta t\,.$$

Knowing the form (5.3.20) of an asymptotic solution, we can calculate the functions $V(x)$, $W(x)$ from the requirements that the asymptotic solution satisfy the D.E. and B.C. of the problem. These requirements lead to a system of ordinary differential equations with boundary conditions for the unknown functions $V(x)$, $W(x)$, which can be solved by elementary methods. The details of this calculation although generally laborious are simple in the cases of greatest interest.

EXERCISES 5b

1. For the problem

D.E.	$u_t - ku_{xx} = e^{-t}$	$0 < x < L,\quad 0 < t$
B.C.	$\left.\begin{aligned} u_x(0,t) &= 0 \\ u(L,t) &= 0 \end{aligned}\right\}$	$0 < t$
I.C.	$u(x,0) = 0$	$0 < x < L:$

(i) Find the solution $u(x,t)$;
(ii) Find an asymptotic solution of simplest possible form.

2. Assume that in problem (5.3.1) $q(x,t) = Q(x)e^{\alpha t}\cos\beta t$. Find the system of ordinary differential equations and the boundary conditions which must be satisfied by $V(x)$ and $W(x)$ in order that (5.3.20) be an asymptotic solution.

3. Using the results of Problem 2, find an asymptotic solution of (5.3.1) when $q(x,t) = Q\cos\beta t$.

4. (i) Find a series solution of the problem

$$\begin{array}{lll} \text{D.E.} & u_t - ku_{xx} = q(x,t) & 0 < x < L\,, \quad 0 < t \\ \text{B.C.} & \left.\begin{array}{l} u(0,t) = 0 \\ u(L,t) = 0 \end{array}\right\} & \qquad\qquad\quad 0 < t \\ \text{I.C.} & u(x,0) = f(x) & 0 < x < L\,. \end{array}$$

(ii) If $q(x,t) = Q(t)$, where

$$Q(t) = \begin{cases} 1 & 0 < t < 3 \\ e^{t-3} & 3 \leq t \end{cases}$$

and $f(x) = 0$, find $u(L/2,t)$.

5. (i) Find the solution of

$$\begin{array}{lll} \text{D.E.} & u_t - ku_{xx} = x & 0 < x < L\,, \quad 0 < t \\ \text{B.C.} & \left.\begin{array}{l} u_x(0,t) = 0 \\ u_x(L,t) = 0 \end{array}\right\} & \qquad\qquad\quad 0 < t \\ \text{I.C.} & u(x,0) = f(x) & 0 < x < L\,. \end{array}$$

(ii) Find an asymptotic solution of the problem of part (i).

6. Find the solution $u(x,t)$ of the problem

$$\begin{array}{lll} \text{D.E.} & u_t - ku_{xx} - ru = q(t) & 0 < x < L\,, \quad 0 < t \\ \text{B.C.} & \left.\begin{array}{l} u(0,t) = 0 \\ u(L,t) = 0 \end{array}\right\} & \qquad\qquad\quad 0 < t \\ \text{I.C.} & u(x,0) = 0 & 0 < x < L \end{array}$$

when

$$q(t) = \begin{cases} 1 & 0 < t < 2 \\ 0 & 2 < t\,. \end{cases}$$

7. (i) Show that the series solution of the problem

$$\begin{array}{lll} \text{D.E.} & u_t = ku_{xx} + q(x,t) & 0 < x < L\,, \quad t > 0\,, \\ \text{B.C.} & u(0,t) = 0 & \qquad\qquad\quad t > 0\,, \\ & u(L,t) = 0 & \qquad\qquad\quad t > 0\,, \\ \text{I.C.} & u(x,0) = 0 & 0 < x < L\,, \end{array}$$

can be formally converted to

$$u(x,t) = \int_0^t \Phi(x,t;\tau)\,d\tau$$

where, for fixed τ, Φ is the solution of

$$\begin{array}{lll} \text{D.E.} & \Phi_t = k\Phi_{xx}\,, & 0 < x < L\,, \quad t > \tau\,, \\ \text{B.C.} & \Phi(0,t;\tau) = 0\,, & \qquad\qquad\quad t > \tau\,, \\ & \Phi(L,t;\tau) = 0\,, & \qquad\qquad\quad t > \tau\,, \\ \text{I.C.} & \Phi(x,\tau;\tau) = q(x,\tau)\,, & 0 < x < L\,. \end{array}$$

This is a special case of *Duhamel's principle.*

(ii) Discuss (formulas *and explanation*) a heuristic derivation of the formula based on consideration of the contribution to the temperature at x at time t due to the heat supplied by the source in the time interval $(\tau,\tau + d\tau)$.

5.4. *FURTHER DISCUSSION OF THE METHOD OF VARIATION OF PARAMETERS*

The discussion in the preceding section of the solution of the general problem (5.1.1) assumed the possibility of the preliminary transformation of the problem into one with homogeneous B.C. This possibility exists if the functions $A(t)$ and $B(t)$ in the B.C. are differentiable for $t > 0$, and even under less restrictive hypotheses which are usually satisfied in problems of interest. It is nevertheless an important proposition, especially in connection with problems we will consider later, that the method of variation of parameters can be applied to problems such as (5.1.1) directly without any preliminary transformation.

Our procedure is exactly that of Section 3 up to Equation (5.3.8); that is, if $u(x,t)$ is the solution of (5.1.1), then

$$u(x,t) = \sum_{n=0}^{\infty} T_n(t)\varphi_n(x) ,$$

where

$$T_n'(t) = \frac{2k}{L}\int_0^L u_{xx}(x,t)\varphi_n(x)\,dx + q_n(t)$$

and

$$\frac{2k}{L}\int_0^L u_{xx}\varphi_n\,dx = \frac{2k}{L}\left[u_x(x,t)\varphi_n(x) - u(x,t)\varphi_n'(x)\right]_0^L + \frac{2k}{L}\int_0^L u(x,t)\varphi_n''(x)\,dx .$$

Now, however, the boundary terms in Equation (5.3.8) above do not all vanish. Instead, since $\varphi_n'(0) = 0$, $\varphi_n(L) = 0$, $u_x(0,t) = A(t)$, $u(L,t) = B(t)$, we have

$$\text{(5.4.1)} \quad \frac{2k}{L}\left[u_x(x,t)\varphi_n(x) - u(x,t)\varphi_n'(x)\right]_0^L = \frac{2k}{L}\left[-B(t)\varphi_n'(L) - A(t)\varphi_n(0)\right] .$$

With the particular eigenfunctions

$$(5.4.2) \qquad \varphi_n(x) = \cos\left(n + \frac{1}{2}\right)\frac{\pi}{L}x\,,$$

we have $\varphi_n(0) = 1$, $\varphi_n'(L) = (-1)^{n+1}(n + \frac{1}{2})\pi/L$, so that (5.4.1) becomes

$$(5.4.3) \qquad \frac{2k}{L}\,[u_x(x,t)\varphi_n(x) - u(x,t)\varphi_n'(x)]_0^L = \frac{2k}{L}\left[\left(n + \frac{1}{2}\right)\frac{\pi}{L}(-1)^n B(t) - A(t)\right],$$

where the right side of (5.4.3) is a known function of t when $A(t)$ and $B(t)$ are given. Substituting (5.4.3) in (5.3.8), we continue the discussion of Section 3, determining $T_n(t)$ and hence $u(x,t)$. The details are left as an exercise.

It should be observed in the discussion of Equation (5.3.8) in this and the preceding section, that a crucial role is played by the fact that the functions $\varphi_n(x)$ are eigenfunctions associated with the *related homogeneous problem*. The argument we have sketched furnishes a series expansion (5.3.4) of the solution $u(x,t)$ of (5.1.1) in terms of the eigenfunctions $\varphi_n(x)$. The reader may object that we have reached a contradiction: the function $u(x,t)$ satisfies inhomogeneous B.C., while each term of the series, and hence its sum, satisfies the related homogeneous B.C. At this point we respond only that there is no contradiction since the eigenfunction expansion is claimed to converge to $u(x,t)$ only for $0 < x < L$. However, further clarification is desirable, and will be furnished in Chapter 7.

EXERCISES 5c

1. Complete the discussion of this section and obtain a formula for the solution $u(x,t)$ of (5.1.1) when $A(t)$, $B(t)$ and $q(x,t)$ are any integrable functions of t, and x and t, respectively.

2. Find the solution $u(x,t)$ of the problem

$$\begin{array}{llll} \text{D.E.} & u_t - ku_{xx} = 0 & 0 < x < L\,, & 0 < t \\ \text{B.C.} & \left.\begin{array}{l} u_x(0,t) = A(t) \\ u(L,t) = 0 \end{array}\right\} & & 0 < t \\ \text{I.C.} & u(x,0) = 0 & 0 < x < L\,, & \end{array}$$

where

$$A(t) = \begin{cases} Q \quad (\text{constant}) & 0 < t < 1 \\ 0 & 1 < t\,. \end{cases}$$

3. Find a series solution of the problem

$$\begin{array}{lll} \text{D.E.} & u_t - ku_{xx} = 0 & 0 < x < L, \quad 0 < t \\ \text{B.C.} & \left.\begin{array}{l} u(0,t) = A(t) \\ u(L,t) = B(t) \end{array}\right\} & 0 < t \\ \text{I.C.} & u(x,0) = 0 & 0 < x < L, \end{array}$$

where $A(t)$, $B(t)$ are any integrable functions.

4. Show that the series solution of Problem 3 can be converted to

$$u(x,t) = \int_0^t \frac{\partial}{\partial t} \Phi(x,t;\tau)\, d\tau\,,$$

where $\Phi(x,t;\tau)$ is the solution of the initial-boundary value problem with *constant* B.C.:

$$\begin{array}{lll} \text{D.E.} & \Phi_t - k\Phi_{xx} = 0 & 0 < x < L, \quad \tau < t, \\ \text{B.C.} & \left.\begin{array}{l} \Phi(0,t;\tau) = A(\tau) \\ \Phi(L,t;\tau) = B(\tau) \end{array}\right\} & \tau < t, \\ \text{I.C.} & \Phi(x,\tau;\tau) = 0 & 0 < x < L. \end{array}$$

(This is *Duhamel's principle* for Problem 3.)

5. (i) Solve the problem

$$\begin{array}{lll} \text{D.E.} & u_t - ku_{xx} = q(x,t) & 0 < x < L, \quad 0 < t \\ \text{B.C.} & \left.\begin{array}{l} u(0,t) = A(t) \\ u(L,t) = B(t) \end{array}\right\} & 0 < t \\ \text{I.C.} & u(x,0) = 0 & 0 < x < L, \end{array}$$

where $A(t)$, $B(t)$ and $q(x,t)$ are any integrable functions of t and x and t, respectively, by the method of variation of parameters.

(ii) Show that the solution of part (i) may be written as

$$u(x,t) = \int_0^t \frac{\partial}{\partial t} \Phi(x,t;\tau)\, d\tau\,,$$

where $\Phi(x,t;\tau)$ is the solution of the *time-independent* inhomogeneous initial-boundary value problem:

$$\begin{array}{lll} \text{D.E.} & \Phi_t - k\Phi_{xx} = q(x,\tau) & 0 < x < L, \quad \tau < t, \\ \text{B.C.} & \left.\begin{array}{l} \Phi(0,t;\tau) = A(\tau) \\ \Phi(L,t;\tau) = B(\tau) \end{array}\right\} & \tau < t, \\ \text{I.C.} & \Phi(x,\tau;\tau) = 0 & 0 < x < L. \end{array}$$

This is the statement of **Duhamel's principle** for the general problem of part (i).

6. Consider a uniform rod of length L, with cross-sectional area S and with thermal constants c, ρ, k. Suppose that the temperature in the rod is initially 0, and that the ends are maintained at temperature 0. Then the solution of the problem

$$\begin{array}{llll} \text{D.E.} & u_t - ku_{xx} = q(x,t) & 0 < x < L\,, & 0 < t \\ \text{B.C.} & \left.\begin{array}{l} u(0,t) = 0 \\ u(L,t) = 0 \end{array}\right\} & & 0 < t \\ \text{I.C.} & u(x,0) = 0 & 0 < x < L & \end{array}$$

is the temperature distribution in the rod when heat is produced by internal sources at the point x and time t at *the rate* $c\rho q(x,t)$ *per unit volume per unit time*. The total heat production by these sources in any interval $[a,b]$ is thus given by

$$\int_0^\infty dt \int_a^b c\rho q(x,t) S\, dx\,.$$

Now suppose that

$$q(x,t;\sigma) = \begin{cases} Q(x)/\sigma & 0 < t < \sigma \\ 0 & \sigma < t \end{cases}$$

where σ is a parameter, and let $u(x,t;\sigma)$ denote the corresponding solution of the initial-boundary value problem. The total heat production in any interval $[a,b]$

$$\begin{aligned} \int_0^\infty dt \int_a^b Sc\rho q(x,t;\sigma) &= \int_0^\sigma dt \int_a^b \frac{c\rho Q(x)}{\sigma} S\, dx \\ &= \int_a^b c\rho Q(x) S\, dx \end{aligned}$$

is independent of σ. If σ is very small, then we may regard $u(x,t;\sigma)$ as roughly the temperature distribution due to sources acting instantaneously at time $t = 0$ and providing heat energy $c\rho Q(x)$ per unit volume. We *define* the temperature distribution $u(x,t)$ due to a distribution of **instantaneous sources** at time $t = 0$ producing at the point x heat energy $c\rho Q(x)$ *per unit volume* by

$$u(x,t) = \lim_{\sigma \to 0} u(x,t;\sigma)\,.$$

(i) Find $u(x,t;\sigma)$ and $u(x,t)$.
(ii) Use the result of part (i) to derive heuristically the result of Exercises 5b, Problem 7.

7. Consider a uniform rod of length L, cross section S and thermal constants c, ρ, κ, k. The solution of the problem

$$\begin{array}{llll} \text{D.E.} & u_t = ku_{xx} & 0 < x < L\,, & 0 < t\,, \\ \text{B.C.} & \left.\begin{array}{l} u_x(0,t) = A(t) \\ u(L,t) = 0 \end{array}\right\} & & 0 < t\,, \\ \text{I.C.} & u(x,0) = 0 & 0 < x < L & \end{array}$$

is the temperature in the rod at time t when the initial temperature is zero, the end $x = L$ is kept at temperature zero, and the flux at the end $x = 0$ is prescribed. The total heat derived from the source at $x = 0$ is

$$Q = \int_0^\infty -\kappa u_x(0,t)S\,dt = -\kappa S \int_0^\infty A(t)\,dt\,.$$

For the special case

$$A(t) = \begin{cases} \dfrac{-1}{\kappa S\sigma}, & 0 < t < \sigma \\ 0\,, & t > \sigma \end{cases}$$

the total heat supplied is $Q = 1$. For this special case let the solution be $v(x,t;\sigma)$.

(i) Find a series formula for $v(x,t;\sigma)$ valid for $t > \sigma$.

(ii) Calculate the limit

$$v(x,t) = \lim_{\sigma \to 0} v(x,t;\sigma)$$

for $0 \le x \le L$ and all $t > 0$. This limit is interpreted as the temperature distribution at time t due to an *initial instantaneous source* of unit heat energy at the end $x = 0$.

(iii) Use $v(x,t)$ to derive a formula for the solution of the problem with general $A(t)$.

8. Let α be a given constant and let $q(x)$ be continuous on $0 \le x \le L$. Consider the inhomogeneous boundary value problem

$$\text{(I)} \qquad \begin{array}{ll} \text{D.E.} & u''(x) + \alpha u(x) = -q(x) \qquad 0 < x < L\,, \\ \text{B.C.} & u(0) = 0\,, \quad u(L) = 0\,, \end{array}$$

and the related homogeneous problem

$$\text{(H)} \qquad \begin{array}{ll} \text{D.E.} & v''(x) + \alpha v(x) = 0 \qquad 0 < x < L\,, \\ \text{B.C.} & v(0) = 0\,, \quad v(L) = 0\,. \end{array}$$

(i) Prove the following propositions (use the principle of superposition). The difference of any two solutions of (I) is a solution of (H). The sum of a solution of (I) and a solution of (H) is a solution of (I).

(ii) Show that if problem (H) has no nontrivial solution, then for every $q(x)$ problem (I) has a solution of series form

$$u(x) = \sum_{n=1}^{\infty} b_n \varphi_n(x)\,,$$

where $\{\varphi_n(x)\}$ are the eigenfunctions of the problem

$$\text{(E)} \qquad \begin{array}{ll} \text{D.E.} & \varphi''(x) + \lambda\varphi(x) = 0\,, \qquad 0 < x < L\,, \\ \text{B.C.} & \varphi(0) = \varphi(L) = 0\,. \end{array}$$

(iii) Show that if problem (H) has a nontrivial solution $v(x)$, then problem (I) has more than one solution for those functions $q(x)$ that satisfy

$$\int_0^L q(x)v(x)\,dx = 0\,,$$

but problem (I) has no solution if $q(x)$ does not satisfy this condition. *Note:* In part (ii) the constant α is not one of the eigenvalues of (E), but in part (iii) α is one of the eigenvalues; that is, $\alpha - \lambda_n = 0$ for a certain value of n.

9. Solve the problem

$$\begin{array}{lll} \text{D.E.} & u_t = u_{xx} + \alpha u + Q & 0 < x < L\,, \quad 0 < t \\ \text{B.C.} & \left.\begin{array}{l} u(0,t) = 0 \\ u(L,t) = 0 \end{array}\right\} & 0 < t \\ \text{I.C.} & u(x,0) = f(x)\,, & 0 < x < L\,, \end{array}$$

where Q is constant and $f(x)$ arbitrary in the following cases:
(i) $\alpha \neq (n\pi/L)^2$ for any integer n;
(ii) $\alpha = (N\pi/L)^2$ for N an integer.
Discuss the differences between the two cases.

6

Fourier series

6.1. *HISTORICAL REMARKS*

In Chapter 4 we obtained a formal solution of the simplest initial-boundary value problem for the heat equation for all initial functions $f(x)$ which can be expanded in a series of the form

$$f(x) = \sum_{n=1}^{\infty} b_n \sin \frac{n\pi x}{L} \qquad 0 < x < L\,. \tag{6.1.1}$$

If every function $f(x)$ can be represented by such a series, then our result furnishes—at least formally—a solution of the initial-boundary value problem in general.

There is no intuitive reason for believing that the expansion (6.1.1) is always possible. On the contrary, it seems quite unlikely. Euler and d'Alembert, among other mathematicians, considered the possibility in similar connections, and rejected it. Nevertheless, Fourier in his study of heat conduction in 1811, boldly undertook to demonstrate the validity of the expansion and to obtain a formula for its coefficients.

Fourier began by considering a function $f(x)$ which could be expanded in a power series. Expanding both sides of (6.1.1) in power series and equating coefficients of equal powers he obtained an infinite system of linear equations for the infinitely many unknowns $b_1, b_2, \ldots$. He solved this system with an elaborate calculation, obtaining infinite series expressions for the coefficients $b_1, b_2, \ldots$. Then he recognized that these infinite series were the values of solutions of certain simple ordinary differential equations which he could solve in closed form. He thus obtained the formula

$$b_n = \frac{2}{L}\int_0^L f(x)\sin\frac{n\pi x}{L}\,dx\,. \tag{6.1.2}$$

From this formula follows

$$\int_0^L \sin\frac{m\pi x}{L}\sin\frac{n\pi x}{L}\,dx = \begin{cases} 0 & n \neq m \\ \dfrac{L}{2} & n = m\,, \end{cases} \tag{6.1.3}$$

since if $f(x) = \sin m\pi x/L$, $b_n = 0$ if $n \neq m$ and $b_m = 1$. Fourier then saw how (6.1.3) could be established directly and used to derive (6.1.2), as we did in Chapter 4.

Finally Fourier announced that the expansion (6.1.1) was valid for any function $f(x)$ for which the formula (6.1.2) made sense. He had not proved this, and indeed he did not claim to have done so, but it is clear that he regarded his calculations as partial justification for his statement. Further justification came from the calculation of some examples for which the convergence of the series to the function could be verified.

Fourier's claim was not correct, but it was very close to the truth. It is not true, for example, that Fourier's expansion is valid for every continuous function. Nevertheless, it is valid for every "reasonable" function—for example, for most of the functions encountered in applied mathematics. The first proof of the convergence of Fourier's series for a large and significant class of functions was provided by Dirichlet in 1829, eighteen years after Fourier's announcement. Further investigations, continuing to the present, have extended this result and revealed that the general question of the convergence of Fourier's series is very difficult. These investigations have led to important developments in all parts of mathematics from mathematical physics to abstract algebra, and have been responsible for the creation of whole new fields of mathematics.

Fourier also considered series of the form

$$f(x) = \frac{a_0}{2} + \sum_{n=1}^{\infty} a_n \cos\frac{n\pi x}{L} \qquad 0 < x < L \tag{6.1.4}$$

and of the form

$$f(x) = \frac{a_0}{2} + \sum_{n=1}^{\infty}\left(a_n\cos\frac{n\pi x}{L} + b_n\sin\frac{n\pi x}{L}\right) \qquad -L < x < L\,. \tag{6.1.5}$$

The series (6.1.1), (6.1.4) and (6.1.5), with the coefficients given by appropriate formulas, are called, respectively, *Fourier sine series*, *Fourier cosine series*, and *Fourier series*. The study of each of these representations can be reduced to the study of the particular case (6.1.5), and this can be specialized further by a change of variables, to the case in which $L = \pi$. We will establish the validity of these Fourier representations for a large class of functions $f(x)$, after an extensive preliminary discussion which includes the justification of the preceding statements.

6.2. *TRIGONOMETRIC SERIES AND PERIODIC FUNCTIONS*

A function $f(x)$ defined for all values of x is said to be **periodic with period** p if

$$f(x + p) = f(x) \tag{6.2.1}$$

for all x. The graphs of such a periodic function in any two successive intervals of length p are identical. Thus the graph of the function for all x is obtained by repeatedly translating through the distance p its graph in any interval $[a,a + p]$ of length p.

If a function is periodic with period p it is also periodic with period kp where k is any integer; that is,

$$f(x + kp) = f(x) \qquad k = \pm 1, \pm 2, \pm 3, \ldots . \tag{6.2.2}$$

For $k = 2$, for example, we get from (6.2.1)

$$f(x + 2p) = f([x + p] + p) = f(x + p) = f(x) .$$

It is easy to verify that the sum, product, quotient, and difference of periodic functions with a common period are periodic with the same period. Another simple property of periodic functions $f(x)$ with period p is

$$\int_a^{a+p} f(x)\,dx = \int_b^{b+p} f(x)\,dx , \tag{6.2.3}$$

that is, if $f(x)$ is integrable, its definite integrals over any two intervals of length p are equal. This is geometrically evident, and can be proved as follows. We have

$$\int_a^{a+p} f(x)\,dx = \int_a^0 f(x)\,dx + \int_0^p f(x)\,dx + \int_p^{a+p} f(x)\,dx .$$

But, considering the sum of the first and third integrals on the right above, and replacing x by $x + p$ in the third integral, we get

$$\int_a^0 f(x)\,dx + \int_p^{a+p} f(x)\,dx = \int_a^0 f(x)\,dx + \int_0^a f(x+p)\,dx$$
$$= \int_a^0 f(x)\,dx + \int_0^a f(x)\,dx = 0\,,$$

so that

$$\int_a^{a+p} f(x)\,dx = \int_0^p f(x)\,dx$$

for any a.

The functions $\sin \omega x$ and $\cos \omega x$ are periodic with period $2\pi/\omega$, for

$$\sin \omega\left(x + \frac{2\pi}{\omega}\right) = \sin(\omega x + 2\pi) = \sin \omega x\,.$$

Since every term of the sum

$$\frac{a_0}{2} + \sum_{k=1}^{n} (a_k \cos k\omega x + b_k \sin k\omega x)\,, \tag{6.2.4}$$

which is called a **trigonometric polynomial,** has period $2\pi/\omega$, the sum also is periodic with this period.

An infinite series of the form

$$\frac{a_0}{2} + \sum_{k=1}^{\infty} (a_k \cos k\omega x + b_k \sin k\omega x) \tag{6.2.5}$$

is called a **trigonometric series.** Since every partial sum of this series has period $2\pi/\omega$, it follows that if the series converges in any interval of length $2\pi/\omega$, then it converges for all x and its sum is periodic with period $2\pi/\omega$.

Given any function $f(x)$ in an interval $a < x \leq a + p$ we can find a periodic function with period p which is equal to $f(x)$ in the given interval. This function, which is called the **periodic extension** of $f(x)$ from the interval $a < x \leq a + p$, is obtained graphically by repeatedly translating the graph of $f(x)$ in $a < x \leq a + p$ to the left and right through the distance p. If we denote the periodic extension of $f(x)$ by $\varphi(x)$, then $\varphi(x)$ is defined by

$$\begin{aligned}
\varphi(x) &= f(x) & a &< x \leq a + p \\
\varphi(x) &= f(x - p) & a + p &< x \leq a + 2p \\
\varphi(x) &= f(x + p) & a - p &< x \leq a \\
\varphi(x) &= f(x - 2p) & a + 2p &< x \leq a + 3p \\
&\cdots & &\cdots\,.
\end{aligned} \tag{6.2.6}$$

The graphs of the periodic extensions of $f(x) = x$, $0 < x \leq 2\pi$, $f(x) = x^2$, $-\pi < x \leq \pi$ and $3 \sin \frac{1}{2}x$, $-\pi < x \leq \pi$ are shown in Figures 6.1, 6.2, and 6.3.

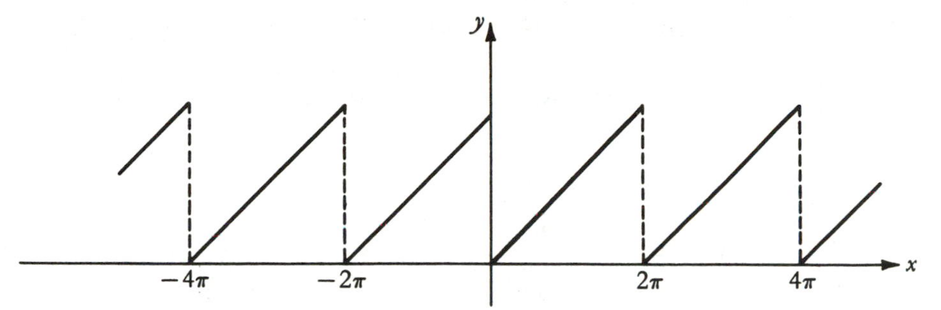

Fig. 6.1. Periodic extension of $f(x) = x \qquad 0 < x \leq 2\pi$.

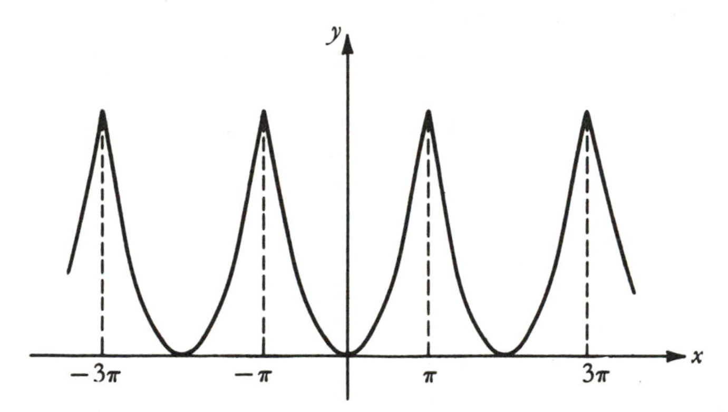

Fig. 6.2. Periodic extension of $f(x) = x^2 \qquad -\pi < x \leq \pi$.

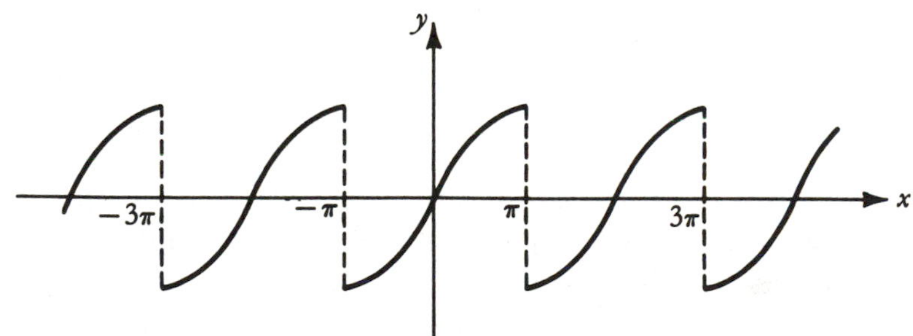

Fig. 6.3. Periodic extension of $f(x) = 3 \sin \frac{x}{2} \qquad -\pi < x \leq \pi$.

6.3. *SOME PROPERTIES OF FUNCTIONS*

A function $f(x)$ is **continuous at a point** x_0 if and only if the limit of the function at the point exists and is equal to the value of the function at the point; that is,

$$\lim_{x \to x_0} f(x) = f(x_0) \; . \tag{6.3.1}$$

A function is **continuous in an interval** if it is continuous at all points in the interval.

A function $f(x)$ is **bounded in an interval** if there is a number M such that

$$|f(x)| \leq M \tag{6.3.2}$$

for all x in the interval. It is known that every function continuous in a closed interval—an interval including its end points—is bounded.

In the study of Fourier series it is necessary, both for the theory and the applications, to consider functions having certain discontinuities but which are *bounded*. The simplest kind of discontinuity of a bounded function is that in which the limit of the function at a point exists, but either the function is not defined at the point or its value is not equal to its limit at the point. Such a discontinuity is called trivial or **removable** because the function can be made continuous just by defining or redefining its value at the point to be the value of its limit at the point. For example, the function

$$f(x) = \frac{\sin x}{x}$$

is not continuous at 0 because it is not defined at $x = 0$. However,

$$\lim_{x \to 0} f(x) = \lim_{x \to 0} \frac{\sin x}{x} = 1 ,$$

so that if we define

$$f(x) = \begin{cases} \dfrac{\sin x}{x} & x \neq 0 \\ 1 & x = 0 \end{cases}$$

the function $f(x)$ is continuous for all x (Fig. 6.4).

If a discontinuity at a point is not removable, then the limit of the function at the point fails to exist. In order to classify disconti-

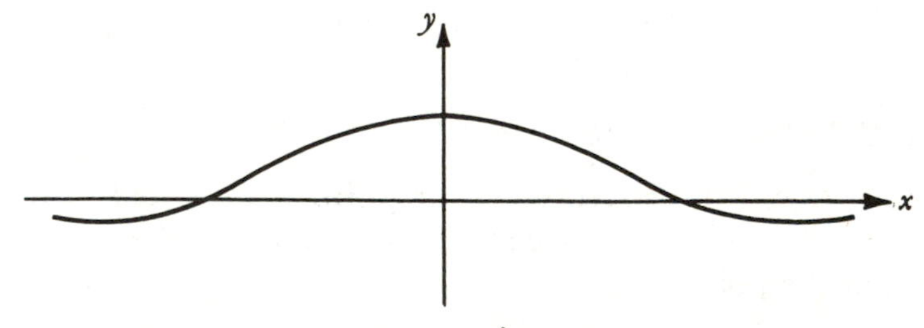

FIG. 6.4. Graph of $f(x) = \dfrac{\sin x}{x}$ $(x \neq 0)$, $f(0) = 1$.

nuities of this kind we introduce the notion of a one-sided limit. The **limit from the left** of $f(x)$ at x_0, denoted by $f(x_0-)$, is the limit of $f(x)$ as x approaches x_0 through values which are less than x_0; that is,

$$f(x_0-) = \lim_{\substack{x \to x_0 \\ x < x_0}} f(x) . \tag{6.3.3}$$

In a similar way we define the **limit from the right,**

$$f(x_0+) = \lim_{\substack{x \to x_0 \\ x > x_0}} f(x) . \tag{6.3.4}$$

If $f(x_0-)$ and $f(x_0+)$ exist and are equal, then clearly $\lim_{x \to x_0} f(x)$ exists and

$$\lim_{x \to x_0} f(x) = f(x_0+) = f(x_0-) ,$$

so that either $f(x)$ is continuous or has a removable discontinuity at x_0.

The next simplest case is that in which $f(x_0+)$ and $f(x_0-)$ both exist but are not equal. Such a point of discontinuity is called a **jump discontinuity,** and the difference $f(x_0+) - f(x_0-)$ is called the *jump* of the function at x_0. An example is

$$f(x) = \begin{cases} \dfrac{\sin x}{|x|} & x \neq 0 \\ 0 & x = 0 \end{cases}$$

whose graph is Figure 6.5. We have

$$f(0+) = \lim_{\substack{x \to 0 \\ x > 0}} \frac{\sin x}{|x|} = \lim_{\substack{x \to 0 \\ x > 0}} \frac{\sin x}{x} = 1$$

$$f(0-) = \lim_{\substack{x \to 0 \\ x < 0}} \frac{\sin x}{|x|} = \lim_{\substack{x \to 0 \\ x < 0}} \frac{\sin x}{-x} = -1$$

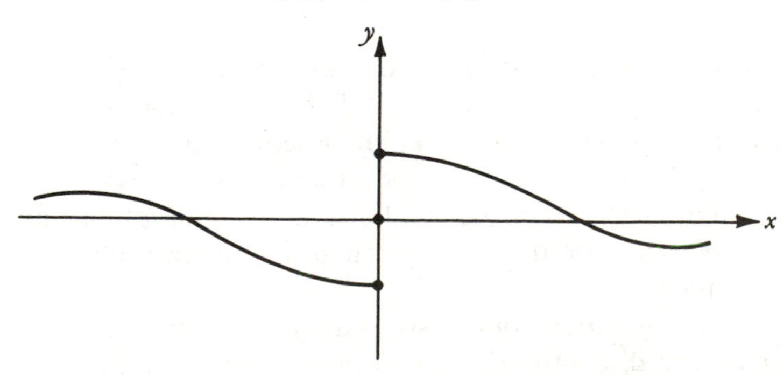

FIG. 6.5. Graph of $f(x) = \dfrac{\sin x}{|x|}$ $(x \neq 0)$, $f(0) = 0$.

so that at $x = 0$, $f(x)$ has a jump of 2. The value of a function at a jump discontinuity is arbitrary, and in our example is not equal to either $f(0+)$ or $f(0-)$. Indeed, a function may be undefined at a jump discontinuity.

Finally, either $f(x_0+)$, $f(x_0-)$, or both, may fail to exist. In this case $f(x)$ is said to have an **oscillatory discontinuity** at x_0. An example is

$$f(x) = \begin{cases} \sin \dfrac{1}{x} & x \neq 0 \\ 0 & x = 0, \end{cases}$$

the graph of which is indicated in Figure 6.6. The graph of a function with an oscillatory discontinuity can only be indicated; it cannot actually be drawn because the function oscillates infinitely often.

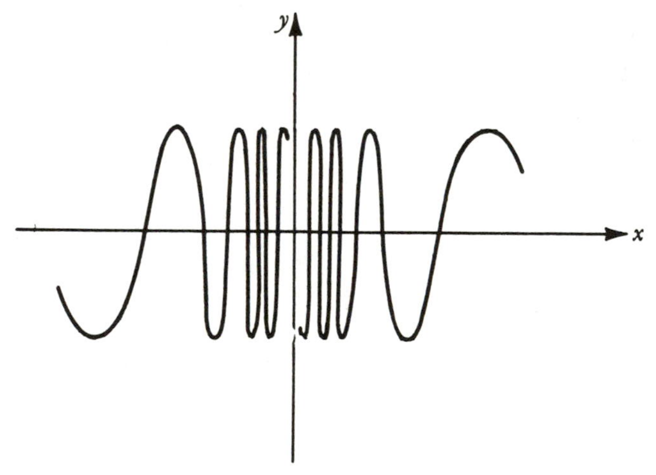

FIG. 6.6. Graph of $f(x) = \sin \frac{1}{x}$ $(x \neq 0)$, $f(0) = 0$.

We are now prepared to define the class of functions whose Fourier series we will study in detail. This class includes most of the functions one expects to encounter in practice. In particular, it includes those functions which can be described picturesquely by the statement that their entire graphs for a finite interval can actually be drawn. We shall use this description as a guide in the formulation of our definition.

First, such functions obviously must be bounded and may not have oscillatory discontinuities. They may have only removable or jump discontinuities, and indeed, only a finite number of these. We shall call a function **piecewise continuous** in a finite interval if it is continuous at all points in the interval, except possibly for a finite number of points at which it has removable or jump discontinuities.

The graphical description suggests a second requirement. At all but a finite number of points, the graph of such a function has a tangent line, and this line turns continuously as we move along the curve between any two exceptional points. Thus the derivative of such a function exists and is continuous except at a finite number of points. Clearly these discontinuities cannot be oscillatory, so that if we require that the derivative be bounded it can have only removable or jump discontinuities. We shall call a function **piecewise smooth** in an interval if it is piecewise continuous and has a piecewise continuous derivative in the interval. A function which is continuous and has a continuous derivative will be called **smooth.**

The class of functions whose Fourier series we will study in detail is the class of *piecewise smooth functions*. Nevertheless much of the theory can be developed for functions which are only piecewise continuous, and even for a somewhat larger class of functions.

We shall call a function $f(x)$ **absolutely integrable** on an interval $a \leq x \leq b$ if it is continuous on the interval except at a finite number of points and

$$\int_a^b |f(x)|\,dx \tag{6.3.5}$$

exists—possibly as an improper integral. It is clear from the geometric interpretation of definite integral that every piecewise continuous function is absolutely integrable, but there are absolutely integrable functions which are not piecewise continuous, for example, $f(x) = x^{-1/2}$ in $0 \leq x \leq 1$. If $f(x)$ is absolutely integrable then it is integrable; that is, if (6.3.5) exists then

$$\int_a^b f(x)\,dx \tag{6.3.6}$$

exists.

The sum, difference, and product of piecewise continuous or piecewise smooth functions are correspondingly piecewise continuous or piecewise smooth. The sum and difference of absolutely integrable functions are absolutely integrable, but the corresponding statement for the product is false. For example, $x^{-1/2}$ is absolutely integrable on $[0,1]$, but $x^{-1} = x^{-1/2} \cdot x^{-1/2}$ is not. However, the product of an absolutely integrable function and a *bounded* integrable function is absolutely integrable.

The preceding definitions refer to functions defined in finite intervals. A function defined in an infinite interval, in particular a periodic function, will be called *piecewise continuous*, or *piecewise smooth*, if it has the property in every finite interval. The definition of absolute integrability in the case of an infinite interval is the same as that for a finite interval.

EXERCISES 6a

1. Sketch the graph of

$$\lim_{n\to\infty} \frac{1}{1 + x^{2n}} \cdot$$

Is this function continuous, smooth, piecewise smooth?

2. For what values of a and b is the function

$$f(x) = \begin{cases} 3x^2 + 5x + 1 & x < 2 \\ ax + b & 2 \leq x \end{cases}$$

(i) Piecewise smooth?
(ii) Continuous and piecewise smooth?
(iii) Smooth (in every interval)?

3. (i) Show that if $a < b$, the function

$$f(x) = \begin{cases} 0 & x < a \\ (x - a)^2(x - b)^2 & a \leq x \leq b \\ 0 & b < x \end{cases}$$

is smooth in every interval.

(ii) Find a function $f(x)$ defined for all x, and not identically zero, such that $f(x) = 0$ for $x \leq a$ and for $x \geq b$, $f(x)$ is a polynomial for $a \leq x \leq b$, and $f(x)$ has a continuous *second* derivative for all x.

4. Let $f(x)$ have a continuous second derivative for $-L \leq x \leq L$. State necessary and sufficient conditions for the periodic extension of $f(x)$ to
(i) be continuous and piecewise smooth,
(ii) be smooth,
(iii) have a continuous second derivative.

5. Show that $f(x) = x \sin(1/x)$ is continuous for all x, but is not piecewise smooth in any interval containing $x = 0$. What can be said of the function $f(x) = x^3 \sin(1/x)$?

6. Show that the formula for integration by parts

$$\int_a^b f(x)g'(x)\,dx = f(x)g(x)\Big|_a^b - \int_a^b f'(x)g(x)\,dx$$

holds if f and g are continuous and piecewise smooth on $[a,b]$. Conclude that Green's formula (4.6.2) holds if f and g are continuous and have continuous piecewise smooth derivatives.

6.4. *FOURIER SERIES*

Let $f(x)$ be a periodic function with period 2π, absolutely integrable on every finite interval, and suppose that it can be expanded in a trigonometric series with period 2π,

$$f(x) = \frac{a_0}{2} + \sum_{k=1}^{\infty} (a_k \cos kx + b_k \sin kx), \tag{6.4.1}$$

where the series on the right converges for all x. If we multiply both sides of Equation (6.4.1) by $\cos nx$ and integrate from $-\pi$ to π, we obtain

$$\int_{-\pi}^{\pi} f(x) \cos nx \, dx = \int_{-\pi}^{\pi} \left(\frac{a_0}{2} + \sum_{k=1}^{\infty} (a_k \cos kx + b_k \sin kx) \right) \cos nx \, dx \, . \tag{6.4.2}$$

If we assume that the series on the right can be integrated term-by-term—this will be true, for example, if the series (6.4.2) converges uniformly in the interval $[-\pi,\pi]$—we get

$$\int_{-\pi}^{\pi} f(x) \cos nx \, dx = \frac{a_0}{2} \int_{-\pi}^{\pi} \cos nx \, dx + \sum_{k=1}^{\infty} \left(a_k \int_{-\pi}^{\pi} \cos kx \cos nx \, dx + b_k \int_{-\pi}^{\pi} \sin kx \cos nx \, dx \right). \tag{6.4.3}$$

Now from the identities

$$\begin{aligned} \cos mx \cos nx &= \frac{\cos (m-n)x + \cos (m+n)x}{2} \\ \sin mx \cos nx &= \frac{\sin (m-n)x + \sin (m+n)x}{2} \\ \sin mx \sin nx &= \frac{\cos (m-n)x - \cos (m+n)x}{2} \end{aligned} \tag{6.4.4}$$

we obtain immediately

$$\begin{aligned} \int_{-\pi}^{\pi} \cos mx \cos nx \, dx &= \begin{cases} 0 & \text{if} \quad m \neq n \\ \pi & \text{if} \quad m = n \neq 0 \\ 2\pi & \text{if} \quad m = n = 0 \end{cases} \\ \int_{-\pi}^{\pi} \sin mx \cos nx \, dx &= 0 \\ \int_{-\pi}^{\pi} \sin mx \sin nx \, dx &= \begin{cases} 0 & \text{if} \quad m \neq n \\ \pi & \text{if} \quad m = n \neq 0 \, . \end{cases} \end{aligned} \tag{6.4.5}$$

From the second formula of (6.4.5) we see that the factor multiplying b_k in (6.4.3) is 0 for all k. From the first of the formulas, the coefficient of a_k in (6.4.3) is 0 unless $k = n$, and the coefficient of a_n is π. Thus the series in (6.4.3) reduces to a single term, and we have

$$\int_{-\pi}^{\pi} f(x) \cos nx \, dx = \pi a_n$$

or

$$a_n = \frac{1}{\pi} \int_{-\pi}^{\pi} f(x) \cos nx \, dx \qquad n = 0, 1, 2, \ldots . \tag{6.4.6}$$

In a similar way, multiplying (6.4.2) by $\sin nx$ and integrating we get

$$b_n = \frac{1}{\pi} \int_{-\pi}^{\pi} f(x) \sin nx \, dx \qquad n = 1, 2, \ldots . \tag{6.4.7}$$

The integrals in (6.4.6) and (6.4.7) are the integrals of products of periodic functions with period 2π over an interval of length 2π, and may therefore be replaced by integrals over any interval $[c, c + 2\pi]$ of length 2π,

$$a_n = \frac{1}{\pi} \int_{c}^{c+2\pi} f(x) \cos nx \, dx \qquad n = 0, 1, 2, \ldots , \tag{6.4.8}$$

$$b_n = \frac{1}{\pi} \int_{c}^{c+2\pi} f(x) \sin nx \, dx \qquad n = 1, 2, \ldots . \tag{6.4.9}$$

Now let $f(x)$ be any periodic function with period 2π, absolutely integrable on any finite interval and *define* a_n and b_n by (6.4.6) and (6.4.7). Both factors of the integrands in these formulas are absolutely integrable and one is bounded, so that their product is absolutely integrable, and hence integrable. Thus these formulas are proper definitions. Our discussion above suggests that if there is any trigonometric series

$$\frac{a_0}{2} + \sum_{k=1}^{\infty} (a_k \cos kx + b_k \sin kx) \tag{6.4.10}$$

which converges to $f(x)$, in some appropriate sense, its coefficients will be those just defined. We shall call the series (6.4.10) with coefficients defined by (6.4.6) and (6.4.7) the *Fourier series of the periodic function* $f(x)$, and write

$$f(x) \sim \frac{a_0}{2} + \sum_{k=1}^{\infty} (a_k \cos kx + b_k \sin kx) . \tag{6.4.11}$$

The symbol $\sim$ is read "is the Fourier series of." It would be incorrect to write $=$ instead of $\sim$, since we do not even know if the series converges, and indeed it may not. Even when the series converges, equality may not hold.

Next, let $f(x)$ be an absolutely integrable function defined on an interval containing the interval $[-\pi,\pi]$. We will again define the coefficients a_n and b_n by (6.4.6) and (6.4.7) and call (6.4.10) the *Fourier series of* $f(x)$ *on the interval* $[-\pi,\pi]$. Symbolically we write

$$f(x) \sim \frac{a_0}{2} + \sum_{k=1}^{\infty} (a_k \cos kx + b_k \sin kx) \qquad [-\pi,\pi]\,.$$

This definition is clearly equivalent to the statement that the Fourier series of a function on the interval $[-\pi,\pi]$ is the Fourier series of its periodic extension (with period 2π) from that interval.

Finally, *consider a function* $f(x)$ *defined and absolutely integrable on an interval* $[-L,L]$ *where* L *is arbitrary. We define the* **Fourier series** *of* $f(x)$ *on this interval by*

$$\text{(6.4.12)}\qquad f(x) \sim \frac{a_0}{2} + \sum_{k=1}^{\infty} \left(a_k \cos \frac{k\pi x}{L} + b_k \sin \frac{k\pi x}{L}\right) \qquad [-L,L]$$

$$\text{(6.4.13)}\qquad a_n = \frac{1}{L}\int_{-L}^{L} f(x) \cos \frac{n\pi x}{L}\, dx \qquad n = 0, 1, 2, \ldots$$

$$\text{(6.4.14)}\qquad b_n = \frac{1}{L}\int_{-L}^{L} f(x) \sin \frac{n\pi x}{L}\, dx \qquad n = 1, 2, \ldots.$$

It is desirable to relate the general Fourier series (6.4.12) to the special case in which $L = \pi$, since then every question concerning (6.4.12) can be reduced to a question concerning this special case. This can be accomplished by a change of scale, that is, by the introduction of a new variable ξ through

$$\text{(6.4.15)}\qquad \xi = \frac{\pi}{L}x\,, \quad x = \frac{L}{\pi}\xi\,.$$

Instead of $f(x)$ we consider the function $f\left(\frac{L}{\pi}\xi\right)$ which is defined and absolutely integrable in the interval $[-\pi,\pi]$. Writing the Fourier series of this function we have

$$f\left(\frac{L}{\pi}\xi\right) \sim \frac{a_0}{2} + \sum_{k=1}^{\infty} (a_k \cos k\xi + b_k \sin k\xi)$$

$$\text{(6.4.16)}\qquad a_n = \frac{1}{\pi}\int_{-\pi}^{\pi} f\left(\frac{L}{\pi}\xi\right) \cos n\xi\, d\xi$$

$$b_n = \frac{1}{\pi}\int_{-\pi}^{\pi} f\left(\frac{L}{\pi}\xi\right) \sin n\xi\, d\xi\,.$$

Substitution of (6.4.15) in Equations (6.4.16) yields (6.4.12)–(6.4.14).

There are two principal questions concerning the convergence of Fourier series. For what values of x does the Fourier series of the function $f(x)$ converge? If the Fourier series of $f(x)$ converges for a particular value x, what is the relation of its sum to the value of the function at x? For piecewise smooth functions complete answers to both questions are furnished by the following theorem which will be proved in Section 7.

Let $f(x)$ be piecewise smooth on the interval $-L \leq x \leq L$, and let $\varphi(x)$ be the periodic extension of $f(x)$. Then the Fourier series of $f(x)$ converges at every value of x, and the sum of the series is the average of the limit from the left and the limit from the right of $\varphi(x)$ at x. In other words,

$$(6.4.17) \quad \frac{a_0}{2} + \sum_{k=1}^{\infty} \left(a_k \cos \frac{k\pi x}{L} + b_k \sin \frac{k\pi x}{L} \right) = \frac{\varphi(x+) + \varphi(x-)}{2},$$

where the series is the Fourier series of $f(x)$.

The most interesting values of x are, of course, in the interval $-L \leq x \leq L$. First suppose that $-L < x < L$. Then $\varphi(x+) = f(x+)$ and $\varphi(x-) = f(x-)$ so (6.4.17) is the same as

$$(6.4.18) \quad \frac{a_0}{2} + \sum_{k=1}^{\infty} \left(a_k \cos \frac{k\pi x}{L} + b_k \sin \frac{k\pi x}{L} \right) = \frac{f(x+) + f(x-)}{2}.$$

Moreover, except for a finite number of values of x, $f(x)$ is continuous and $f(x+) = f(x-) = f(x)$. Hence if $-L < x < L$ and $f(x)$ is continuous at x, then

$$(6.4.19) \quad \frac{a_0}{2} + \sum_{k=1}^{\infty} \left(a_k \cos \frac{k\pi x}{L} + b_k \sin \frac{k\pi x}{L} \right) = f(x).$$

We return to Equation (6.4.17) and consider the case $x = L$. Since

$$\varphi(L-) = \lim_{\substack{x \to L \\ x < L}} \varphi(x) = \lim_{\substack{x \to L \\ x < L}} f(x) = f(L-)$$

and

$$\varphi(L+) = \lim_{\substack{x \to L \\ x > L}} \varphi(x) = \lim_{\substack{x \to -L \\ x > -L}} f(x) = f(-L+),$$

we have from (6.4.17)

$$(6.4.20) \quad \left[\frac{a_0}{2} + \sum_{k=1}^{\infty} \left(a_k \cos \frac{k\pi x}{L} + b_k \sin \frac{k\pi x}{L} \right) \right]_{x=L} = \frac{f(L-) + f(-L+)}{2}.$$

Since the sum of the series is periodic with period $2L$, it has the same value at $x = -L$ as at $x = L$.

As a first example we will calculate the Fourier series of $f(x) = 1 + x$ on the interval $[-\pi,\pi]$. From (6.4.13) we have

$$a_n = \frac{1}{\pi}\int_{-\pi}^{\pi} (1 + x) \cos nx \, dx = \frac{1}{\pi}\left[\frac{(1 + x) \sin nx}{n} + \frac{\cos nx}{n^2}\right]_{-\pi}^{\pi} = 0 .$$

Here we must assume $n \neq 0$, since otherwise the integration formula does not make sense. It is true that the formula for a_0 is a special case of that for a_n. Indeed it was to achieve this that the constant term in the Fourier series was written as $a_0/2$ rather than a_0. Nevertheless, as this example shows, the calculation of a_0 should be performed separately. We have

$$a_0 = \frac{1}{\pi}\int_{-\pi}^{\pi} (1 + x) \, dx = \frac{1}{\pi}\left[\frac{x^2}{2} + x\right]_{-\pi}^{\pi} = 2 .$$

From (6.4.14) we obtain

$$\begin{aligned} b_n = \frac{1}{\pi}\int_{-\pi}^{\pi} (1 + x) \sin nx \, dx &= -\frac{1}{\pi}\left[\frac{(1 + x) \cos nx}{n} + \frac{\sin nx}{n^2}\right]_{-\pi}^{\pi} \\ &= \frac{2(-1)^{n+1}}{n} . \end{aligned}$$

Thus

$$1 + x \sim 1 + \sum_{k=1}^{\infty} \frac{2(-1)^{k+1}}{k} \sin kx \qquad [-\pi,\pi] .$$

Since $f(x) = 1 + x$ is continuous for $-\pi < x < \pi$, we have from the convergence theorem stated above

$$\begin{aligned} 1 + x &= 1 + \sum_{k=1}^{\infty} \frac{2(-1)^{k+1}}{k} \sin kx \\ &= 1 + 2\left(\sin x - \frac{\sin 2x}{2} + \frac{\sin 3x}{3} - \cdots\right), \qquad -\pi < x < \pi . \end{aligned}$$

In particular, setting $x = \pi/2$ we obtain a remarkable series for $\pi/4$,

$$\frac{\pi}{4} = 1 - \frac{1}{3} + \frac{1}{5} - \frac{1}{7} + \cdots .$$

At the end-points π and $-\pi$ of the interval the Fourier series converges to 1, in accordance with the convergence theorem. Outside the interval $-\pi \leq x \leq \pi$ the Fourier series converges to the periodic extension of $1 + x$ from $[-\pi,\pi]$ sketched in Figure 6.7.

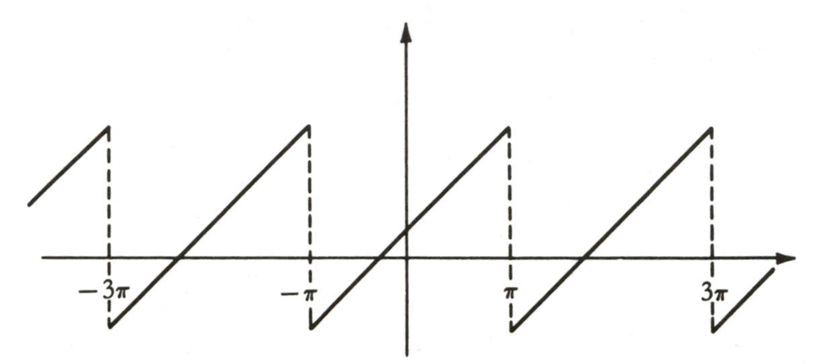

FIG. 6.7. The graph of $1 + \sum_{k=1}^{\infty} \frac{2(-1)^{k+1}}{k} \sin kx$.

As a second example we write the Fourier series of the function $f(x) = x$ on the interval $[-\pi,\pi]$, namely

$$x \sim \sum_{k=1}^{\infty} \frac{2(-1)^{k+1}}{k} \sin kx \qquad [-\pi,\pi] .$$

This can be inferred from the preceding example using the simple but useful proposition that the Fourier coefficients of sums, differences, or constant multiples of functions are the corresponding sums, differences, or constant multiples of the Fourier coefficients of the functions, together with the easily verified statement that the Fourier series of the constant function 1 in an interval is the series consisting of the single term 1.

As a final example consider the function

$$f(x) = \begin{cases} 1 & -1 \leq x < 0 \\ \frac{1}{2} & x = 0 \\ x & 0 < x \leq 1 \end{cases}$$

in the interval $[-1,1]$. Its Fourier coefficients are

$$\begin{aligned} a_n &= \int_{-1}^{1} f(x) \cos n\pi x \, dx = \int_{-1}^{0} f(x) \cos n\pi x \, dx + \int_{0}^{1} f(x) \cos n\pi x \, dx \\ &= \int_{-1}^{0} \cos n\pi x \, dx + \int_{0}^{1} x \cos n\pi x \, dx \\ &= \left[\frac{\sin n\pi x}{n\pi}\right]_{-1}^{0} + \left[\frac{x \sin n\pi x}{n\pi} + \frac{\cos n\pi x}{n^2\pi^2}\right]_{0}^{1} \\ &= \frac{\cos n\pi - 1}{n^2\pi^2} = \frac{(-1)^n - 1}{n^2\pi^2} \qquad n = 1, 2, \ldots \end{aligned}$$

$$b_n = \int_{-1}^{1} f(x) \sin n\pi x \, dx = \int_{-1}^{0} f(x) \sin n\pi x \, dx + \int_{0}^{1} f(x) \sin n\pi x \, dx$$
$$= \int_{-1}^{0} \sin n\pi x \, dx + \int_{0}^{1} x \sin n\pi x \, dx$$
$$= \left[-\frac{\cos n\pi x}{n\pi} \right]_{-1}^{0} + \left[-\frac{x \cos n\pi x}{n\pi} + \frac{\sin n\pi x}{n^2\pi^2} \right]_{0}^{1}$$
$$= -\frac{1}{n\pi}$$
$$a_0 = \int_{-1}^{1} f(x) \, dx = \int_{-1}^{0} f(x) \, dx + \int_{0}^{1} f(x) \, dx = \int_{-1}^{0} 1dx + \int_{0}^{1} x \, dx$$
$$= \frac{3}{2} \cdot$$

Since at its only point of discontinuity, $x = 0$, the function f has been defined so that $f(0) = \frac{1}{2}[f(0+) + f(0-)]$, and further since $f(-1) = f(1) = f(1-) = f(-1+)$, the Fourier series of f converges to $f(x)$ for all x in the interval $-1 \leq x \leq 1$; that is,

$$\begin{Bmatrix} 1 & -1 \leq x < 0 \\ \frac{1}{2} & x = 0 \\ x & 0 < x \leq 1 \end{Bmatrix} = \frac{3}{4} + \sum_{n=1}^{\infty} \left[\frac{(-1)^n - 1}{n^2\pi^2} \cos n\pi x - \frac{1}{n\pi} \sin n\pi x \right] \cdot$$

If, in particular, we set $x = 0$ we obtain the formula

$$\frac{\pi^2}{8} = 1 + \frac{1}{3^2} + \frac{1}{5^2} + \frac{1}{7^2} + \cdots.$$

The graph of the periodic function to which the series converges is shown in Figure 6.8. The realization that an "artificial" function

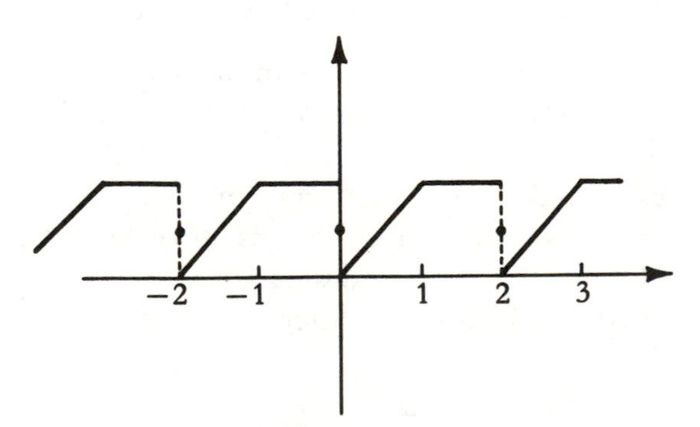

FIG. 6.8. Graph of $\frac{3}{4} + \sum_{n=1}^{\infty} \left[\frac{(-1)^n - 1}{n^2\pi^2} \cos n\pi x - \frac{1}{n\pi} \sin n\pi x \right]$.

such as this could be represented by such a "natural" series resulted in a profound change in the concept of function in Fourier's time.

In practice Fourier series are employed in two ways: to represent a periodic function $f(x)$ for all x; to represent a function $f(x)$ defined in a finite interval in that interval alone. In the latter case we may still consider the Fourier series as that of a periodic function, namely the appropriate periodic extension of the given function, and it will often be useful to do so.

EXERCISES 6b

1. Calculate the Fourier series of the function

$$f(x) = \begin{cases} -1 & -3 < x < 0 \\ +1 & 0 < x < 3 \end{cases}$$

on the interval $[-3,3]$. Sketch the graph of the sum of the first three terms of this series. (Graph each term and use graphical addition.)

2. Calculate the Fourier series of the function $f(x) = x^2$ on the interval $[-\pi,\pi]$. Find the sum of the series $1 - \frac{1}{2^2} + \frac{1}{3^2} - \cdots$.

3. Calculate the Fourier series of the function $f(x) = 2x - 1$ on the interval $[-1,1]$. Sketch the graph of the periodic function to which the Fourier series converges.

4. The function $f(x)$ is periodic with period 2, and on the interval $1 \leq x \leq 3$ is given by $f(x) = (x - 1)(x - 3)$. Find its Fourier series.

5. Show that the convergence theorem of Section 4 holds for arbitrary L if it holds for the special case $L = \pi$.

6.5. *FOURIER SINE AND COSINE SERIES*

It may happen, as in the second example in the preceding section, that in the Fourier series of a function all of the cosine terms or all of the sine terms are absent. Such series are called *sine series* and *cosine series*, respectively. We shall determine when and why they occur, and this knowledge will enable us to represent any function in an interval by either a sine or cosine series.

A function $f(x)$ defined on an interval $-L \leq x \leq L$ is said to be **even** if

$$f(-x) = f(x) \tag{6.5.1}$$

and **odd** if

$$f(-x) = -f(x) \,. \tag{6.5.2}$$

The functions x^2, $\cos ax$ are examples of even functions and x, $\sin ax$ are examples of odd functions. A function is even if its graph is symmetric with respect to the vertical axis and odd if its graph is symmetric with respect to the origin (Fig. 6.9). The graph of an

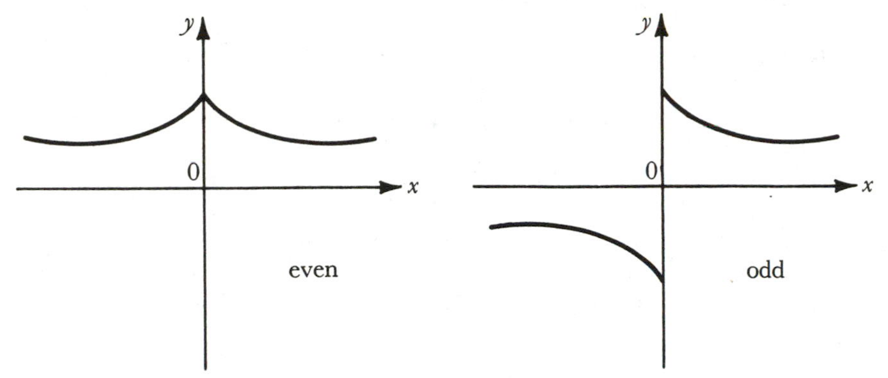

Fig. 6.9. Graphs of even and odd functions.

even function can be obtained from its graph to the right of the vertical axis by reflection in that axis; that of an odd function by first reflecting in the vertical axis and then reflecting in the horizontal axis.

It is easy to verify that: the sum of even functions is even and the sum of odd functions is odd; the product of two even functions or of two odd functions is even; the product of an even function and an odd function is odd. For example, if $f(x)$ and $g(x)$ are odd and $h(x) = f(x)g(x)$, we have

$$f(-x) = -f(x),\ g(-x) = -g(x)$$
$$h(-x) = f(-x)g(-x) = [-f(x)][-g(x)] = f(x)g(x) = h(x)$$

so that $h(x)$ is even.

Since each of the functions $\sin(n\pi x/L)$ is odd, a periodic function whose Fourier series is a sine series must be odd, and similarly a function whose Fourier series is a cosine series must be even. We shall establish the converse of this statement. The Fourier series on $[-L,L]$ of an odd function is a sine series, and that of an even function is a cosine series.

The additional properties of even and odd functions that we need for this are the following. If $f(x)$ is odd on $[-L,L]$, then

$$\int_{-L}^{L} f(x)\, dx = 0 \qquad (6.5.3)$$

and if $f(x)$ is even on $[-L,L]$, then

(6.5.4) $$\int_{-L}^{L} f(x)\,dx = 2\int_0^L f(x)\,dx\,.$$

These statements are geometrically obvious (see Fig. 6.9), and are easily proved. Suppose $f(x)$ is odd. Then

$$\begin{aligned}\int_{-L}^{L} f(x)\,dx &= \int_{-L}^{0} f(x)\,dx + \int_0^L f(x)\,dx = -\int_L^0 f(-\xi)\,d\xi + \int_0^L f(x)\,dx\\ &= \int_0^L f(-\xi)\,d\xi + \int_0^L f(x)\,dx\\ &= \int_0^L -f(\xi)\,d\xi + \int_0^L f(x)\,dx = 0\,.\end{aligned}$$

The statement for even functions is proved similarly.

Now let $f(x)$ be an even function on $[-L,L]$ and consider its Fourier series on this interval. Since $\sin(n\pi x/L)$ is odd for all n, the products $f(x)\sin(n\pi x/L)$ are odd, and hence

$$b_n = \frac{1}{L}\int_{-L}^{L} f(x)\sin\frac{n\pi x}{L}\,dx = 0\,.$$

Furthermore $f(x)\cos n\pi x/L$ is even for all n, so that

$$a_n = \frac{1}{L}\int_{-L}^{L} f(x)\cos\frac{n\pi x}{L}\,dx = \frac{2}{L}\int_0^L f(x)\cos\frac{n\pi x}{L}\,dx\,.$$

Thus *if $f(x)$ is even, its Fourier series is a cosine series,*

(6.5.5) $$f(x) \sim \frac{a_0}{2} + \sum_{k=1}^{\infty} a_k\cos\frac{k\pi x}{L} \qquad [-L,L]$$

(6.5.6) $$a_n = \frac{2}{L}\int_0^L f(x)\cos\frac{n\pi x}{L}\,dx \qquad n = 0, 1, 2, \ldots.$$

In a similar way we can show that *if $f(x)$ is odd on $[-L,L]$, its Fourier series is a sine series*

(6.5.7) $$f(x) \sim \sum_{k=1}^{\infty} b_k\sin\frac{k\pi x}{L} \qquad [-L,L]$$

(6.5.8) $$b_n = \frac{2}{L}\int_0^L f(x)\sin\frac{n\pi x}{L}\,dx\,.$$

Notice that the integrals in (6.5.6) and (6.5.8) depend only on the values of $f(x)$ in the interval $0 \le x \le L$.

Suppose $f(x)$ is any function on the interval $0 \le x \le L$. There is an even function on the interval $-L \le x \le L$, called the **even extension** of $f(x)$, which is identical with $f(x)$ in $0 \le x \le L$. Graphically, the even extension of $f(x)$ is obtained simply by reflecting the graph of $f(x)$ in the vertical axis. The Fourier series of the even extension of $f(x)$ is a cosine series, which is called the *Fourier cosine series* of $f(x)$ on the interval $[0,L]$. The formulas for the coefficients of this series involve only $f(x)$ since the integration is performed over the interval $0 \leq x \leq L$, and in this interval $f(x)$ and its even extension are identical. Thus the **Fourier cosine series** *of* $f(x)$ *on* $[0,L]$ *is defined by*

$$f(x) \sim \frac{a_0}{2} + \sum_{k=1}^{\infty} a_k \cos\frac{k\pi x}{L} \qquad [0,L], \tag{6.5.9}$$

where

$$a_n = \frac{2}{L}\int_0^L f(x)\cos\frac{n\pi x}{L}\,dx \qquad n = 0, 1, 2, \ldots. \tag{6.5.10}$$

If $f(x)$ is piecewise smooth, then its even extension will also be piecewise smooth. Hence the Fourier cosine series of $f(x)$ will converge at each point to the average of the limit from the left and the limit from the right at the point, of the periodic extension of the even extension of $f(x)$. In particular, at all points interior to the interval $0 < x < L$ at which $f(x)$ is continuous, the series will converge to $f(x)$. In this manner every piecewise smooth function in $[0,L]$ can be expanded in a cosine series.

In a similar way we define the **odd extension** of a function $f(x)$, and the **Fourier sine series** of $f(x)$ on $[0,L]$,

$$f(x) \sim \sum_{k=1}^{\infty} b_k \sin\frac{k\pi x}{L} \qquad [0,L] \tag{6.5.11}$$

$$b_n = \frac{2}{L}\int_0^L f(x)\sin\frac{n\pi x}{L}\,dx \qquad n = 1, 2, \ldots, \tag{6.5.12}$$

as the Fourier series of the odd extension of $f(x)$. If $f(x)$ is piecewise smooth, then its Fourier sine series will converge to $f(x)$ at all points in the interval $0 < x < L$ at which $f(x)$ is continuous, and to $[f(x+) + f(x-)]/2$ at discontinuities. At $x = 0$ and $x = L$ the series converges to zero.

As examples we will calculate the Fourier cosine series, and the Fourier sine series of the function

$$f(x) = \begin{cases} 1 & 0 \leq x \leq \dfrac{\pi}{2} \\ 0 & \dfrac{\pi}{2} < x \leq \pi \end{cases}$$

on the interval $[0,\pi]$.

The coefficients of the Fourier cosine series of $f(x)$ on $[0,\pi]$ are

$$a_0 = \frac{2}{\pi}\int_0^\pi f(x)\,dx = \frac{2}{\pi}\int_0^{\pi/2} 1\,dx = 1$$

$$a_n = \frac{2}{\pi}\int_0^\pi f(x)\cos nx\,dx = \frac{2}{\pi}\int_0^{\pi/2}\cos nx\,dx = \frac{2\sin\frac{n\pi}{2}}{n\pi}, \qquad n = 1, 2, \ldots$$

so that the Fourier cosine series of $f(x)$ in $[0,\pi]$ is

$$f(x) \sim \frac{1}{2} + \frac{2}{\pi}\cos x - \frac{2}{3\pi}\cos 3x + \frac{2}{5\pi}\cos 5x - \cdots \qquad [0,\pi]\,.$$

This series converges to an *even* function with period 2π whose graph is sketched in Figure 6.10. The series converges to $f(x)$ at

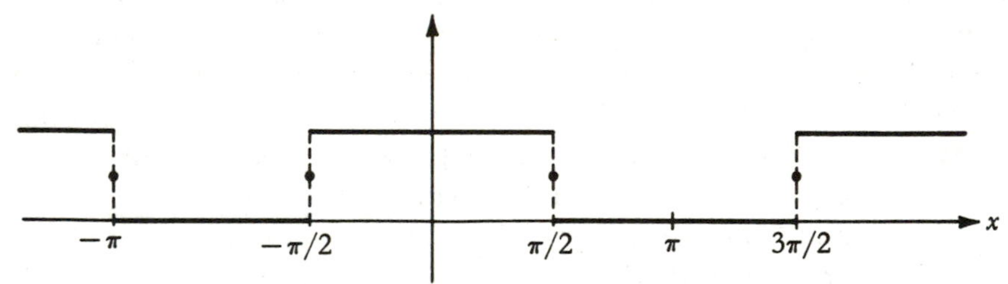

FIG. 6.10. Cosine series of $f(x) = 1 \quad 0 \leq x < \pi/2,\ f(x) = 0 \quad \pi/2 \leq x \leq \pi$.

all points on $0 \leq x \leq \pi$ except $\pi/2$, and at $\pi/2$ converges to $1/2$.

The coefficients of the Fourier sine series of $f(x)$ on $[0,\pi]$ are given by

$$b_n = \frac{2}{\pi}\int_0^\pi f(x)\sin nx\,dx = \frac{2}{\pi}\int_0^{\pi/2}\sin nx\,dx = \frac{2\left(1 - \cos\frac{n\pi}{2}\right)}{n\pi}, \qquad n = 1, 2, \ldots$$

so that the Fourier sine series of the function is

$$f(x) \sim \frac{2}{\pi}\sin x + \frac{4}{2\pi}\sin 2x + \frac{2}{3\pi}\sin 3x + \frac{2}{5\pi}\sin 5x + \cdots \qquad [0,\pi]\,.$$

This series converges to an *odd* function with period 2π which is sketched in Figure 6.11. At all points of the interval $0 \leq x \leq \pi$ other than 0, $\pi/2$, and π the series converges to $f(x)$. At $\pi/2$ it converges to $1/2$, as does the cosine series of f, but at 0 and π it converges to zero.

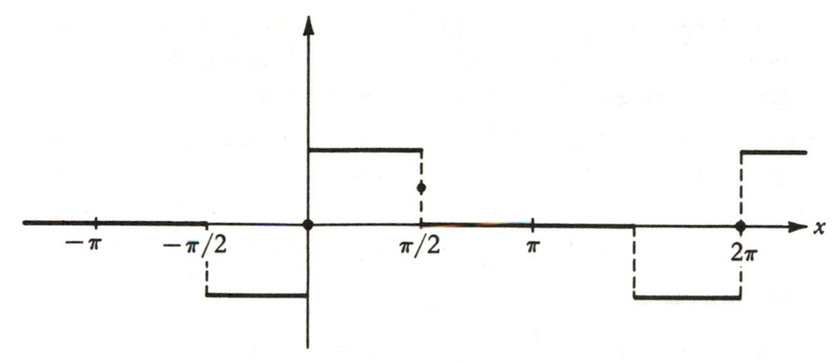

FIG. 6.11. Sine series of $f(x) = 1 \quad 0 \leq x < \pi/2$, $f(x) = 0 \quad \pi/2 \leq x \leq \pi$.

It may seem strange that a function in a finite interval can be represented by several different trigonometric series. In fact, a little reflection will show that there are infinitely many different trigonometric series which can represent a given function in a finite interval. The most important in practice, however, are the Fourier series, Fourier sine and cosine series, and series whose relation to these is similar to that between Fourier sine and cosine series and Fourier series (see Exercises 6c, Problems 3 and 4). Even though all of these series represent the given function in the given interval, they represent *different periodic* functions.

EXERCISES 6c

1. Calculate the Fourier sine series and cosine series of the function $f(x) = 2x + 1$ on the interval $[0,1]$. Sketch the graphs of the periodic functions to which each of these series converge.

2. Calculate the Fourier cosine series of the function $f(x) = ax^2 + bx$ on the interval $[0,1]$. Find the sum of the series $\sum_{n=1}^{\infty} 1/n^2$.

3. (i) Show that if a function $f(x)$ on the interval $[0,\pi]$ is symmetric with respect to the line $x = \pi/2$, that is, if $f(x) = f(\pi - x)$ then the Fourier sine series of $f(x)$ on $[0,\pi]$ has the form

$$f(x) \sim b_1 \sin x + b_3 \sin 3x + b_5 \sin 5x + \cdots + b_{2n+1} \sin (2n + 1)x + \cdots$$

where

$$b_{2n+1} = \frac{4}{\pi} \int_0^{\pi/2} f(x) \sin (2n + 1)x \, dx \, .$$

(ii) Use the result of part (i) to find a series of the stated form which converges to x^2 on the interval $[0,\pi/2]$, and sketch the graph of the periodic function to which the series converges.

4. Generalize the discussion of Problem 3 to show that if $f(x)$ is continuous and piecewise smooth on $0 \leq x \leq L$, then

$$f(x) = \sum_{n=0}^{\infty} b_n \sin\left(n + \frac{1}{2}\right)\frac{\pi}{L}x\,,$$

where

$$b_n = \frac{2}{L}\int_0^L f(x) \sin\left[\left(n + \frac{1}{2}\right)\frac{\pi}{L}x\right] dx\,,$$

and the series converges and the equation holds for $0 < x < L$.

5. (i) Assuming that α is not an integer (positive, negative, or zero), find the Fourier cosine series on $0 \leq x \leq \pi$ of the function $\cos \alpha x$.

(ii) By setting $x = 0$ in your answer to part (i) show that for nonintegral α

$$\frac{\pi}{\sin \pi\alpha} = \frac{1}{\alpha} + \sum_{m=1}^{\infty} (-1)^m \left(\frac{1}{\alpha + m} + \frac{1}{\alpha - m}\right).$$

This is called the partial fraction expansion of $f(\alpha) = \pi/\sin \pi\alpha$.

(iii) By setting $x = \pi$ in (i) show that for nonintegral α

$$\pi \cot \pi\alpha = \frac{1}{\alpha} + \sum_{m=1}^{\infty} \left(\frac{1}{\alpha + m} + \frac{1}{\alpha - m}\right).$$

(iv) The result in part (iii) can be written in the form

$$\pi \cot \pi\alpha - \frac{1}{\alpha} = \sum_{m=1}^{\infty} \left(\frac{1}{\alpha + m} + \frac{1}{\alpha - m}\right).$$

In this form it holds even when $\alpha = 0$ in the sense that

$$\lim_{\alpha\to 0} (\pi \cot \pi\alpha - 1/\alpha) = 0$$

and when $\alpha = 0$ the series on the right is $0 + 0 + 0 + \cdots$.

(v) Show that if $0 < a < 1$, then the series in (iv), considered as a series of functions of α, converges uniformly on the interval $-a \leq \alpha \leq a$. *Hint:* Find a constant M such that whenever $m \geq 1$

$$\left|\frac{1}{\alpha + m} + \frac{1}{\alpha - m}\right| \leq \frac{M}{m^2} \qquad \text{for } -a \leq \alpha \leq a\,.$$

(vi) Because of the uniform convergence, the series in (iv) can be integrated termwise to yield a series expansion of

$$\int_0^x \left(\pi \cot \pi\alpha - \frac{1}{\alpha}\right) d\alpha \qquad -1 < x < 1\,.$$

Show that

$$\log \frac{\sin \pi x}{\pi x} = \sum_{m=1}^{\infty} \log \left(1 - \frac{x^2}{m^2}\right) \qquad -1 < x < 1\,.$$

(vii) If $s_n \to s$, then $e^{s_n} \to e^s$. Use (vi) to obtain

$$\frac{\sin \pi x}{\pi x} = \prod_{m=1}^{\infty} \left(1 - \frac{x^2}{m^2}\right) \qquad -1 < x < 1\,.$$

(In texts on functions of a complex variable this last formula is shown to be valid for all real or complex values of x. It furnishes a representation of the function $\sin \pi x$ analogous to the representation of a polynomial as a product of linear factors in terms of its roots.)

6. Let $f(x)$ and its first and second derivatives be continuous for $0 \leq x \leq L$.

(i) State conditions on f and its derivatives at $x = 0$ for the odd extension of f to be continuous together with its first and second derivatives for $-L \leq x \leq L$.

(ii) Repeat (i) for the even extension of f.

6.6. *COMPLEX FORM OF FOURIER SERIES*

In discussions involving the trigonometric functions $\sin \omega x$ and $\cos \omega x$ it is often convenient to employ the complex trigonometric function $e^{i\omega x}$. First, it is easier to operate with a single complex-valued function than with a pair of real-valued functions. Second, although the relations between $\sin n\omega x$, $\cos n\omega x$ and $\sin \omega x$, $\cos \omega x$ are complicated, that between $e^{in\omega x}$ and $e^{i\omega x}$ is simply expressed by

$$e^{in\omega x} = (e^{i\omega x})^n\,.$$

As an illustration of these remarks we establish the formula

$$\text{(6.6.1)} \quad \frac{1}{2} + \cos u + \cos 2u + \cdots + \cos nu = \frac{\sin (n + \frac{1}{2})u}{2 \sin \frac{1}{2}u}$$

which we will need later. We start with the formula for the sum of a finite geometric series

$$1 + z + z^2 + \cdots + z^n = \frac{1 - z^{n+1}}{1 - z}, \qquad z \neq 1\,,$$

which is easily verified by multiplying both sides by $1 - z$. If we set $z = e^{iu}$, we obtain

$$1 + e^{iu} + e^{2iu} + \cdots + e^{niu} = \frac{1 - e^{i(n+1)u}}{1 - e^{iu}}\,.$$

The left side of this equation is

$$(6.6.2)\quad 1 + e^{iu} + e^{2iu} + \cdots + e^{inu} = (1 + \cos u + \cos 2u + \cdots + \cos nu) + i(\sin u + \sin 2u + \cdots + \sin nu)\,.$$

The right side can be simplified by multiplying numerator and denominator by $e^{-1/2iu}$. We get

$$\frac{1 - e^{i(n+1)u}}{1 - e^{iu}} = \frac{e^{-1/2iu} - e^{i(n+1/2)u}}{e^{-1/2iu} - e^{1/2iu}} = i\,\frac{e^{-1/2iu} - e^{i(n+1/2)u}}{2 \sin \frac{u}{2}}$$

$$(6.6.3)\qquad = \left(\tfrac{1}{2} + \frac{\sin (n + \frac{1}{2})u}{2 \sin \frac{1}{2}u}\right) + i\left(\frac{\cos \frac{1}{2}u - \cos (n + \frac{1}{2})u}{2 \sin \frac{1}{2}u}\right).$$

Equating the real parts of (6.6.2) and (6.6.3), we obtain (6.6.1).

We have spoken so far of Fourier series of real-valued functions of a real variable, but it is convenient, and not difficult, to extend our considerations to complex-valued functions of a real variable. If $f(x)$ is any absolutely integrable complex-valued function on the interval $-\pi \leq x \leq \pi$, then we define its Fourier coefficients and Fourier series on $[-\pi,\pi]$ by (6.4.12), (6.4.13) and (6.4.14). There is nothing remarkable about this. Indeed the Fourier series of the complex-valued function $f(x) = f_1(x) + if_2(x)$ is just the Fourier series of $f_1(x)$ plus i times the Fourier series of $f_2(x)$ (see Section 1.4).

It is useful, however, to write the Fourier series of a function in terms of the complex trigonometric functions. If we substitute (1.4.2) and (1.4.3) in (6.4.11), we obtain

$$f(x) \sim \frac{a_0}{2} + \sum_{k=1}^{\infty} a_k \left(\frac{e^{ikx} + e^{-ikx}}{2}\right) - ib_k \left(\frac{e^{ikx} - e^{-ikx}}{2}\right)$$

$$\sim \frac{a_0}{2} + \sum_{k=1}^{\infty} \frac{a_k - ib_k}{2}\, e^{ikx} + \sum_{k=1}^{\infty} \frac{a_k + ib_k}{2}\, e^{-ikx}\,,$$

so that if we set

$$(6.6.4)\qquad \begin{aligned} c_n &= \frac{a_n - ib_n}{2} \qquad n = 1, 2, \ldots \\ c_0 &= \frac{a_0}{2} \\ c_{-n} &= \frac{a_n + ib_n}{2} \qquad n = 1, 2, \ldots, \end{aligned}$$

we have

$$f(x) \sim c_0 + \sum_{k=1}^{\infty} c_k e^{ikx} + \sum_{k=-1}^{-\infty} c_k e^{ikx},$$

or

$$f(x) \sim \sum_{k=-\infty}^{\infty} c_k e^{ikx} \qquad [-\pi,\pi]. \tag{6.6.5}$$

Substituting (6.4.13) and (6.4.14) in (6.6.4), we obtain the single formula

$$c_n = \frac{1}{2\pi}\int_{-\pi}^{\pi} f(x)e^{-inx}\,dx \qquad n = \dots -2, -1, 0, 1, 2, \dots. \tag{6.6.6}$$

Formulas (6.6.5) and (6.6.6) define the **complex form of the Fourier series** of $f(x)$.

We emphasize that the Fourier series of a function may be written in complex form even when the function is real-valued. In this case, however, the coefficients are still in general complex-valued. If a function is real-valued then its Fourier coefficients a_n and b_n are real and from (6.6.4) we see that c_{-n} and c_n are conjugate. This property is characteristic for real-valued functions.

The formula (6.6.6) for the complex Fourier coefficients of a function can be derived directly, as those for the real Fourier coefficients were derived in Section 6.4. The argument is entirely similar. The property of the complex trigonometric functions e^{inx} analogous to—in fact, equivalent to—the orthogonality of the trigonometric functions sin nx, cos nx is

$$\int_{-\pi}^{\pi} e^{imx}e^{-inx}\,dx = \begin{cases} 0 & n \neq m \\ 2\pi & n = m, \end{cases} \tag{6.6.7}$$

which follows immediately from

$$\int e^{imx}e^{-inx}\,dx = \int e^{i(m-n)x}\,dx = \frac{e^{i(m-n)x}}{i(m-n)} \qquad m \neq n.$$

Now, assume that equality holds in (6.6.5) and multiply both sides of the equation by e^{-inx}. Integrate both sides of the resulting equation over the interval $[-\pi,\pi]$, integrating the series term-by-term. Using (6.6.7) we obtain the formula (6.6.6).

The formula for the complex form of the Fourier series of a function on the interval $[-L,L]$ can be obtained either by changing variables in the formulas above, or by repeating the preceding arguments with obvious modifications.

The complex representation of Fourier series permits us to establish an important connection between the theory of Fourier series and the theory of analytic functions of a complex variable. Let $F(x)$ be a real-valued function of the real-variable x which is analytic in the neighborhood of the origin, that is, which can be expanded in a power series,

$$F(x) = \sum_{k=0}^{\infty} c_k x^k , \tag{6.6.8}$$

which converges for all x such that $|x| < C$, where C is some positive constant. The coefficients c_n are given in terms of F by Taylor's formula

$$c_n = \frac{F^{(n)}(0)}{n!} .$$

It is shown in the theory of power series that (6.6.8) will also converge when x is replaced by the complex variable $z = x + iy$ if $|z| < C$. Thus by means of

$$F(z) = \sum_{k=0}^{\infty} c_k z^k , \qquad |z| < C \tag{6.6.9}$$

the definition of the function F can be extended to complex values of its argument. The resulting function is an analytic function of the complex variable z. For example, the exponential e^z is defined for all complex z by

$$e^z = 1 + z + \frac{z^2}{2!} + \frac{z^3}{3!} + \cdots .$$

If, in particular, we set $z = i\beta x$ in the above equation and separate the right side into real and imaginary parts, we obtain Euler's formula (1.4.4).

Now let $F(z)$ be an analytic function of z whose power series converges for $|z| < C$ where $C > 1$. Since $|e^{it}| = 1$, for t real, we may set $z = e^{it}$ in (6.6.9) to obtain

$$F(e^{it}) = \sum_{k=0}^{\infty} c_k e^{ikt} . \tag{6.6.10}$$

Thus the sum of the trigonometric series in (6.6.10) is $F(e^{it})$—or to put it another way—this series is the Fourier series of $F(e^{it})$ (since the series converges uniformly). Through this relation every theorem about Fourier series becomes a theorem about the behavior of analytic functions of z on the unit circle in the complex z-plane. Conversely, such theorems about analytic functions become theorems about certain classes of Fourier series.

We will treat only a very simple application of the relationship we have described. In certain simple cases it can be used to find the sum of a trigonometric series. Consider the problem of finding the sum of the series

$$f(t) = 1 + a \cos t + a^2 \cos 2t + a^3 \cos 3t + \cdots$$

where a is some real constant. We combine this series with the companion series

$$g(t) = a \sin t + a^2 \sin 2t + a^3 \sin 3t + \cdots$$

to obtain

$$f(t) + ig(t) = 1 + ae^{it} + a^2e^{i2t} + a^3e^{i3t} + \cdots$$
$$= \sum_{n=0}^{\infty} (ae^{it})^n .$$

We recognize this as the familiar geometric series

$$\sum_{n=0}^{\infty} u^n = \frac{1}{1-u}, \qquad |u| < 1 .$$

If $-1 < a < 1$, then $|ae^{it}| < 1$ for all real t, and we can put $u = ae^{it}$ in the geometric series, which leads to

$$f(t) + ig(t) = \frac{1}{1 - ae^{it}} .$$

To find the real part of the right-hand side we write it in the form

$$\frac{1}{1-ae^{it}} = \frac{1}{1 - a\cos t - ia \sin t}$$
$$= \frac{(1 - a\cos t) + ia\sin t}{(1 - a\cos t)^2 + a^2 \sin^2 t}$$
$$= \frac{1 - a\cos t + ia\sin t}{1 - 2a\cos t + a^2} .$$

The real part of this expression is $f(t)$, that is

$$\frac{1 - a\cos t}{1 - 2a\cos t + a^2} = \sum_{n=0}^{\infty} a^n \cos nt \tag{6.6.11}$$

valid for all real t if $-1 < a < 1$. This is the desired result. As a by-product we also obtain, since the imaginary part is $g(t)$,

$$\frac{a\sin t}{1 - 2a\cos t + a^2} = \sum_{n=1}^{\infty} a^n \sin nt , \tag{6.6.12}$$

valid for all real t if $-1 < a < 1$.

EXERCISES 6d

1. Find the complex form of the Fourier series of e^x in the interval $[-3,3]$.

2. Find the function $f(x)$, $-\pi \leq x \leq \pi$ whose Fourier series is

$$1 + \cos x + \frac{\cos 2x}{2!} + \frac{\cos 3x}{3!} + \cdots + \frac{\cos nx}{n!} + \cdots .$$

3. Let $f(x)$ be complex-valued with Fourier series in real and complex form

$$f(x) \sim \frac{a_0}{2} + \sum_{k=1}^{\infty} a_k \cos kx + b_k \sin kx \qquad [-\pi,\pi],$$

$$f(x) \sim \sum_{-\infty}^{\infty} c_k e^{ikx} \qquad [-\pi,\pi],$$

respectively. Show that the corresponding Fourier series of $\bar{f}(x)$ (the bar denotes complex conjugate) are given by

$$\bar{f}(x) \sim \frac{\bar{a}_0}{2} + \sum_{k=1}^{\infty} \bar{a}_k \cos kx + \bar{b}_k \sin kx \qquad [-\pi,\pi]$$

$$\bar{f}(x) \sim \sum_{-\infty}^{\infty} \bar{c}_{-k} e^{ikx} \qquad [-\pi,\pi].$$

6.7. *CONVERGENCE PROOF*

On the basis of the discussion in preceding sections, all the statements we have made about the convergence of various trigonometric series will be established if we prove the **convergence theorem:**

Let $f(x)$ be a piecewise-smooth function with period 2π and let

$$\text{(6.7.1)} \qquad \begin{aligned} a_k &= \frac{1}{\pi}\int_{-\pi}^{\pi} f(x) \cos kx \, dx \qquad k = 0, 1, 2, \ldots \\ b_k &= \frac{1}{\pi}\int_{-\pi}^{\pi} f(x) \sin kx \, dx \qquad k = 1, 2, \ldots . \end{aligned}$$

Then, for all x,

$$\text{(6.7.2)} \qquad \frac{f(x+) + f(x-)}{2} = \frac{a_0}{2} + \sum_{k=1}^{\infty} (a_k \cos kx + b_k \sin kx) .$$

A statement that a series has a certain sum is a statement about the limit of a sequence, the sequence of partial sums of the series. Let $s_n(x)$ denote the nth partial sum of the series in (6.7.2), that is,

$$\text{(6.7.3)} \qquad s_n(x) = \frac{a_0}{2} + \sum_{k=1}^{n} (a_k \cos kx + b_k \sin kx) .$$

Then we have to show that

$$\lim_{n\to\infty} s_n(x) = \frac{f(x+) + f(x-)}{2}$$

or, equivalently,

$$\lim_{n\to\infty}\left[s_n(x) - \frac{f(x+) + f(x-)}{2}\right] = 0 . \tag{6.7.4}$$

Our first step will be to find a simple integral representation for the sequence of (6.7.4). For this we shall need the formulas

$$\frac{1}{2} + \cos u + \cos 2u + \cdots + \cos nu = \frac{\sin (n + \frac{1}{2})u}{2 \sin \frac{1}{2}u} \tag{6.7.5}$$

$$\int_{-\pi}^{0} \frac{\sin (n + \frac{1}{2})u}{2 \sin \frac{1}{2}u}\, du = \int_{0}^{\pi} \frac{\sin (n + \frac{1}{2})u}{2 \sin \frac{1}{2}u}\, du = \frac{\pi}{2}. \tag{6.7.6}$$

Equation (6.7.5) is just Equation (6.6.1), which was established in Section 6. If we integrate (6.7.5) from $-\pi$ to π the integral of each term on the left, other than the first, vanishes, so that the integral from $-\pi$ to π of the function on the right has the value π. Since this function is even, its integrals from $-\pi$ to 0 and from 0 to π are equal and each is equal to half its integral from $-\pi$ to π. This proves (6.7.6). Notice also that since each term on the left of (6.7.5) is periodic with period 2π the sum on the left and hence the function on the right is periodic with period 2π.

The integral representation we are going to establish is known as **Dirichlet's formula,** and is one of the basic tools of the theory of Fourier series. The statement of this theorem follows.

If $f(x)$ is periodic with period 2π and absolutely integrable, and if $s_n(x)$ is the nth partial sum of its Fourier series, then

$$s_n(x) = \frac{1}{\pi}\int_{-\pi}^{\pi} f(x + t)\, \frac{\sin (n + \frac{1}{2})t}{2 \sin \frac{1}{2}t}\, dt . \tag{6.7.7}$$

If $f(x+)$ and $f(x-)$ exist for some x, then

$$\begin{aligned} s_n(x) &- \frac{f(x+) + f(x-)}{2} \\ &= \frac{1}{\pi}\int_{-\pi}^{0} [f(x + t) - f(x-)]\, \frac{\sin (n + \frac{1}{2})t}{2 \sin \frac{1}{2}t}\, dt \\ &\quad + \frac{1}{\pi}\int_{0}^{\pi} [f(x + t) - f(x+)]\, \frac{\sin (n + \frac{1}{2})t}{2 \sin \frac{1}{2}t}\, dt . \end{aligned} \tag{6.7.8}$$

For the proof we substitute (6.7.1) in (6.7.3). In doing this we must use a variable of integration different from x in (6.7.1), since otherwise x would appear both as a variable and a variable of integration in the same equation. We obtain

$$s_n(x) = \frac{1}{2\pi}\int_{-\pi}^{\pi} f(\xi)\,d\xi + \frac{1}{\pi}\sum_{k=1}^{n}\left[\cos kx\int_{-\pi}^{\pi} f(\xi)\cos k\xi\,d\xi + \sin kx\int_{-\pi}^{\pi} f(\xi)\sin k\xi\,d\xi\right]\cdot \tag{6.7.9}$$

Since $\cos kx$ and $\sin kx$ are constant with respect to the integration variable ξ, we may multiply under the integral signs. Further, since the sum in (6.7.9) is a finite sum we may interchange summation and integration. Thus we have

$$s_n(x) = \frac{1}{\pi}\int_{-\pi}^{\pi} f(\xi)\left[\frac{1}{2} + \sum_{k=1}^{n}(\cos kx\cos k\xi + \sin kx\sin k\xi)\right]d\xi\,.$$

Using the addition theorem for the cosine this becomes

$$s_n(x) = \frac{1}{\pi}\int_{-\pi}^{\pi} f(\xi)\left[\frac{1}{2} + \sum_{k=1}^{n}\cos k(x-\xi)\right]d\xi\,. \tag{6.7.10}$$

The sum in (6.7.10) is that in (6.7.5) if we set $u = x - \xi$. Hence

$$s_n(x) = \frac{1}{\pi}\int_{-\pi}^{\pi} f(\xi)\,\frac{\sin(n+\frac{1}{2})(x-\xi)}{2\sin\frac{1}{2}(x-\xi)}\,d\xi\,. \tag{6.7.11}$$

In (6.7.11) we make the change of variable $t = \xi - x$, obtaining

$$s_n(x) = \frac{1}{\pi}\int_{x-\pi}^{x+\pi} f(x+t)\,\frac{\sin(n+\frac{1}{2})t}{2\sin\frac{1}{2}t}\,dt\,. \tag{6.7.12}$$

Finally, we observe that both factors of the integrand in (6.7.12), and hence their product, have period 2π, and the integral is taken over an interval of length 2π. The integral may therefore be replaced by an integral over the interval $-\pi$ to π which also has length 2π, so that

$$s_n(x) = \frac{1}{\pi}\int_{-\pi}^{\pi} f(x+t)\,\frac{\sin(n+\frac{1}{2})t}{2\sin\frac{1}{2}t}\,dt\,.$$

This is the first equation, (6.7.7), of our statement.

To establish the second equation, (6.7.8), observe that from Equation (6.7.6) we have

$$\begin{aligned}\frac{f(x-)}{2} &= f(x-)\,\frac{1}{\pi}\int_{-\pi}^{0}\frac{\sin(n+\frac{1}{2})t}{2\sin\frac{1}{2}t}\,dt = \frac{1}{\pi}\int_{-\pi}^{0} f(x-)\,\frac{\sin(n+\frac{1}{2})t}{2\sin\frac{1}{2}t}\,dt\\ \frac{f(x+)}{2} &= f(x+)\,\frac{1}{\pi}\int_{0}^{\pi}\frac{\sin(n+\frac{1}{2})t}{2\sin\frac{1}{2}t}\,dt = \frac{1}{\pi}\int_{0}^{\pi} f(x+)\,\frac{\sin(n+\frac{1}{2})t}{2\sin\frac{1}{2}t}\,dt\,.\end{aligned} \tag{6.7.13}$$

Writing the integral in (6.7.7) as the sum of integrals from $-\pi$ to 0 and from 0 to π and subtracting Equation (6.7.13) we obtain (6.7.8), completing the proof of Dirichlet's formula.

To prove the *convergence theorem* of this section it suffices to show that each of the integrals in (6.7.8) converges to 0 as n becomes infinite. For this purpose we use the following special form of a famous **lemma of Riemann.**

Let $g(t)$ be piecewise continuous in the interval $a \leq t \leq b$, and suppose that, except for a finite number of points, $g(t)$ has a continuous derivative at each point in this interval. Then

$$\lim_{m\to\infty} \int_a^b g(t) \sin mt\, dt = 0\,. \tag{6.7.14}$$

We shall establish this statement in three steps. Suppose first that $g(t)$ and $g'(t)$ are continuous on $a \leq t \leq b$. Then, integrating by parts we obtain

$$\int_a^b g(t) \sin mt\, dt = -\left.\frac{g(t)\cos mt}{m}\right|_a^b + \int_a^b \frac{g'(t)\cos mt}{m}\,dt\,. \tag{6.7.15}$$

Since g and g' are continuous on $[a,b]$, they are bounded, that is, there are constants A, B such that

$$|g(t)| \leq A\,, \quad |g'(t)| \leq B \qquad a \leq t \leq b\,. \tag{6.7.16}$$

Taking absolute values in (6.7.15) and using (6.7.16) we obtain

$$\left|\int_a^b g(t) \sin mt\, dt\right| \leq \frac{2A}{m} + \frac{B(b-a)}{m}\,. \tag{6.7.17}$$

Since the right side of (6.7.17) converges to 0 as m becomes infinite, the left does also, and the statement is proved in this case. It is important to notice that a and b are any two numbers here.

Next, suppose that $g(t)$ is continuous for $a \leq t \leq b$, and that $g'(t)$ is continuous for $a < t < b$, but possibly discontinuous at a or b. Let ϵ be any small positive number. Then

$$\begin{aligned}\int_a^b g(t) \sin mt\, dt = \int_a^{a+\epsilon} g(t)\sin mt\, dt + \int_{a+\epsilon}^{b-\epsilon} g(t) \sin mt\, dt \\ + \int_{b-\epsilon}^b g(t) \sin mt\, dt\,.\end{aligned} \tag{6.7.18}$$

Since $g(t)$ is continuous on the closed interval $a \leq t \leq b$, it is bounded there, say by A. Taking absolute values in (6.7.18) we get therefore

$$\left|\int_a^b g(t) \sin mt\, dt\right| \leq 2A\epsilon + \left|\int_{a+\epsilon}^{b-\epsilon} g(t) \sin mt\, dt\right|. \tag{6.7.19}$$

The first term on the right in this last inequality can be made as small as we wish by choosing ϵ sufficiently small. In the interval

$a + \epsilon \leq t \leq b - \epsilon$, $g(t)$ satisfies the conditions previously considered, so that the integral on the right converges to 0 as m becomes infinite. The statement that a function of m converges to 0 as m becomes infinite means precisely that the absolute value of the function can be made as small as we wish by taking m sufficiently large. Thus, the first term on the right can be made as small as we wish and then the second term can be made as small as we wish by taking m sufficiently large. Since the right side of the inequality is therefore as small as we wish if m is sufficiently large, the integral on the left side of the inequality converges to 0 and our statement is proved in this second case.

Finally, consider the general case. Let $t_1 < t_2 < \cdots t_r$ be the points at which $g'(t)$ is discontinuous. Then

$$\text{(6.7.20)} \qquad \int_a^b g(t) \sin mt\, dt = \int_a^{t_1} + \int_{t_1}^{t_2} + \cdots + \int_{t_r}^b g(t) \sin mt\, dt\,.$$

In each of the subintervals $[a,t_1]$, $[t_1,t_2]$, . . . , $[t_r,b]$, $g(t)$ satisfies the conditions of the preceding case. Hence each of the integrals on the right in (6.7.20), and therefore their sum, converges to 0. This completes the proof of Riemann's lemma.

Now we can complete the proof of the *convergence theorem* for Fourier series. Consider the first integral in Equation (6.7.8). It has the form of the integral in Equation (6.7.14) with

$$\text{(6.7.21)} \qquad g(t) = \frac{f(x+t) - f(x-)}{2 \sin \frac{1}{2}t}$$

and $m = n + (1/2)$. Hence if we can show that the function (6.7.21) is piecewise continuous and has a continuous derivative at all but a finite number of points in $-\pi \leq t \leq 0$, it will follow from Riemann's lemma that the first term of (6.7.8) converges to 0. At all points $t \neq 0$ at which f has a continuous derivative, g will have a continuous derivative. Since f is piecewise smooth, *g has a continuous derivative except at a finite number of points possibly including $t = 0$.* Again, g is continuous where f is continuous, except possibly at $t = 0$, and therefore g is continuous except at a finite number of points. If f has a jump discontinuity for a value $t \neq 0$, then g also has a jump discontinuity. If we can show that g has a removable discontinuity at $t = 0$, it will follow that g is *piecewise continuous* on $[-\pi,0]$. Hence, g will satisfy the hypotheses of Riemann's lemma. But,

$$\begin{aligned}\lim_{\substack{t\to 0\\ t<0}} \frac{f(x+t)-f(x-)}{2\sin\frac{1}{2}t} &= \lim_{\substack{t\to 0\\ t<0}} \frac{f(x+t)-f(x-)}{t}\,\frac{t}{2\sin\frac{1}{2}t}\\ &= \lim_{\substack{t\to 0\\ t<0}} \frac{f(x+t)-f(x-)}{t}\lim_{\substack{t\to 0\\ t<0}} \frac{t}{2\sin\frac{1}{2}t}\\ &= \lim_{\substack{t\to 0\\ t<0}} \frac{f(x+t)-f(x-)}{t} \qquad (6.7.22)\\ &= \lim_{\substack{t\to 0\\ t<0\\ t<\tau<0}} f'(x+\tau)\\ &= \lim_{\substack{\tau\to 0\\ \tau<0}} f'(x+\tau) = f'(x-),\end{aligned}$$

where we have used the mean-value theorem and the fact that $f'(x-)$ exists since f is piecewise smooth. Thus g has a removable discontinuity at $t = 0$, and we may use Riemann's lemma to conclude that the first integral in (6.7.8) converges to 0. In the same way we show that the second integral in this equation converges to 0. Equation (6.7.4) and therefore the convergence theorem stated at the beginning of this section are proved.

6.8. *UNIFORM CONVERGENCE*

The convergence theorem of Section 7, and those convergence theorems which follow from it in virtue of the discussions in preceding sections, deal with what is called **pointwise convergence.** Each asserts that if $f(x)$ is one of a rather large class of functions, the partial sums of one of its Fourier series converge to $f(x)$ at each point; that is, for each value of x, in an interval. Although this seems the simplest and most natural kind of convergence—and it is indeed an important kind of convergence—it is inadequate for many applications. A principal deficiency is that term-by-term integration of a pointwise convergent series is not in general possible. Thus, for example, we cannot conclude that if a trigonometric series converges pointwise to a periodic function it must be the Fourier series of the function (compare Section 4).

The simplest notion of convergence which implies pointwise convergence and also justifies term-by-term integration is that of *uniform convergence*. We begin by recalling the definition of this notion.

We say that $g(x)$ approximates $f(x)$ uniformly on the interval $[a,b]$ with error $\leq \epsilon$ if

$$|f(x) - g(x)| \leq \epsilon \qquad \text{for all } x \text{ in } [a,b] .$$

The corresponding notion of convergence is **uniform convergence.**

The functions $s_n(x)$ are said to converge to $f(x)$ uniformly on $[a,b]$ if for every positive ϵ, no matter how small, there is an integer N such that whenever $n \geq N$,

$$|f(x) - s_n(x)| \leq \epsilon$$

for all x in $[a,b]$. In other words, the functions $s_n(x)$ approximate $f(x)$ uniformly on $[a,b]$ with arbitrarily small error for n sufficiently large.

A series

$$\sum_{k=0}^{\infty} u_k(x)$$

of functions defined on $[a,b]$ is said to converge to $f(x)$ uniformly on $[a,b]$ if its sequence of partial sums

$$s_n(x) = \sum_{k=0}^{n} u_k(x)$$

converges to $f(x)$ uniformly on $[a,b]$.

Uniform convergence of a series clearly implies pointwise convergence of the series, and more. Thus it is to be expected that the uniform convergence to a function of a Fourier series of the function can be established only for functions satisfying conditions more stringent than those imposed in the preceding sections. We shall prove the following typical theorem.

Let $f(x)$ be defined and continuous with continuous first and second derivatives on $[0,\pi]$, and assume

$$f'(0) = 0 , \quad f'(\pi) = 0 . \tag{6.8.1}$$

Then the Fourier cosine series of $f(x)$

$$f(x) \sim \frac{a_0}{2} + \sum_{k=1}^{\infty} a_k \cos kx$$

converges to $f(x)$ uniformly on $[0,\pi]$.

We observe first that since $f(x)$ is continuous on $[0,\pi]$, its even periodic extension is continuous for all real x, and since $f(x)$ has a continuous derivative on $[0,\pi]$, its even periodic extension is piecewise smooth. It follows from the convergence theorem of Section 7 that the Fourier cosine series converges to $f(x)$ for all x in $[0,\pi]$; that is,

(6.8.2) $$f(x) = \frac{a_0}{2} + \sum_{k=1}^{\infty} a_k \cos kx \qquad 0 \leq x \leq \pi \, .$$

The proof that the convergence in (6.8.2) is uniform will be based on the estimate

(6.8.3) $$|a_k \cos kx| \leq |a_k| \leq \frac{A}{k^2} \quad \text{for} \quad k \geq 1 \, ,$$

where A is a constant, i.e., independent of k. To establish (6.8.3) we write $\cos kx = \varphi_k(x)$ in the formula

$$a_k = \frac{2}{\pi} \int_0^{\pi} f(x) \cos kx \, dx$$

and note that $\varphi_k(x) = -\varphi_k''/k^2$, thus obtaining

$$a_k = -\frac{2}{k^2\pi} \int_0^{\pi} f(x)\varphi_k''(x) \, dx \qquad k \geq 1 \, .$$

Applying Green's formula to the preceding equation we get

$$a_k = -\frac{2}{k^2\pi} \left\{ [f(x)\varphi_k'(x) - f'(x)\varphi_k(x)]_0^{\pi} + \int_0^{\pi} f''(x)\varphi_k(x) \, dx \right\} \cdot$$

Since $f'(0)$ and $f'(\pi) = \varphi_k'(0) = \varphi_k'(\pi) = 0$, the boundary terms above vanish, yielding

$$a_k = -\frac{2}{k^2\pi} \int_0^{\pi} f''(x)\varphi_k(x) \, dx \, .$$

Therefore, using the fact that $|\varphi_k(x)| \leq 1$,

$$|a_k| \leq \frac{2}{k^2\pi} \int_0^{\pi} |f''(x)| \, dx = \frac{A}{k^2} \, ,$$

where

$$A = \frac{2}{\pi} \int_0^{\pi} |f''(x)| \, dx \, ,$$

and this proves (6.8.3).

From (6.8.2)

$$f(x) - s_n(x) = \sum_{k=n+1}^{\infty} a_k \cos kx \, ,$$

and using (6.8.3) we have for all x in $[0,\pi]$

(6.8.4) $$\begin{aligned} |f(x) - s_n(x)| &\leq \sum_{k=n+1}^{\infty} |a_k| \\ &\leq \sum_{k=n+1}^{\infty} \frac{A}{k^2} \cdot \end{aligned}$$

The last expression is the remainder, after n terms, of the series $\sum_{k=1}^{\infty} A/k^2$. Since this series is convergent, the remainder converges to 0 as n becomes infinite. Hence for any positive number ϵ, no matter how small, there is an integer N such that whenever $n \geq N$,

$$\sum_{k=n+1}^{\infty} \frac{A}{k^2} \leq \epsilon \,. \tag{6.8.5}$$

It follows from (6.8.4) and (6.8.5) that whenever $n \geq N$,

$$|f(x) - s_n(x)| \leq \epsilon \,,$$

for all x in $[0,\pi]$. This proves the theorem.

Instead of going through the argument in the preceding paragraph we could have inferred the theorem directly from the estimate (6.8.3) and the simple and widely applicable **"Weierstrass M-test"** for uniform convergence. We state this theorem below. The reader should recognize that the argument above, which we presented because of its simplicity, is a portion of the proof of Weierstrass's test for a special case.

Let $\sum_{k=1}^{\infty} u_k(x)$ *be an infinite series of functions defined on the interval* $[a,b]$, *and suppose that for* $k = 1, 2, 3, \ldots$ *and all* x *in* $[a,b]$

$$|u_k(x)| \leq M_k \,,$$

where $\sum_{k=1}^{\infty} M_k$ *is a convergent series of constants. Then the series* $\sum_{k=1}^{\infty} u_k(x)$ *is uniformly convergent on* $[a,b]$.

The theorem above on the uniform convergence of Fourier cosine series may be regarded as a uniform convergence theorem for the eigenvalue problem

$$\begin{array}{lll} \text{D.E.} & \varphi'' + \lambda\varphi = 0 & 0 < x < \pi \\ \text{B.C.} & \varphi'(0) = 0 \,, & \varphi'(\pi) = 0 \,. \end{array} \tag{6.8.6}$$

The theorem could be briefly stated as follows. *If* $f(x)$ *is continuous together with its first and second derivatives and if* $f(x)$ *satisfies the boundary conditions, then the eigenfunction expansion of* $f(x)$ *converges to* $f(x)$ *uniformly.* The assumption (6.8.1) is just the assumption that $f(x)$ satisfies the boundary conditions.

By the same method of proof we can establish similar statements for several other eigenvalue problems. Thus, for the eigenvalue problem

$$\text{(6.8.7)} \qquad \begin{array}{lll} \text{D.E.} & \varphi'' + \lambda\varphi = 0 & -\pi < x < \pi \\ \text{B.C.} & \varphi(-\pi) = \varphi(\pi)\,, & \varphi'(-\pi) = \varphi'(\pi) \end{array}$$

the corresponding theorem may be stated as follows.

Let $f(x)$ together with its first two derivatives $f'(x)$, $f''(x)$ be defined and continuous on $[-\pi,\pi]$ and assume

$$f(-\pi) = f(\pi)\,, \quad f'(-\pi) = f'(\pi)\,.$$

Then the Fourier series of $f(x)$ converges to $f(x)$ uniformly on $[-\pi,\pi]$.

We observed at the beginning of this section that an important advantage of uniform convergence over pointwise convergence is the possibility of **term-by-term integration.** This possibility persists after multiplication by any piecewise continuous function.

If the series $\sum_{k=1}^{\infty} u_k(x)$ of functions continuous on $[a,b]$ converges uniformly on $[a,b]$ to the continuous function $f(x)$, and if $g(x)$ is piecewise continuous on $[a,b]$, then

$$\text{(6.8.8)} \qquad \int_a^b f(x)g(x)\,dx = \sum_{k=1}^{\infty} \int_a^b u_k(x)g(x)\,dx\,.$$

This theorem follows immediately from the definition of uniform convergence and the inequality

$$\left|\int_a^b [f(x)g(x) - s_n(x)g(x)]\,dx\right| \leq G \int_a^b |f(x) - s_n(x)|\,dx\,,$$

where

$$s_n(x) = \sum_{k=1}^{n} u_k(x)\,,$$

and G is a constant such that

$$|g(x)| \leq G \qquad a \leq x \leq b\,.$$

Now, let $g(x)$ be piecewise continuous on $[0,\pi]$ with Fourier cosine series given by

$$g(x) \sim \frac{b_0}{2} + \sum_{k=1}^{\infty} b_k \cos kx\,.$$

If $f(x)$ is defined and continuous with two continuous derivatives on $[0,\pi]$ and satisfies (6.8.1), then as we have shown, its Fourier cosine series

$$f(x) \sim \frac{a_0}{2} + \sum_{k=1}^{\infty} a_k \cos kx$$

converges uniformly. Thus the hypothesis of the preceding theorem on term-by-term integration holds. We have therefore,

$$\int_0^\pi f(x)g(x)\,dx = \int_0^\pi \left[\frac{a_0}{2} + \sum_{k=1}^{\infty} a_k \cos kx\right] g(x)\,dx$$
$$= \frac{a_0}{2}\int_0^\pi g(x)\,dx + \sum_{k=1}^{\infty} a_k \int_0^\pi g(x)\cos kx\,dx\ ;$$

that is,

$$\int_0^\pi f(x)g(x)\,dx = \frac{\pi}{4}a_0b_0 + \frac{\pi}{2}\sum_{k=1}^{\infty} a_kb_k\ . \tag{6.8.9}$$

Equation (6.8.9) is known as *Parseval's equation* for the Fourier cosine series [or, for the eigenvalue problem (6.8.6)]. The right side of this equation is symmetric in f and g, although the restrictions imposed on f and g for the proof of the equation were very different. We shall see that (6.8.9), and analogous equations for other Fourier series hold, in fact, when f satisfies only the restriction imposed on g. This result is one of the central theorems in the theory of Fourier series.

EXERCISES 6e

1. State and prove a uniform convergence theorem for Fourier sine series.
2. Supply the proof of the uniform convergence theorem for Fourier series on $[-\pi,\pi]$ stated in Section 8.
3. State and prove a uniform convergence theorem for series of the form

$$f(x) = \sum_{k=0}^{\infty} c_k \sin\left(k + \frac{1}{2}\right)\frac{\pi x}{L}\cdot$$

4. Prove the theorem on term-by-term integration stated in Section 8.
5. Let $f(x)$ be continuous and periodic with period 2π and suppose that the trigonometric series

$$\frac{a_0}{2} + \sum_{k=1}^{\infty} a_k \cos kx + b_k \sin kx$$

converges uniformly to $f(x)$ on $[-\pi,\pi]$. Show that the trigonometric series must be the Fourier series of $f(x)$.

6. Let $f(x)$ together with its first two derivatives be continuous on $[0,\pi]$ and assume

$$f(0) = 0\ , \quad f(\pi) = 0\ .$$

Let $g(x)$ be any piecewise continuous function on $[0,\pi]$. Show that if the Fourier sine series of $f(x)$ and $g(x)$ are

$$f(x) \sim \sum_{k=1}^{\infty} b_k \sin kx$$

$$g(x) \sim \sum_{k=1}^{\infty} \beta_k \sin kx,$$

then the series

$$\sum_{k=1}^{\infty} b_k \beta_k$$

is convergent, and

$$\int_0^{\pi} f(x)g(x)\,dx = \frac{\pi}{2} \sum_{k=1}^{\infty} b_k \beta_k .$$

(This is *Parseval's equation* for Fourier sine series.)

7. Let $f(x)$ and $g(x)$ together with their first two derivatives be continuous on $[-\pi,\pi]$, and satisfy

$$f(-\pi) = f(\pi)\,, \quad f'(-\pi) = f'(\pi)$$
$$g(-\pi) = g(\pi)\,, \quad g'(-\pi) = g'(\pi)\,.$$

Let the Fourier series of $f(x)$ and $g(x)$ be, respectively,

$$f(x) \sim \frac{a_0}{2} + \sum_{k=1}^{\infty} a_k \cos kx + b_k \sin kx \qquad [-\pi,\pi]$$

$$g(x) \sim \frac{\alpha_0}{2} + \sum_{k=1}^{\infty} \alpha_k \cos kx + \beta_k \sin kx \qquad [-\pi,\pi]\,.$$

(i) Show that

$$\int_{-\pi}^{\pi} [f(x)]^2\,dx = \pi\left[\frac{a_0^2}{2} + \sum_{k=1}^{\infty} a_k^2 + b_k^2\right].$$

(The corresponding result for g also holds, of course.)

(ii) Show that $f(x) + g(x)$ satisfies the same conditions as $f(x)$, and by replacing $f(x)$ by $f(x) + g(x)$ in the result of part (i), show that

$$\int_{-\pi}^{\pi} f(x)g(x)\,dx = \pi\left[\frac{a_0\alpha_0}{2} + \sum_{k=1}^{\infty} a_k\alpha_k + b_k\beta_k\right].$$

(Either of the equations in this exercise may be called *Parseval's equation* for Fourier series in real form. They are equivalent, since that of part (ii) followed from that of part (i), and, conversely, the former is a special case of the latter.)

8. Let $f(x)$ be a complex-valued function which satisfies the conditions of Problem 7, and let the real and complex form of the Fourier series of $f(x)$ be, respectively,

$$f(x) \sim \frac{a_0}{2} + \sum_{k=1}^{\infty} a_k \cos kx + b_k \sin kx \qquad [-\pi,\pi]$$

$$f(x) \sim \sum_{k=-\infty}^{\infty} c_k e^{ikx} \qquad [-\pi,\pi] .$$

By setting $g(x) = \bar{f}(x)$ in the result of Exercise 7(ii), and using Equations (6.6.4) and the result of Exercises 6d, Problem 3, show that

$$\int_{-\pi}^{\pi} |f(x)|^2 \, dx = \int_{-\pi}^{\pi} f(x)\bar{f}(x) \, dx = 2\pi \sum_{k=-\infty}^{\infty} c_k \bar{c}_k = 2\pi \sum_{k=-\infty}^{\infty} |c_k|^2 .$$

(This is *Parseval's equation* for Fourier series in complex form.)

6.9. *CONVERGENCE THEOREMS FOR MORE GENERAL EIGENFUNCTION EXPANSIONS*

The Fourier series we have discussed in preceding sections are the prototypes of more general eigenfunction expansions. We recall from Section (4.11) that for a *self-adjoint regular Sturm-Liouville problem* (4.11.4) the totality of its real-valued eigenfunctions can be arranged in a sequence which forms an orthogonal system relative to the weight function $\sigma(x)$ on the interval $[a,b]$. In the following when we refer to the eigenfunctions of such a problem we shall assume that they have been so chosen. We consider real-valued functions $f(x)$, $g(x)$, and first make the following definition.

Definition. *Let $\varphi_0, \varphi_1, \ldots, \varphi_n, \ldots$ be an infinite system of functions orthogonal relative to a weight function $\sigma(x)$ on an interval $[a,b]$. If $f(x)$ is piecewise continuous, then the infinite series*

$$\sum_{k=0}^{\infty} a_k \varphi_k(x) \tag{6.9.1}$$

with

$$a_n = \frac{\int_a^b f(x)\varphi_n(x)\sigma(x) \, dx}{\int_a^b \varphi_n^2(x)\sigma(x) \, dx} \tag{6.9.2}$$

is called the **Fourier series** *of $f(x)$ relative to the system $\{\varphi_n(x)\}$, and we write*

$$f(x) \sim \sum_{k=0}^{\infty} a_k \varphi_k(x) . \tag{6.9.3}$$

The coefficient a_n is called the nth **Fourier coefficient** *of f relative to the system $\{\varphi_n\}$.*

Thus, for example, the Fourier sine series of a function on the interval $[0,\pi]$ may also be called the Fourier series of the function relative to the system $\{\sin nx\}$. A consequence of the definition above is that in a general context the term "Fourier series" is undefined without further statement.

The basic general **pointwise convergence** theorem for Fourier series relative to a system of eigenfunctions is the following.

Theorem. *Let $f(x)$ be piecewise smooth on the interval $[a,b]$ and let $\{\varphi_n(x)\}$ be the eigenfunctions of a self-adjoint regular Sturm–Liouville problem* (4.11.4). *Then for each value of x in $[a,b]$, the Fourier series of $f(x)$ relative to $\{\varphi_n(x)\}$ converges, and*

$$\frac{f(x+) + f(x-)}{2} = \sum_{k=0}^{\infty} a_k\varphi_k(x) \qquad a < x < b\,. \tag{6.9.4}$$

This theorem has been proved, in the text and exercises preceding, for the case in which the operator is $S = -D^2$, and the boundary conditions are either unmixed and of the first and second kind, or are the periodic boundary conditions. The proof in more general cases, even for the case in which $S = -D^2$ and one of the boundary conditions is of the third kind, is far more difficult, and beyond the scope of this book.

The basic general theorem on **uniform convergence** is the following.

Theorem. *Let $f(x)$ be defined and continuous with two continuous derivatives on $[a,b]$ and assume that $f(x)$ satisfies the boundary conditions of* (4.11.4). *Then the Fourier series of $f(x)$ relative to the eigenfunctions $\{\varphi_n(x)\}$ of* (4.11.4) *converges uniformly to $f(x)$ on $[a,b]$.*

We have proved this theorem for the same problems for which we have proved the pointwise convergence theorem. As exercises the reader can show that, assuming the theorem on pointwise convergence, the uniform convergence theorem can be proved for the operator $S = -D^2$ and all boundary conditions.

We have finally **Parseval's theorem.**

Theorem. *Let $f(x)$ and $g(x)$ be piecewise continuous on $[a,b]$ and let a_n, b_n $n = 0, 1, 2, \ldots$ be the Fourier coefficients of f and g, respectively, relative to the eigenfunctions $\{\varphi_n(x)\}$ of* (4.11.4). *Then*

$$\int_a^b f(x)g(x)\sigma(x)\,dx = \sum_{k=0}^{\infty} a_k b_k \int_a^b \varphi_k^2(x)\sigma(x)\,dx\,. \tag{6.9.5}$$

We have not yet proved this theorem in any case. In the next section we shall present a proof based partly on the uniform convergence theorem of this section. In the course of the proof we will see that the validity of Equation (6.9.5) is equivalent to the convergence to the function of the Fourier series of the function in a sense we have not yet considered. Thus the three theorems of this section are all convergence theorems.

EXERCISES 6f

1. Assuming that the pointwise convergence theorem of Section 9 holds for Fourier series relative to the eigenfunctions of

$$\text{D.E.}\qquad \varphi'' + \lambda\varphi = 0 \qquad\qquad 0 < x < L$$
$$\text{B.C.}\qquad \varphi'(0) = 0\,,\quad \varphi'(L) + h\varphi(L) = 0\,,$$

show that the uniform convergence theorem of the section also holds for such series.

2. Assuming that the uniform convergence theorem of Section 9 holds, prove Equation (6.9.5) when f satisfies the hypothesis of the uniform convergence theorem and g is piecewise continuous.

6.10. PARSEVAL'S THEOREM AND MEAN SQUARE CONVERGENCE

Our principal object in this section is an indication of the proof of Parseval's theorem, the last theorem of Section 9. We observe first that it is sufficient to establish the theorem in the special case $f = g$; that is, to show that if $h(x)$ is piecewise continuous on $[a,b]$ and if

$$h(x) \sim \sum_{k=0}^{\infty} c_k\varphi_k(x)\,, \tag{6.10.1}$$

then

$$\int_a^b [h(x)]^2\sigma(x)\,dx = \sum_{k=0}^{\infty} c_k^2 \int_a^b \varphi_k^2(x)\sigma(x)\,dx\,. \tag{6.10.2}$$

For, if f and g are piecewise continuous on $[a,b]$, then so are $f + g$ and $f - g$. Setting $h = f + g$ in (6.10.1), (6.10.2), we get

$$\int_a^b [f(x) + g(x)]^2\sigma(x)\,dx = \sum_{k=0}^{\infty} (a_k + b_k)^2 \int_a^b \varphi_k^2(x)\sigma(x)\,dx\,,$$

and similarly, setting $h = f - g$,

$$\int_a^b [f(x) - g(x)]^2\sigma(x)\,dx = \sum_{k=0}^{\infty} (a_k - b_k)^2 \int_a^b \varphi_k^2(x)\sigma(x)\,dx\,.$$

Subtracting the second of this pair of equations from the first and dividing by 4, we obtain (6.9.5). Our argument shows that (6.9.5), and (6.9.5) with $f = g$ are equivalent: either of these is called **Parseval's equation.**

We next establish an identity which has two important consequences for our discussion.

Let

$$t_n(x) = \sum_{k=0}^{n} \alpha_k \varphi_k(x) \tag{6.10.3}$$

be a linear combination of the functions $\varphi_0(x), \varphi_1(x), \ldots, \varphi_n(x)$ with any real coefficients $\alpha_0, \ldots, \alpha_n$. Then

$$\begin{aligned}\int_a^b [f(x) - t_n(x)]^2\sigma(x)\,dx &= \int_a^b [f(x)]^2\sigma(x)\,dx - 2\sum_{k=0}^{n} \alpha_k a_k \int_a^b \varphi_k^2(x)\sigma(x)\,dx \\ &\quad + \sum_{k=0}^{n} \alpha_k^2 \int_a^b \varphi_k^2(x)\sigma(x)\,dx\end{aligned} \tag{6.10.4}$$

where $a_0, a_1, \ldots, a_n$ are the Fourier coefficients of f relative to $\{\varphi_n\}$.

The proof of the identity is a simple calculation. We have

$$\begin{aligned}\int_a^b [f(x) - \sum_{k=0}^{n} \alpha_k\varphi_k(x)]^2\sigma(x)\,dx &= \int_a^b \left[f(x) - \sum_{k=0}^{n} \alpha_k\varphi_k(x)\right]\left[f(x) - \sum_{j=0}^{n} \alpha_j\varphi_j(x)\right]\sigma(x)\,dx \\ &= \int_a^b [f(x)]^2\sigma(x)\,dx - 2\sum_{k=0}^{n} \alpha_k \int_a^b f(x)\varphi_k(x)\sigma(x)\,dx \\ &\quad + \sum_{j=0}^{n}\sum_{k=0}^{n} \alpha_j\alpha_k \int_a^b \varphi_j(x)\varphi_k(x)\sigma(x)\,dx\,,\end{aligned}$$

from which (6.10.4) follows in virtue of the orthogonality of the system $\{\varphi_n(x)\}$ relative to $\sigma(x)$ and the definition (6.9.2) of the Fourier coefficients of f relative to $\{\varphi_n\}$.

The most important consequence of (6.10.4) is obtained by setting $\alpha_k = a_k$, $k = 0, 1, 2, \ldots, n$. Then $t_n(x)$ becomes $s_n(x)$, the partial sum of the Fourier series of f,

$$s_n(x) = \sum_{k=0}^{n} a_k\varphi_k(x) \tag{6.10.5}$$

and (6.10.4) becomes

$$\int_a^b [f(x) - s_n(x)]^2\sigma(x)\,dx = \int_a^b [f(x)]^2\sigma(x)\,dx - \sum_{k=0}^{n} a_k^2 \int_a^b \varphi_k^2(x)\sigma(x)\,dx\,. \tag{6.10.6}$$

Setting $g = f$ in (6.9.5) and comparing the result with (6.10.6), we see that *Parseval's equation holds for the piecewise continuous function* $f(x)$ *if and only if*

$$\lim_{n\to\infty} \int_a^b [f(x) - s_n(x)]^2\sigma(x)\,dx = 0\,. \tag{6.10.7}$$

A second consequence of the identity (6.10.4) is obtained by completing the square in the sum on the right side. The equation then becomes

$$\begin{aligned}\int_a^b [f(x) - t_n(x)]^2\sigma(x)\,dx &= \int_a^b [f(x)]^2\sigma(x)\,dx - \sum_{k=0}^{n} a_k^2 \int_a^b \varphi_k^2(x)\sigma(x)\,dx \\ &\quad + \sum_{k=0}^{n} [a_k - \alpha_k]^2 \int_a^b \varphi_k^2(x)\sigma(x)\,dx\,.\end{aligned} \tag{6.10.8}$$

Substituting (6.10.6) in (6.10.8) and observing that the remaining sum in (6.10.8) is nonnegative we conclude that *if* $t_n(x)$ *is any linear combination of* $\varphi_0, \ldots, \varphi_n$ *and* $s_n(x)$ *is the partial sum of the Fourier series of* $f(x)$ *relative to* $\{\varphi_n(x)\}$, *then*

$$\int_a^b [f(x) - s_n(x)]^2\sigma(x)\,dx \le \int_a^b [f(x) - t_n(x)]^2\sigma(x)\,dx\,. \tag{6.10.9}$$

These two consequences of Equation (6.10.4) can be stated succinctly if we introduce the following definitions.

If $f(x)$ *and* $g(x)$ *are piecewise continuous functions on* $[a,b]$, $g(x)$ *is said to approximate* $f(x)$ *on* $[a,b]$ *with mean square error relative to* $\sigma(x)$ *on* $[a,b]$ *not greater than* ϵ *if*

$$\int_a^b [f(x) - g(x)]^2\sigma(x)\,dx \le \epsilon\,.$$

The corresponding notion of convergence is **mean square convergence.**

The sequence $s_n(x)$ *of piecewise continuous functions is said to converge to* $f(x)$ *in mean square relative to* $\sigma(x)$ *on* $[a,b]$ *provided that*

$$\lim_{n\to\infty} \int_a^b [f(x) - s_n(x)]^2\sigma(x)\,dx = 0$$

or, what is the same, *provided that for every positive number* ϵ, *no matter how small, there is an integer* N *such that whenever* $n \geq N$,

$$\int_a^b [f(x) - s_n(x)]^2 \sigma(x)\, dx \leq \epsilon .$$

A series of functions is said to converge to $f(x)$ *in mean square relative to* $\sigma(x)$ *on* $[a,b]$ *if its sequence of partial sums converges to* $f(x)$ *in mean square relative to* $\sigma(x)$ *on* $[a,b]$.

Our two propositions can now be stated briefly: Parseval's equation for a function is equivalent to the mean square convergence to the function of its Fourier series; the best mean square approximations to a function by linear combinations of eigenfunctions are given by the partial sums of its Fourier series.

The reader should compare the definitions above with the corresponding definitions in Section 8 on uniform convergence and observe the analogy. Each refers to approximation over an entire interval and to a related notion of convergence. However, to say that $g(x)$ is close to $f(x)$ in the mean square sense does not imply that $[f(x) - g(x)]^2$ is small at every value of x but only that $[f(x) - g(x)]^2$ is small on the average. This averaging of the error makes it possible to approximate with small mean square error a rather arbitrary function $f(x)$ by a function $g(x)$ satisfying quite stringent conditions. In fact, the following holds.

Let $f(x)$ *be piecewise continuous on* $[a,b]$ *and let* δ *be a positive number. Then there is a function* $g(x)$ *which is continuous on* $[a,b]$ *together with its derivatives* $g'(x)$, $g''(x)$ *and for which*

$$g(a) = g'(a) = g(b) = g'(b) = 0 , \tag{6.10.10}$$

such that

$$\int_a^b [f(x) - g(x)]^2 \sigma(x)\, dx \leq \delta .$$

The reader can persuade himself of the plausibility of this statement by geometric considerations. A proof of the statement can be based on the *Weierstrass approximation theorem* which asserts that every function continuous on an interval $[a,b]$ can be approximated uniformly on the interval, with arbitrarily small prescribed error, by a polynomial. An outline of such a proof, in which the reader can complete the details, is presented in Exercises 6g, Problem 7. The Weierstrass approximation theorem is proved in many textbooks, some of which are cited at the end of this chapter.

We can now complete the proof of Parseval's theorem. Given f

choose g in the last proposition corresponding to the value $\delta = \epsilon/4$, that is, so that

$$\int_a^b [f(x) - g(x)]^2\sigma(x)\,dx \le \frac{\epsilon}{4}\cdot \tag{6.10.11}$$

Because of (6.10.10), g satisfies any homogeneous boundary conditions and hence the uniform convergence theorem of Section 9 holds. It is easy to show that uniform convergence on $[a,b]$ implies mean square convergence relative to any $\sigma(x)$ on $[a,b]$. Thus, if we denote the partial sums of the Fourier series of g by t_n, there is an N such that

$$\int_a^b [g(x) - t_n(x)]^2\sigma(x)\,dx \le \frac{\epsilon}{4} \tag{6.10.12}$$

for $n \ge N$. In the easily verified algebraic inequality, valid for any real u, v

$$(u + v)^2 \le 2(u^2 + v^2) \tag{6.10.13}$$

set $u = f - g$, $v = g - t_n$. From this we get, upon integration,

$$\int_a^b [f - t_n]^2\sigma\,dx \le 2\left[\int_a^b [f - g]^2\sigma\,dx + \int_a^b [g - t_n]^2\sigma\,dx\right]$$

and hence, from (6.10.11) and (6.10.12),

$$\int_a^b [f - t_n]^2\sigma\,dx \le \epsilon \tag{6.10.14}$$

for $n \ge N$. Finally, from (6.10.9) and (6.10.14) we have

$$\int_a^b [f - s_n]^2\sigma\,dx \le \epsilon \tag{6.10.15}$$

for $n \ge N$, which is to say that

$$\lim_{n\to\infty} \int_a^b [f - s_n]^2\sigma\,dx = 0\,.$$

This completes the proof.

As an example, consider the function

$$f(x) = \begin{cases} 1\,, & 0 \le x \le \dfrac{\pi}{2} \\ 0\,, & \dfrac{\pi}{2} < x \le \pi\,. \end{cases}$$

In Section 5 we found the series

$$f(x) \sim \frac{1}{2} + \frac{2}{\pi}\cos x - \frac{2}{3\pi}\cos 3x + \frac{2}{5\pi}\cos 5x - \cdots$$

which gives, by Parseval's equation,

$$1 = \frac{2}{\pi}\int_0^\pi [f(x)]^2\,dx = \frac{1}{2} + \frac{4}{\pi^2}\left[1 + \frac{1}{3^2} + \frac{1}{5^2} + \cdots\right].$$

This agrees with our previous evaluation

$$\frac{\pi^2}{8} = 1 + \frac{1}{3^2} + \frac{1}{5^2} + \cdots.$$

The Parseval equation for the Fourier sine series

$$x \sim \sum_{n=1}^{\infty} \frac{2(-1)^{n+1}}{n} \sin nx \qquad [0, \pi]$$

is

$$\frac{2\pi^2}{3} = \frac{2}{\pi}\int_0^\pi x^2\,dx = \sum_{n=1}^{\infty} \frac{4}{n^2},$$

which is equivalent to

$$\frac{\pi^2}{6} = 1 + \frac{1}{2^2} + \frac{1}{3^2} + \cdots.$$

EXERCISES 6g

1. Find the numerical equation obtained from Parseval's equation in each of the following cases:

(i) $f(x) = x\,,\quad 0 \le x \le \pi\,,$ cosine series

(ii) $f(x) = 1\,,\quad 0 \le x \le \pi\,,$ sine series .

2. Evaluate $\sum_{n=1}^{\infty} 1/n^4$.

3. Show that if $f(x)$ is continuous on $[0,\pi]$ and

$$\int_0^\pi f(x)\cos kx\,dx = 0 \qquad k = 0, 1, 2, \ldots,$$

then $f(x) = 0$ at every point of $[0,\pi]$.

4. Use identity (6.10.4) to prove the following statement, called **Bessel's inequality:** Let $\varphi_0, \varphi_1, \ldots, \varphi_n, \ldots$ be any sequence of functions orthogonal relative to $\sigma(x)$ on $[a,b]$, and let $f(x)$ be any piecewise continuous function on $[a,b]$. Then if

$$c_n = \frac{\int_a^b f(x)\varphi_n(x)\sigma(x)\,dx}{\int_a^b \varphi_n^2(x)\sigma(x)\,dx},$$

$$\int_a^b [f(x)]^2\sigma(x)\,dx \ge \sum_{n=0}^{\infty} c_n^2 \int_a^b \varphi_n^2(x)\sigma(x)\,dx\,.$$

5. Show that if $s_n(x)$ converges uniformly to $f(x)$ on $[a,b]$ and $\sigma(x)$ is any piecewise continuous positive function, then $s_n(x)$ converges to $f(x)$ in mean square relative to $\sigma(x)$ on $[a,b]$.

6. Show that, for any real numbers u, v, w:
(i) $2uv \leq u^2 + v^2$
(ii) $(u + v)^2 \leq 2(u^2 + v^2)$
(iii) $(u + v + w)^2 \leq 4(u^2 + v^2 + w^2)$.

7. Let $f(x)$ be piecewise continuous and $\sigma(x)$ continuous and positive on the interval $[0,L]$, and let F and K be constants such that $|f(x)| \leq F$, $0 \leq \sigma(x) \leq K$ for $0 \leq x \leq L$. Let δ be any positive real number.

(i) Suppose that the jump discontinuities of $f(x)$ occur at $x_1 < x_2 < \cdots < x_q$. For $\alpha > 0$ arbitrarily (and sufficiently) small, an approximating continuous function $g_1(x)$ can be constructed according to the following graphical description: the graph of $g_1(x)$ is obtained from the graph of $f(x)$ by replacing the graph of $f(x)$ in each interval $[x_j - \alpha, x_j + \alpha]$, $j = 1, 2, \ldots, q$ by the straight line segment joining the points $[x_j - \alpha, f(x_j - \alpha)]$ and $[x_j + \alpha, f(x_j + \alpha)]$.

Sketch the graph of a typical function $f(x)$ and the corresponding function $g_1(x)$, with convenient choices of q and α.

Show that

$$\int_0^L [f(x) - g_1(x)]^2\sigma(x)\,dx \leq 8F^2Kq\alpha\,,$$

and that, in particular, α and $g_1(x)$ can be chosen so that

$$\int_0^L [f(x) - g_1(x)]^2\sigma(x)\,dx \leq \frac{\delta}{4}\cdot$$

(ii) Let $g_1(x)$ be the continuous function chosen in (i). By the Weierstrass approximation theorem, given arbitrary $\beta > 0$, there is a polynomial $g_2(x)$ such that

$$|g_1(x) - g_2(x)| \leq \beta \qquad 0 \leq x \leq L\,.$$

Show that

$$\int_0^L [g_1(x) - g_2(x)]^2\sigma(x)\,dx \leq LK\beta^2\,,$$

and that, in particular, the number β satisfying $0 < \beta < F$ can be chosen so that

$$\int_0^L [g_1(x) - g_2(x)]^2\sigma(x)\,dx \leq \frac{\delta}{4}$$

and

$$|g_2(x)| \leq 2F \qquad 0 \leq x \leq L\,.$$

(iii) Let $g_2(x)$ be the function chosen in (ii). Let γ be a positive number, less than $L/2$, but otherwise arbitrary. Let $\psi(x) = \theta(x)\,\theta(L - x)$, where

$$\theta(x) = \begin{cases} \dfrac{12}{\gamma^4} \displaystyle\int_0^x \xi(\gamma - \xi)^2 \, d\xi & 0 \le x \le \gamma \\ 1 & \gamma \le x \le L \, . \end{cases}$$

Define $g(x)$ by $g(x) = \psi(x) g_2(x)$.

Show that $\psi(x)$ together with its first and second derivatives are continuous on $[0,L]$, and that ψ and its first derivative vanish at $x = 0$ and $x = L$. Show that $g(x)$ has the same stated properties. Show also that

$$0 \le \psi(x) \le 1 \qquad 0 \le x \le L \, .$$

Sketch the graph of $\psi(x)$, and using a sketch of a typical graph of $g_2(x)$, sketch the graph of the corresponding function $g(x)$.

Show that

$$\int_0^L [g_2(x) - g(x)]^2 \sigma(x) \, dx \le 8F^2 K\gamma \, ,$$

and, in particular, that γ and $g(x)$ can be chosen so that

$$\int_0^L [g_2(x) - g(x)]^2 \sigma(x) \, dx \le \frac{\delta}{4} \, .$$

(iv) Use (i), (ii) and (iii) and the inequality of Problem 6(iii) to conclude that $g(x)$ is continuous together with its first and second derivatives on $[0,L]$,

$$g(0) = g'(0) = g(L) = g'(L) = 0 \, ,$$

and

$$\int_0^L [f(x) - g(x)]^2 \sigma(x) \, dx \le \delta \, .$$

REFERENCES

Apostol, Tom M., *Mathematical Analysis*, Addison-Wesley, Reading, 1957, pp. 478–482.

Bartle, Robert G., *The Elements of Real Analysis*, Wiley, New York, 1964, pp. 177–183.

Rudin, Walter, *Principles of Mathematical Analysis*, 2nd ed., McGraw-Hill, New York, 1964, pp. 146–148.

7

Existence, uniqueness, and representation of solutions

7.1. *INTRODUCTION*

The formal solution of the initial-boundary value problem

$$\text{(7.1.1)}\quad \begin{array}{lll} \text{D.E.} & u_t = u_{xx} & 0 < x < \pi, \quad 0 < t \\ \text{B.C.} & u(0,t) = 0, \quad u(\pi,t) = 0 & 0 < t \\ \text{I.C.} & u(x,0) = f(x) & 0 < x < \pi, \end{array}$$

which we found in Chapter 4, was

$$u(x,t) = \sum_{n=1}^{\infty} b_n e^{-n^2 t} \sin nx, \tag{7.1.2}$$

where

$$b_n = \frac{2}{\pi} \int_0^{\pi} f(x) \sin nx \, dx . \tag{7.1.3}$$

The statement that a solution of problem (7.1.1) is given by (7.1.2) and (7.1.3) apparently rested on two assumptions:

(a) The series (7.1.2), (7.1.3), each term of which satisfies the D.E. and B.C., converges, and indeed so that its sum is a solution of the D.E.

(b) The function $f(x)$ in $0 < x < \pi$ can be represented by a series,

$$f(x) = \sum_{n=1}^{\infty} b_n \sin nx ,$$

where the coefficients b_n are given by (7.1.3).

In our discussion of Fourier series we showed that statement (b) holds if $f(x)$ is piecewise smooth on $0 \leq x \leq \pi$. To complete our solution of the initial-boundary value problem apparently we have now only to prove statement (a).

Suppose this were done. Then we would have established that problem (7.1.1) has a solution, and that (7.1.2) and (7.1.3) furnish a formula for a solution. A statement that a problem has a solution is called an **existence** statement for the problem; a statement of a formula for a solution is a **representation** statement. Thus, a proof for (a) would apparently settle the questions of existence and representation for the initial-boundary value problem. However, an important question would be left open, the question of **uniqueness,** that is, the question: Does the problem have only one solution?

Of the three statements—the initial-boundary value problem has a solution; it has only one solution; the solution is given by Equations (7.1.2) and (7.1.3)—the first is the most plausible and the last the least plausible. This suggests that instead of trying to prove the last, thereby proving the first, and attacking the second independently, which is what the program we have outlined requires, we assume the existence of a solution, try to establish that it must have the stated representation, and thus establish its uniqueness. This procedure is indeed the one usually followed in the solution of algebraic and ordinary differential equations. The first program proposed corresponds to checking the solution of such an equation.

We are going to follow the second program because for this we can present a discussion which is generally applicable—to problems we shall consider later as well as those we have already considered—and for which we need nothing that we have not already established. This program will leave open the question of existence, which we will discuss separately.

First, however, we must recognize that for a discussion following either program, more is needed than we have so far indicated.

7.2. PRECISE FORMULATION OF THE PROBLEM

Consider the function $u(x,t)$ defined by:

$$u(x,t) = \begin{cases} 1 & 0 < x < \pi\,, \qquad t > 0 \\ 0 & x = 0\,,\ x = \pi\,, \quad t > 0 \\ f(x) & 0 < x < \pi\,, \qquad t = 0\,. \end{cases} \tag{7.2.1}$$

The function $u(x,t)$ satisfies the D.E., B.C. and I.C. of the initial-boundary value problem of Section 1; that is, it is a solution of the problem. It will be agreed that it is not a satisfactory solution of the problem, but the question is, why is it not? It is a perfectly well-defined function for $0 \leq x \leq \pi$, $0 \leq t$; that is, Equation (7.2.1) is a rule which associates a definite value with every pair (x,t) for which $0 \leq x \leq \pi$, $0 \leq t$. To an objection that $u(x,t)$ is not defined by a single formula it can be answered that in our discussion of Fourier series we have seen that functions similarly defined by more than one formula can also be given by a single formula: indeed it is a priori possible (although not true) that Equation (7.1.2) represents the same function as Equation (7.2.1).

The reason the function $u(x,t)$ defined by equation (7.2.1) is unsatisfactory as a solution is that we have tacitly expected a solution to have certain continuity and smoothness properties. For example, although $u(x,t)$ satisfies the B.C.

$$u(0,t) = 0\,, \tag{7.2.2}$$

it is not true that

$$\lim_{\substack{x\to 0\\ x>0}} u(x,t) = u(0,t) = 0\,, \tag{7.2.3}$$

and we tacitly expected that (7.2.2) would imply (7.2.3). Our example shows that this need not be so, and that we must explicitly require that (7.2.3) hold. In fact, it is not important that (7.2.2) hold, or even that $u(0,t)$ be defined, but only that

$$\lim_{\substack{x\to 0\\ x>0}} u(x,t) = 0\,. \tag{7.2.4}$$

If (7.2.4) holds, then we can define or redefine $u(0,t)$ to be 0 and the resulting function will be continuous when $x = 0$. Equation (7.2.4) is the precise formulation of the B.C. of problem (7.1.1) at $x = 0$. A corresponding statement holds for the B.C. at $x = \pi$.

It might seem natural to take as a precise formulation of the I.C. of problem (7.1.1), the equation

$$\lim_{\substack{t\to 0\\ t>0}} u(x,t) = f(x)\,. \tag{7.2.5}$$

This is not so, if by "natural" we mean *appropriate* in the sense that with the stated auxiliary conditions the problem has one and only one solution. In fact, if (7.2.5) is taken as the precise formulation of the I.C., the resulting problem will have more than one solution. A demonstration of this statement is too difficult to be presented here, but a corresponding statement in a closely related context will be demonstrated in Chapter 9, Section 6.

Instead of (7.2.5) we shall require that, for every piecewise continuous function $\psi(x)$ on $[0,\pi]$,

$$\lim_{\substack{t\to 0\\ t>0}} \int_0^\pi u(x,t)\psi(x)\,dx = \int_0^\pi f(x)\psi(x)\,dx\,. \tag{7.2.6}$$

This condition arises mathematically in a natural way in the proof of the uniqueness and representation theorem that we will present in Section 4. The replacement of (7.2.5) by (7.2.6) also has a natural physical interpretation: instead of requiring that at each point the temperature $u(x,t)$ approach the initial temperature $f(x)$ as $t \to 0$, we require that every *average value* of $u(x,t)$ approach the corresponding average value of $f(x)$. From the physical point of view this is entirely reasonable, since in fact, any measurement of a quantity at a point is actually a determination of a certain average of the corresponding function. We shall briefly discuss the notion of average value of a function and the interpretation of (7.2.6) in the next section.

In the precise formulation of problem (7.1.1) which we shall consider, we shall include, in addition to the conditions mentioned above, the requirements that $u_x(0+,t)$ and $u_x(\pi-,t)$ exist for all $t > 0$, and that, for each t_1, t_2 with $0 < t_1 < t_2$, $u_t(x,t)$ be bounded for $0 < x < \pi$, $t_1 < t < t_2$. Although the need for these requirements is not a priori evident, the interpretation of u_x as the heat flux (except for a constant factor), and of u_t as the time rate of change of the temperature, makes them physically reasonable. The full statement of the precise formulation of problem (7.1.1) will be given in Section 4, after the following brief digression.

7.3. *AVERAGE VALUES AND FUNCTION VALUES*

Consider a piecewise continuous function $f(x)$ on an interval, say $[0,\pi]$, and let $[a,b]$ be a subinterval of $[0,\pi]$. In the calculus, the average of $f(x)$ on $[a,b]$ is defined to be

$$\frac{1}{b-a}\int_a^b f(\xi)\,d\xi\,. \tag{7.3.1}$$

Suppose that f is continuous at x. Then the average of f over a small interval centered at x is an approximation to $f(x)$. More precisely, if f is continuous at x, then

$$f(x) = \lim_{\epsilon \to 0} \frac{1}{\epsilon} \int_{x-\epsilon/2}^{x+\epsilon/2} f(\xi)\, d\xi\,. \tag{7.3.2}$$

For, if ϵ is sufficiently small, then f is continuous in the interval $[x - \epsilon/2, x + \epsilon/2]$. Hence by the mean-value theorem

$$\frac{1}{\epsilon} \int_{x-\epsilon/2}^{x+\epsilon/2} f(\xi)\, d\xi = f(\tilde{x})\,,$$

where $x - \epsilon/2 \leq \tilde{x} \leq x + \epsilon/2$. Letting $\epsilon \to 0$ we see that $\tilde{x} \to x$, so that because of the continuity of f at x, $f(\tilde{x}) \to f(x)$. This proves (7.3.2).

Now, we can reformulate the definition (7.3.1) in a way which suggests a generalization. Assuming $0 < a < b < \pi$, we set

$$\psi(x;a,b) = \begin{cases} 0 & x < a \\ \dfrac{1}{b-a} & a < x < b \\ 0 & b < x\,. \end{cases} \tag{7.3.3}$$

Then (7.3.1) may be written

$$\int_0^\pi f(\xi)\psi(\xi;a,b)\, d\xi\,.$$

The function $\psi(\xi;a,b)$ has the properties

$$\psi(\xi;a,b) \geq 0$$

$$\int_0^\pi \psi(\xi;a,b) = 1\,.$$

We generalize the notion of average, on the basis of the preceding paragraph as follows. *Let $\psi(x)$ be a piecewise continuous function on $[0,\pi]$ with the properties*

$$\psi(x) \geq 0 \qquad 0 \leq x \leq \pi \tag{7.3.4}$$

$$\int_0^\pi \psi(\xi)\, d\xi = 1\,. \tag{7.3.5}$$

If $f(x)$ is piecewise continuous on $[0,\pi]$, then the **average** *of $f(x)$ relative to the density $\psi(x)$ on $[0,\pi]$ is given by*

$$\int_0^\pi f(\xi)\psi(\xi)\, d\xi\,. \tag{7.3.6}$$

We shall refer to any function $\psi(x)$ satisfying the conditions above as a *density function*, or a *density*, on $[0,\pi]$.

It is easy to verify that the average relative to a density has the following properties: (i) the average of a linear combination $c_1f_1(x) + c_2f_2(x)$ is the corresponding linear combination of the averages of f_1

and f_2; (ii) if $f_1(x) \geq f_2(x)$ for all x in $[a,b]$, then the same relation holds for their averages; (iii) the average of a constant function is the value of the constant. These are indeed the properties which we would expect to be characteristic of an average.

The relation between the average values of a function and the values of the function, which was established at the beginning of this section, can be reformulated as follows.

If $f_1(x)$ and $f_2(x)$ are piecewise continuous functions on $[0,\pi]$ whose averages relative to all densities are equal, then $f_1(x) = f_2(x)$ at every point x at which f_1 and f_2 are continuous.

For referring to (7.3.3), (7.3.2), and the hypothesis of the statement, we have

$$f_1(x) = \lim_{\epsilon \to 0} \int_0^\pi f_1(\xi)\psi\left(\xi; x - \frac{\epsilon}{2}, x + \frac{\epsilon}{2}\right) d\xi$$
$$= \lim_{\epsilon \to 0} \int_0^\pi f_2(\xi)\psi\left(\xi; x - \frac{\epsilon}{2}, x + \frac{\epsilon}{2}\right) d\xi = f_2(x) .$$

In Section 2 we made the statement that the requirement

$$\lim_{t \to 0} \int_0^\pi u(x,t)\psi(x)\, dx = \int_0^\pi f(x)\psi(x)\, dx$$

for every piecewise continuous function $\psi(x)$ was equivalent to a statement about averages, that is, to the same requirement when $\psi(x)$ is a function satisfying (7.3.4) and (7.3.5). The verification of this statement is easy and is left as an exercise.

EXERCISES 7a

1. Verify the properties (i)–(iii) of averages relative to an arbitrary density.

2. Let $\psi(x)$ be an arbitrary piecewise continuous function on $[0,\pi]$. Show that

$$\psi_1(x) = \frac{|\psi(x)| + \psi(x)}{\displaystyle\int_0^\pi [|\psi(\xi)| + \psi(\xi)]\, d\xi}$$

$$\psi_2(x) = \frac{|\psi(x)| - \psi(x)}{\displaystyle\int_0^\pi [|\psi(\xi)| - \psi(\xi)]\, d\xi}$$

are densities on $[0,\pi]$ and that with appropriate constants c_1, c_2

$$\psi(x) = c_1\psi_1(x) + c_2\psi_2(x)$$

(if either ψ_1 or ψ_2 is undefined because of a vanishing denominator, the corresponding term in the preceding equation is to be omitted). Conclude that the statement

$$\lim_{t\to 0} \int_0^\pi u(x,t)\psi(x)\,dx = \int_0^\pi f(x)\psi(x)\,dx$$

holds for all piecewise continuous functions $\psi(x)$ on $[0,\pi]$ if it holds for all such functions which are densities on $[0,\pi]$.

7.4. *REPRESENTATION AND UNIQUENESS OF SOLUTIONS*

We shall now show that if our problem, precisely formulated as in Section 2, has a solution, then it has only one solution, which must have the representation given by Equations (7.1.2) and (7.1.3). We give the explicit statement below, and call attention to the fact that we require of the initial function $f(x)$ only that it be piecewise continuous.

Suppose there is a function $u(x,t)$ *such that* (i) u, u_t, u_x, *and* u_{xx} *are defined and continuous for* $0 < x < \pi$, $0 < t$ *and satisfy*

$$u_t = ku_{xx} \qquad 0 < x < \pi\,, \quad 0 < t\,, \tag{7.4.1}$$

and further, for each t_1, t_2 *with* $0 < t_1 < t_2$, $u_t(x,t)$ *is bounded for* $0 < x < L$, $t_1 < t < t_2$; (ii) $u(0+,t)$, $u(\pi-,t)$, $u_x(0+,t)$ *and* $u_x(\pi-,t)$ *exist for all* $t > 0$, *and*

$$u(0+,t) = 0\,, \quad u(\pi-,t) = 0 \qquad 0 < t\,; \tag{7.4.2}$$

(iii) *for every piecewise continuous function* $\psi(x)$ *on* $[0,\pi]$

$$\lim_{\substack{t\to 0\\ t>0}} \int_0^\pi u(x,t)\psi(x)\,dx = \int_0^\pi f(x)\psi(x)\,dx\,, \tag{7.4.3}$$

where $f(x)$ *is piecewise continuous on* $[0,\pi]$. *Then there is only one such function and it is given for* $0 < x < \pi$, $0 < t$ *by*

$$u(x,t) = \sum_{n=1}^{\infty} b_n e^{-n^2 t} \sin nx \tag{7.4.4}$$

with

$$b_n = \frac{2}{\pi}\int_0^\pi f(x)\sin nx\,dx\,. \tag{7.4.5}$$

We have only to demonstrate the last part of this statement, that is, that $u(x,t)$ is given by (7.4.4) and (7.4.5) for $0 < x < \pi$, $0 < t$. The *uniqueness* of the solution of our problem is a particular and immediate consequence of this result. If there are two solutions, they are each equal to the sum of the series in (7.4.4), and hence equal to each other, for all $0 < x < \pi$, $0 < t$.

For our proof we shall need a slight generalization of a standard theorem on the **differentiation under the integral sign** of integrals depending on a parameter. A precise statement of this theorem, together with its generalization, is given below. For the proof of the theorem, and a discussion of the essential features of its generalization, see the references cited at the end of this chapter.

If in the closed rectangle $a \leq x \leq b$, $t_1 \leq t \leq t_2$ the function $g(x,t)$ is continuous and has a continuous partial derivative $g_t(x,t)$, then the function

$$G(t) = \int_a^b g(x,t)\,dx$$

is continuous and has a derivative for $t_1 < t < t_2$, with

$$\frac{d}{dt}G(t) = \int_a^b g_t(x,t)\,dx\,.$$

More generally, the conclusion above holds if $g(x,t)$ and $g_t(x,t)$ are continuous in the open rectangle $a < x < b$, $t_1 < t < t_2$, the derivative $g_t(x,t)$ is bounded there, and the limits $g(a+,t)$, $g(b-,t)$ exist for $t_1 < t < t_2$.

In addition to the theorem just quoted, our argument will require only the following three propositions, each of which has already been proved. In these propositions the statement that a function $\chi(x)$ is continuous for $0 \leq x \leq \pi$ means that $\chi(x)$ is continuous for $0 < x < \pi$ and $\chi(0+)$, $\chi(\pi-)$ exist.

(A) The eigenvalues and eigenfunctions of the problem

$$\text{D.E.} \qquad \frac{d^2\varphi}{dx^2} + \lambda\varphi = 0 \qquad 0 < x < \pi$$

$$\text{B.C.} \qquad \varphi(0+) = 0\,, \quad \varphi(\pi-) = 0$$

are $\lambda_n = n^2$, $\varphi_n(x) = \sin nx$, $n = 1, 2, \ldots$.

(B) If $p(x)$ and $p'(x)$ are continuous for $0 \leq x \leq \pi$, then

$$p(x) = \sum_{n=1}^{\infty} b_n \sin nx \qquad 0 < x < \pi$$

where

$$b_n = \frac{2}{\pi}\int_0^{\pi} p(x)\sin nx\,dx\,.$$

(C) If $\varphi(x)$, $\varphi'(x)$, $\psi(x)$, $\psi'(x)$ are continuous for $0 \leq x \leq \pi$, and $\varphi''(x)$, $\psi''(x)$ are continuous for $0 < x < \pi$, then Green's formula,

$$\int_0^{\pi}\left[\varphi\frac{d^2\psi}{dx^2} - \psi\frac{d^2\varphi}{dx^2}\right]dx = [\varphi\psi' - \psi\varphi']_{0+}^{\pi-}\,,$$

holds, where the integral on the left is an improper integral. If φ

and ψ satisfy the B.C. of (A), then the right side of Green's formula vanishes, and if either $\varphi''(x)$ or $\psi''(x)$ is continuous for $0 \leq x \leq \pi$, we have

$$\int_0^\pi \varphi \frac{d^2\psi}{dx^2}\, dx = \int_0^\pi \psi \frac{d^2\varphi}{dx^2}\, dx .$$

The argument proceeds as follows. Let $u(x,t)$ be a solution of our problem. Then for each fixed $t > 0$, $u(x,t)$ and $u_x(x,t)$ are continuous functions of x for $0 \leq x \leq \pi$ because of (i) and (ii). According to (B), therefore, we have

$$u(x,t) = \sum_{n=1}^{\infty} B_n(t) \sin nx \qquad 0 < x < \pi , \tag{7.4.6}$$

where the coefficients $B_n(t)$ are functions of t given by

$$B_n(t) = \frac{2}{\pi} \int_0^\pi u(x,t) \sin nx\, dx . \tag{7.4.7}$$

For $t > 0$, $B_n(t)$ is continuous and differentiable. Differentiating Equation (7.4.7), we obtain

$$\frac{d}{dt} B_n(t) = \frac{2}{\pi} \int_0^\pi u_t(x,t) \sin nx\, dx , \tag{7.4.8}$$

the differentiation under the integral sign being justified by the theorem quoted above, and (i) and (ii). After substituting (7.4.1), (7.4.8) becomes

$$B_n'(t) = \frac{2}{\pi} \int_0^\pi u_{xx}(x,t) \sin nx\, dx . \tag{7.4.9}$$

The function $\sin nx$ satisfies the conditions of (C), and again because of (i) and (ii), so does $u(x,t)$ for $t > 0$. Using (C) to replace the integral in (7.4.9) we get

$$B_n'(t) = \frac{2}{\pi} \int_0^\pi u(x,t) \frac{d^2}{dx^2} \sin nx\, dx . \tag{7.4.10}$$

However, because of (A),

$$\frac{d^2}{dx^2} \sin nx = -n^2 \sin nx ,$$

so that

$$B_n'(t) = -n^2 \cdot \frac{2}{\pi} \int_0^\pi u(x,t) \sin nx\, dx , \tag{7.4.11}$$

and referring to Equation (7.4.7) we find that, for $t > 0$, $B_n(t)$ satisfies the ordinary differential equation

(7.4.12) $$B_n'(t) + n^2 B_n(t) = 0 .$$

Solving Equation (7.4.12) we get

(7.4.13) $$B_n(t) = b_n e^{-n^2 t} \qquad t > 0 ,$$

where b_n is constant. Substituting (7.4.13) in (7.4.6) we have for $u(x,t)$ the representation

$$u(x,t) = \sum_{n=1}^{\infty} b_n e^{-n^2 t} \sin nx \qquad 0 < x < \pi , \quad 0 < t .$$

To determine the coefficients b_n we observe that, from (7.4.13),

(7.4.14) $$\lim_{t \to 0} B_n(t) = B_n(0+) = b_n ,$$

while from (7.4.3) with $\psi(x) = \sin nx$, and (7.4.7), we have

(7.4.15) $$B_n(0+) = \lim_{\substack{t \to 0 \\ t > 0}} \frac{2}{\pi} \int_0^{\pi} u(x,t) \sin nx \, dx = \frac{2}{\pi} \int_0^{\pi} f(x) \sin nx \, dx .$$

Thus

$$b_n = \frac{2}{\pi} \int_0^{\pi} f(x) \sin nx \, dx .$$

This completes the proof.

Although we have presented a theorem and its proof for a particular initial-boundary value problem for a special linear homogeneous equation of heat conduction, we have formulated the problem so that the theorem can be generalized to other problems that we have solved formally, with changes in the proof which are, so to speak, only typographical. In Exercises 7b it will be seen that similar statements hold for a large class of inhomogeneous problems for the heat equation and related equations. In the next and later chapters we will see that corresponding statements hold for problems involving differential equations of other types.

We remark, finally, that the preceding discussion furnishes the resolution of the apparent contradiction, observed in Section 5.4, in the representation of the solution of a problem with inhomogeneous boundary conditions by a series of eigenfunctions of the related homogeneous problem. If $u(x,t)$ is the solution of such a problem, for which the B.C. precisely formulated, are

(7.4.16) $$u_x(0+,t) = A(t) , \quad u(L-,t) = B(t)$$

and if $\Phi(x,t)$ denotes the sum of the formally calculated series representation, then it can be shown, following the model above, that

(7.4.17) $$u(x,t) = \Phi(x,t) \qquad 0 < x < L , \quad 0 < t .$$

All that is required for the argument is that, for each $t > 0$, the expansion of $u(x,t)$ in terms of the eigenfunctions associated with the related homogeneous problem *converges pointwise* to $u(x,t)$ in $0 < x < L$. According to the results of Chapter 6, this will hold if u is smooth, on $[0,L]$, regardless of the values of $u_x(0,t)$ and $u(L,t)$. From (7.4.17) it follows that

$$u_x(0+,t) = \Phi_x(0+,t)\,, \quad u(L-,t) = \Phi(L-,t) \tag{7.4.18}$$

so that the B.C. (7.4.16) are satisfied by Φ. The fact that, for example, $\Phi_x(0,t) \neq \Phi_x(0+,t)$ is irrelevant.

7.5. EXISTENCE OF SOLUTIONS OF INITIAL-BOUNDARY VALUE PROBLEMS FOR THE HEAT EQUATION AND RELATED EQUATIONS

We shall now show that the series determined in Section 4 is actually a solution of the problem of that section. We shall need the following results from the theory of infinite series of functions of two variables, each of which is strictly analogous, in statement and proof, to a corresponding theorem on functions of one variable.

If the terms of a series of functions are majorized on a rectangle by the terms of a convergent numerical series, then the series of functions is absolutely and uniformly convergent on the rectangle (Weierstrass's M-test).

If an infinite series of continuous functions converges uniformly on a rectangle, then its sum is continuous on the rectangle.

If the series obtained from a given convergent series by formal term-by-term partial differentiation is a uniformly convergent series of continuous functions on a closed rectangle, then the given series has a continuous derivative which on the rectangle is the sum of the series obtained by term-by-term differentiation.

Now consider the series (7.4.4), (7.4.5). We have, from (7.4.5),

$$|b_n| = \left|\int_0^\pi f(x)\sin nx\,dx\right| \leq \int_0^\pi |f(x)|\,dx = B \tag{7.5.1}$$

so that, for the nth term of (7.4.4) we obtain the majorization, for $t \geq t_0 > 0$,

$$|b_n e^{-n^2t}\sin nx| \leq Be^{-n^2t} \leq Be^{-n^2t_0}\,. \tag{7.5.2}$$

It follows from the first of the theorems quoted above, since the numerical series

$$\sum_{n=1}^{\infty} Be^{-n^2t_0}$$

is convergent, that the series (7.4.4) is uniformly convergent for $0 \leq x \leq \pi$, $t \geq t_0 > 0$, for $t_0 > 0$ arbitrary. From the second theorem quoted, and the fact that t_0 is any positive constant, we infer that the sum $u(x,t)$ of the series is continuous for $0 \leq x \leq \pi$, $0 < t$.

If we formally differentiate the series (7.4.4) term-by-term with respect to x and t, we obtain

$$\sum_{n=1}^{\infty} nb_n e^{-n^2 t} \cos nx$$

$$\sum_{n=1}^{\infty} -n^2 b_n e^{-n^2 t} \sin nx \, ,$$

respectively. These series have as majorizing series the convergent series

$$\sum_{n=1} Bne^{-n^2 t_0}$$

$$\sum_{n=1}^{\infty} Bn^2 e^{-n^2 t_0} \, ,$$

respectively, for $0 \leq x \leq \pi$, $t \geq t_0 > 0$. It follows as above that the formally differentiated series are uniformly convergent for $0 \leq x \leq \pi$, $t \geq t_0 > 0$, and hence by the third quoted theorem that the sum, $u(x,t)$, of the series (7.4.4) has continuous partial derivatives with respect to x and t for $0 \leq x \leq \pi$, $t > 0$, and that these derivatives can be calculated term-by-term. In a similar way it can be shown that the second partial derivative of u with respect to x exists, is continuous, and can be calculated term-by-term.

Thus, we have shown that u, u_t and u_{xx} are continuous for $0 < x < \pi$, $0 < t$, and that

$$u_t(x,t) = \sum_{n=1}^{\infty} b_n \frac{\partial}{\partial t} (e^{-n^2 t} \sin nx) \tag{7.5.3}$$

$$u_{xx}(x,t) = \sum_{n=1}^{\infty} b_n \frac{\partial^2}{\partial x^2} (e^{-n^2 t} \sin nx) \, . \tag{7.5.4}$$

Subtracting (7.5.4) from (7.5.3) we get

$$u_t(x,t) - u_{xx}(x,t) = \sum_{n=1}^{\infty} b_n \left[\frac{\partial}{\partial t} (e^{-n^2 t} \sin nx) - \frac{\partial^2}{\partial x^2} (e^{-n^2 t} \sin nx) \right] . \tag{7.5.5}$$

Since, by construction, each term of the series (7.5.5) satisfies

$$\frac{\partial}{\partial t} (e^{-n^2 t} \sin nx) - \frac{\partial^2}{\partial x^2} (e^{-n^2 t} \sin nx) = 0 \, ,$$

it follows that $u(x,t)$ satisfies the differential equation (7.4.1), and the main part of statement (i) of the problem of Section 4 has been established.

The reader should now refer to the discussion of the principle of superposition for infinite linear combinations in Section 1.3 and observe that the preceding discussion can be formulated in two statements, as follows. First, if A is a linear differential operator, the statement on p. 8 holds if "suitable convergence" is understood to mean "for each derivative appearing in A, the term-by-term calculation of the derivative yields a uniformly convergent series." Second, the series (7.4.4) "converges suitably," in the preceding sense.

We return to the problem of Section 4. To show that the remaining condition of (i) is satisfied we use the statement, proved above, that $u_t(x,t)$ is continuous for $0 \leq x \leq \pi$, $0 < t$, and hence, for arbitrary t_1, t_2 with $0 < t_1 < t_2$, it is continuous in the closed rectangle $0 \leq x \leq \pi$, $t_1 \leq t \leq t_2$. Since a function continuous in a closed rectangle is bounded there, it is true a fortiori, that $u_t(x,t)$ is bounded for $0 < x < \pi$, $t_1 < x < t_2$. The statements, also proved above, that $u(x,t)$ and $u_x(x,t)$ are continuous for $0 \leq x \leq \pi$, $0 < t$, imply that $u(0+,t)$, $u(\pi-,t)$, $u_x(0+,t)$, $u_x(\pi-,t)$ exist for all $t > 0$, and further, since $u(0,t) = u(\pi,t) = 0$ by inspection,

$$u(0+,t) = u(0,t) = 0\,, \quad u(\pi-,t) = u(\pi,t) = 0\,.$$

Thus, all the conditions of (ii) are satisfied.

It remains to show that condition (iii) is satisfied. Let $\psi(x)$ be any piecewise continuous function and let its Fourier sine series be

$$\psi(x) \sim \sum_{n=1}^{\infty} \beta_n \sin nx\,.$$

Then, by Parseval's theorem,

$$\int_0^{\pi} u(x,t)\psi(x)\,dx = \frac{\pi}{2}\sum_{n=1}^{\infty} b_n\beta_n e^{-n^2t}\,, \quad 0 < t \tag{7.5.6}$$

and

$$\int_0^{\pi} f(x)\psi(x)\,dx = \frac{\pi}{2}\sum_{n=1}^{\infty} b_n\beta_n\,. \tag{7.5.7}$$

Also by Parseval's theorem,

$$\int_0^{\pi} [f(x)]^2\,dx = \frac{\pi}{2}\sum_{n=1}^{\infty} b_n^2 \tag{7.5.8}$$

$$\int_0^\pi [\psi(x)]^2 \, dx = \frac{\pi}{2} \sum_{n=1}^{\infty} \beta_n^2 . \tag{7.5.9}$$

Now from the easily verified inequality

$$|b_n \beta_n| \leq \tfrac{1}{2}(b_n^2 + \beta_n^2) ,$$

it follows that the terms of the series in (7.5.6) are majorized for all $t \geq 0$ by those of the numerical series

$$\frac{\pi}{4} \sum_{n=1}^{\infty} (b_n^2 + \beta_n^2) ,$$

which is a convergent series by (7.5.8) and (7.5.9). Hence by the one-variable analogues of the first and second theorems on infinite series cited above, (7.5.6) is a continuous function of t for $t \geq 0$. In particular then, the limit of (7.5.6) as $t \to 0$ is equal to its value at $t = 0$, and noting (7.5.7), this proves statement (iii). Thus the existence proof is completed.

It is worth observing again that we have assumed only that the initial function is piecewise continuous, and have shown that for arbitrary $t > 0$, $u(x,t)$ has continuous derivatives of second order. In fact by repeating the argument with which this last statement was established we can show that $u(x,t)$ has continuous partial derivatives *of all orders* for *all* positive t. Thus any discontinuity in the initial function or its derivatives is, so to speak, immediately smoothed out. We shall contrast this later with the behavior of solutions of other initial-boundary value problems.

EXERCISES 7b

1. Following the discussion in Section 4, establish a representation and uniqueness theorem for the initial-boundary value problem

D.E.	$u_t = u_{xx} + ru$	$0 < x < \pi ,$	$0 < t$
B.C.	$u(0,t) = 0 , \quad u(\pi,t) = 0$		$0 < t$
I.C.	$u(x,0) = f(x)$	$0 < x < \pi ,$	

where r is a constant.

2. Formulate and prove a representation and uniqueness theorem for the problem

D.E.	$u_t = u_{xx} + q(x,t)$	$0 < x < \pi ,$	$0 < t$
B.C.	$u(0,t) = 0 , \quad u(\pi,t) = 0$		
I.C.	$u(x,0) = f(x) ,$		

where $q(x,t)$ is continuous for $0 \leq x \leq \pi$, $0 \leq t$, and $f(x)$ is piecewise continuous for $0 \leq x \leq \pi$.

3. Formulate and prove a representation and uniqueness theorem for the problem

D.E.	$u_t = u_{xx}$	$0 < x < L, \quad 0 < t$
B.C.	$u_x(0,t) = A(t), \quad u(L,t) = B(t)$	
I.C.	$u(x,0) = 0$	$0 < x < L,$

where $A(t)$ and $B(t)$ are continuous for $0 \leq t$.

4. Show that if $f(x)$ has continuous first and second derivatives and $f(0) = f(\pi) = 0$, then the solution $u(x,t)$ of the problem of Section 4 satisfies

$$\lim_{t \to 0} u(x,t) = f(x) \qquad 0 \leq x \leq \pi .$$

(Use the estimates obtained in Exercises 6e, Problem 1.)

5. Assume that in Problem 2, $f(x) = 0$, and the functions

$$q_n(t) = \frac{2}{\pi} \int_0^\pi q(x,t) \sin nx \, dx , \qquad n = 1, 2, \ldots,$$

satisfy the inequality

$$|q_n(t)| \leq \frac{Q}{n^{1+\alpha}} \qquad 0 \leq t, \quad n = 1, 2, \ldots,$$

where Q and α are positive constants. Show that the series solution found for Problem 2 converges to a function $u(x,t)$ which is continuous together with its first- and second-order partial derivatives and satisfies the D.E. of the problem for $0 < x < \pi$, $0 < t$.

REFERENCES

Protter, Murray H. and Charles B. Morrey, Jr., *Modern Mathematical Analysis*, Addison-Wesley, Reading, 1964, pp. 521–525, 542–546.

Taylor, Angus E., *Advanced Calculus*, Ginn, Boston, 1955, pp. 522–525, 666–668.

8

The wave equation and related equations

8.1. *THE VIBRATING STRING*

Consider a uniform string stretched between fixed end points. Suppose that no external forces act on it and that it is at rest. The linear density (mass per unit length) of the string is a constant ρ_0. Each segment of the string on one side of a point exerts a force on the segment on the other side whose direction lies along that of the string, and whose magnitude is a constant T_0.

Suppose that the string is displaced from rest in a plane. The density at each point will assume a new value because of the change in length in the neighborhood of the point. The force at each point exerted by the segment of string to one side of the point on the segment on the other side will also change both in direction and magnitude.

In an actual string the change in force referred to above depends on the amount of stretching and bending of the string. It is a matter of common experience that a string has considerable resistance to stretching, but very little resistance to bending—in fact, this is one property that we use to distinguish "strings" from "wires" or "ropes." We idealize this property by introducing the notion of a *perfectly flexible* string, which we define by the statement that the force at a point exerted by the segment of the string to one side of the point on the segment on the other side has direction lying along the tangent line to the string at the point, and magnitude which depends only on the relative change in length of the string in the neighborhood of the point. We call this force the *tension* in the string at the point.

Let us return to the consideration of the plane displacement of the string. The displacement of each point of the string can be resolved into two components, one in the direction of the equilibrium position of the string, called *longitudinal*, and one perpendicular to this direction, called *transverse*. As idealizations of displacements in which one of these components is small relative to the other we take the small component to be zero, and speak of *purely longitudinal* and *purely transverse* displacements.

Now suppose that the string is released. Then it will vibrate in the plane. We shall find the conditions determining the **purely transverse plane vibrations of a perfectly flexible uniform string,** at first under the assumption that no external forces, such as gravity, act on the string.

Let the equilibrium position of the string lie along the x-axis with the fixed ends at $x = 0$ and $x = L$. Choose the u-axis perpendicular to the x-axis and in the plane of vibration. The position of the string at any time t will then coincide with the graph of a function $u = u(x,t)$ called the **displacement** function (see Fig. 8.1). Let $s(x,t)$,

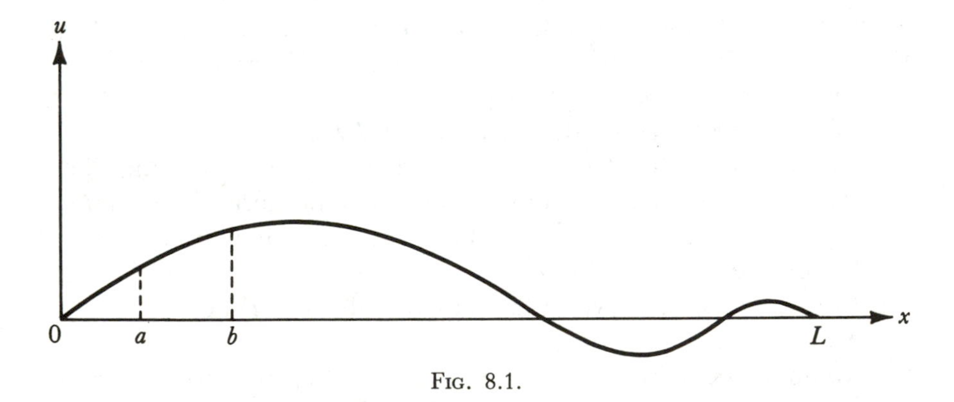

FIG. 8.1.

$\rho(x,t)$, $T(x,t)$ be, respectively, the arc length, density and tension in the string, and $\theta(x,t)$ the angle of inclination of the graph of the displacement function.

We first observe that, since the mass of the segment above the interval $[a,b]$ is constant, we have

$$\frac{d}{dt}\int_a^b \rho\, ds = \frac{d}{dt}\int_a^b \rho\,\frac{\partial s}{\partial x}\,dx = \int_a^b \frac{\partial}{\partial t}\left[\rho\,\frac{\partial s}{\partial x}\right]dx = 0\,,$$

which implies, since a and b are arbitrary,

$$\frac{\partial}{\partial t}\left[\rho \frac{\partial s}{\partial x}\right] = 0 .$$

This equation implies, in turn, assuming that at some time (possibly in the very distant past or future) the string is in its equilibrium position, that

$$\rho \frac{\partial s}{\partial x} = \rho_0 . \tag{8.1.1}$$

To find the conditions determining $u(x,t)$ we shall use the following proposition of mechanics. The forces acting on the masses of a system may be divided into two classes: *internal* forces, which are forces acting between particles of the system, and *external* forces. The center of mass of the system moves as if it were a single particle of mass equal to the total mass of the system acted on by the total of the external forces acting on the system.

We apply this proposition to the segment of string lying above the interval $[a,b]$ (Fig. 8.2). The mass of this segment is, by (8.1.1),

$$\int_{s(a,t)}^{s(b,t)} \rho \, ds = \rho_0(b - a) . \tag{8.1.2}$$

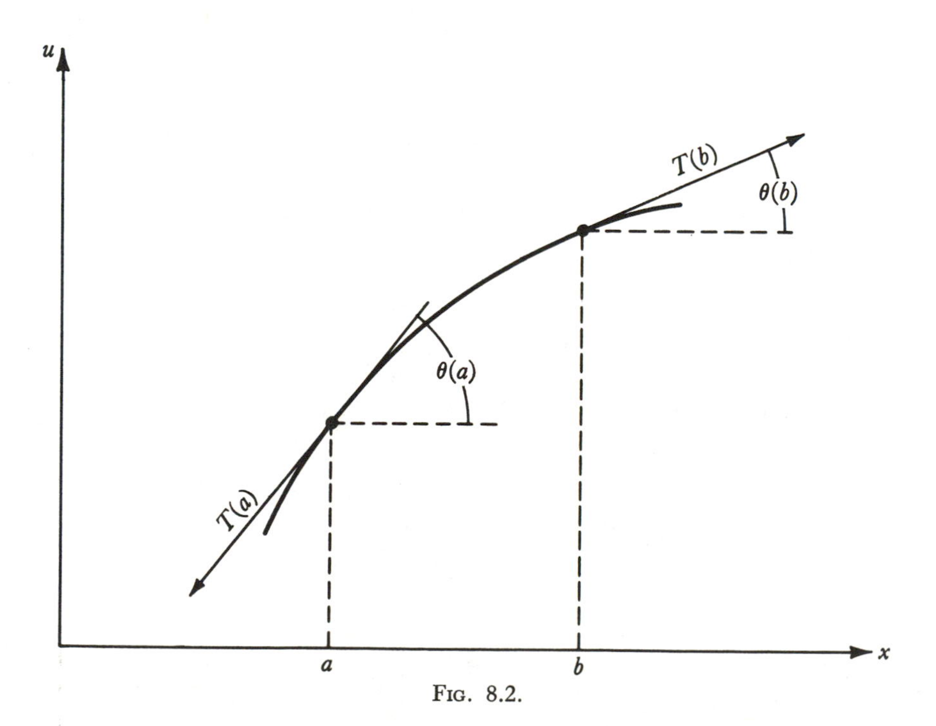

Fig. 8.2.

The u-coordinate of the center of mass of the segment is

$$\frac{1}{\rho_0(b-a)}\int_{s(a,t)}^{s(b,t)} \rho u\, ds = \frac{1}{\rho_0(b-a)}\int_a^b \rho_0 u\, dx = \frac{1}{b-a}\int_a^b u\, dx \tag{8.1.3}$$

and the u-component of the acceleration of the center of mass is

$$\frac{d^2}{dt^2}\frac{1}{(b-a)}\int_a^b u\, dx = \frac{1}{b-a}\int_a^b u_{tt}\, dx\,. \tag{8.1.4}$$

The total of the external forces acting on the segment is determined as follows (see Fig. 8.2). The part of the string to the right of b exerts on the segment a force whose component in the positive u-direction is $+T(b,t)\sin\theta(b,t)$. Similarly the u-component of the force exerted by the part of the string to the left of a is $-T(a,t)\sin\theta(a,t)$. The sum of these forces is

$$T(b,t)\sin\theta(b,t) - T(a,t)\sin\theta(a,t) = \int_a^b \frac{\partial}{\partial x}[T\sin\theta]\, dx\,. \tag{8.1.5}$$

Equating the product of the mass (8.1.2) and the acceleration (8.1.4) to the force (8.1.5), we obtain

$$\int_a^b \rho_0 u_{tt}\, dx = \int_a^b \frac{\partial}{\partial x}[T\sin\theta]\, dx$$

or, since a and b are arbitrary,

$$\rho_0 u_{tt} = \frac{\partial}{\partial x}[T\sin\theta]\,. \tag{8.1.6}$$

In a similar way, since the x-component of the acceleration of the center of mass is zero, we get

$$0 = \frac{\partial}{\partial x}[T\cos\theta]\,. \tag{8.1.7}$$

Since for fixed t, $\theta(x,t)$ is the angle of inclination of the curve $u = u(x,t)$ whose slope is $u_x(x,t)$, we have $u_x = \tan\theta$, and hence

$$\sin\theta = \frac{u_x}{\sqrt{1+u_x^2}}, \quad \cos\theta = \frac{1}{\sqrt{1+u_x^2}}$$

so that (8.1.6) and (8.1.7) become

$$\begin{aligned} \rho_0 u_{tt} &= \frac{\partial}{\partial x}\left[T\frac{u_x}{\sqrt{1+u_x^2}}\right] \\ &= \frac{T}{\sqrt{1+u_x^2}}u_{xx} + u_x\frac{\partial}{\partial x}\left[\frac{T}{\sqrt{1+u_x^2}}\right] \end{aligned} \tag{8.1.8}$$

and

$$0 = \frac{\partial}{\partial x}\left[\frac{T}{\sqrt{1 + u_x^2}}\right]. \tag{8.1.9}$$

Substituting (8.1.9) in (8.1.8), we obtain

$$\rho_0 u_{tt} = \frac{T}{\sqrt{1 + u_x^2}} u_{xx} . \tag{8.1.10}$$

Now, the dependence of the tension T on u and its derivatives must be determined empirically. In any event, (8.1.10) is a *nonlinear* equation. To avoid these difficulties we restrict our consideration to vibrations which are *small* in the sense that the relative change in arc length of the string at each point is small; that is,

$$\frac{\partial s}{\partial x} = \sqrt{1 + u_x^2}$$

is approximately equal to 1, with small error. Since the change in the magnitude T of the tension from its equilibrium value T_0 depends on the relative change in arc length this implies that T varies little from T_0. Our two assumptions combined yield the approximation T_0 for

$$\frac{T}{\sqrt{1 + u_x^2}},$$

the coefficient of u_{xx} in (8.1.10). It is worth remarking that (8.1.9) shows that this coefficient is independent of x, so that our approximation is equivalent to the assumption that it is also independent of t.

Making this approximation we obtain the equation

$$u_{tt} = c^2 u_{xx} \qquad 0 < x < L, \quad 0 < t \tag{8.1.11}$$

where the constant c is given by

$$c = \sqrt{T_0/\rho_0} . \tag{8.1.12}$$

Equation (8.1.11), which is the differential equation governing the small plane transverse vibrations of a perfectly flexible string, is called the one-dimensional **wave equation.**

The displacement function must also satisfy the *boundary conditions*

$$u(0,t) = 0 , \quad u(L,t) = 0 \qquad 0 < t . \tag{8.1.13}$$

To find the additional auxiliary conditions that must be imposed we return to the consideration of the motion of the center of mass of a segment. To specify that motion, we must, in addition to the differential equation of motion of the particle, specify its initial position and initial velocity, which means, referring to equation (8.1.3) and considering that a and b are arbitrary, that we must specify $u(x,0)$ and $u_t(x,0)$; that is, we must specify the *initial conditions*

$$u(x,0) = f(x)\,, \quad u_t(x,0) = g(x) \qquad 0 < x < L\,. \tag{8.1.14}$$

Thus we arrive at the initial-boundary value problem

$$\begin{array}{lll} \text{D.E.} & u_{tt}(x,t) = c^2 u_{xx}(x,t)\,, & 0 < x < L\,,\ 0 < t \\ \text{B.C.} & \left.\begin{array}{l} u(0,t) = 0 \\ u(L,t) = 0 \end{array}\right\} & 0 < t \\ \text{I.C.} & \left.\begin{array}{l} u(x,0) = f(x) \\ u_t(x,0) = g(x) \end{array}\right\} & 0 < x < L\,. \end{array} \tag{8.1.15}$$

It seems plausible, and in fact it is true, that the stated set of auxiliary conditions is appropriate for the wave equation.

If *distributed external forces* such as gravity, frictional damping due to air resistance, etc., act on the string it is necessary to modify equation (8.1.11). Let $q = q(x,t,u,u_t,u_x)$ be the distributed external force per unit mass in the positive u-direction. Then the total external force (8.1.5) must be replaced by

$$\int_a^b \frac{\partial}{\partial x}(T \sin \theta)\, dx + \int_a^b \rho_0 q\, dx$$

and this leads to the modified differential equation

$$\rho_0 \frac{\partial^2 u}{\partial t^2} = T_0 \frac{\partial^2 u}{\partial x^2} + \rho_0 q\,. \tag{8.1.16}$$

Suppose for example that the string vibrates in a vertical plane and we wish to investigate the effect of gravity. Then, assuming the u-axis points upward we have $q = -g$, where g is the gravitational constant.

In general, Equation (8.1.16) is nonlinear. However, the equation is linear in special cases of interest. We cite in particular the equation

$$u_{tt} = c^2 u_{xx} - r u_t - k u + q(x,t)\,, \tag{8.1.17}$$

where $c > 0$, $r \geq 0$, $k \geq 0$, (r and k not both 0), are constants, which is sometimes called the **telegraph equation,** and which can be interpreted as the equation governing the small vibrations of the string under the action of: a **damping force,** $-ru_t$, proportional to the velocity and oppositely directed; a **restoring force,** $-ku$, proportional to the displacement and oppositely directed; and an **applied force,** $q(x,t)$, all per unit mass.

The reader should compare the statement above, and the statements concerning boundary and jump conditions in the next section, with the corresponding discussions for the heat equation in Chapter 3.

8.2. *GENERAL BOUNDARY AND JUMP CONDITIONS*

The discussion of the vibrating string in Section 1, leading to the wave equation, remains valid if the ends are not fixed at $x = 0$, $x = L$, but are constrained to move on these lines. For such problems the D.E. and I.C. are unchanged, but other B.C. arise.

For the interpretation of these boundary conditions, and of jump conditions when they occur, we need the fact, based on the approximations in Section 1, that for small vibrations the u-component of the tension in the string

$$\pm T \sin \theta = \pm T u_x/(1 + u_x^2)^{1/2}$$

is given approximately by

$$\pm T_0 u_x \,. \tag{8.2.1}$$

Consider the other forms of boundary conditions which may arise. While the condition $u(0,t) = 0$ corresponds to a **fixed end,** we say the condition $u_x(0,t) = 0$ corresponds to a **free end.** The latter condition would be appropriate if the end of the string were attached to a massless runner free to slide along a frictionless groove on the u-axis, since then the u-component of the tension, $T_0 u_x$, must vanish at the end. The boundary condition $u_x(0,t) = hu(0,t)$ arises when the end of the string is attached to the origin by a spring, and for this reason is often called an **elastic boundary condition.** The corresponding inhomogeneous boundary conditions $u(0,t) = A(t)$, $u_x(0,t) = B(t)$ describe, respectively, a prescribed motion at the end $x = 0$, and a prescribed force in the direction of the u-axis acting at $x = 0$, while $u_x(0,t) = hu(0,t) + C(t)$ describes the situation when the force acting at the end at $x = 0$ is the sum of a spring force and a given force.

For examples of **jump conditions,** suppose that two strings, one of density ρ_1 and length L_1, the second of density ρ_2 and length L_2, are joined so that the first extends from $x = 0$ to $x = L_1$ and the second from $x = L_1$ to $x = L_1 + L_2$. Then the jump conditions

$$
\left.\begin{aligned} u(L_1-,t) &= u(L_1+,t) \\ u_x(L_1-,t) &= u_x(L_1+,t) \end{aligned}\right\} \qquad 0 < t \tag{8.2.2}
$$

express the requirements that the string be continuous and that the force per unit mass acting at L_1 be finite. If, however, there is a finite mass, m, at the point L_1, acted on by a given force of magnitude $P(t)$ in the positive u-direction, the second equation of (8.2.2) must be replaced by

$$
T_0[u_x(L_1+,t) - u_x(L_1-,t)] = -P(t) + mu_{tt}(L_1,t) \qquad 0 < t\,, \tag{8.2.3}
$$

which is the equation of motion of the mass under the action of the given force and the tension forces in the string at the point. [Here $u_{tt}(L_1,t)$ may be either $u_{tt}(L_1+,t)$ or $u_{tt}(L_1-,t)$ because of the first equation of (8.2.2).] Of particular interest is the case in which the mass m is negligibly small and may be replaced in approximation by 0. In this case the jump condition (8.2.3) becomes

$$
T_0[u_x(L_1+,t) - u_x(L_1-,t)] = -P(t)\,,
$$

and we call $P(t)$ an **external point force** applied at L_1.

8.3. *THE ENERGY EQUATION*

From the homogeneous or inhomogeneous wave equation we can derive an equation which expresses for the vibrating string the principle of conservation of energy, and which has, in addition to its physical interest, considerable importance for the mathematical theory of the wave equation.

Let $u(x,t)$ be a solution of

$$
\rho_0 u_{tt} = T_0 u_{xx} + \rho_0 q(x,t) \tag{8.3.1}
$$

which is continuous together with its first- and second-order partial derivatives for $a \le x \le b$ and $0 < t$.

We multiply Equation (8.3.1) by u_t, and integrate the result with respect to x from a to b, obtaining

$$
\int_a^b \rho_0 u_{tt} u_t\, dx = \int_a^b T_0 u_{xx} u_t\, dx + \int_a^b \rho_0 q u_t\, dx\,. \tag{8.3.2}
$$

Now we observe that the integrand on the left is

$$
\frac{1}{2}\rho_0 \frac{\partial}{\partial t} u_t^2\,,
$$

so that (8.3.2) can be written

$$
\frac{d}{dt}\int_a^b \frac{1}{2}\rho_0 u_t^2\, dx = \int_a^b T_0 u_{xx} u_t\, dx + \int_a^b \rho_0 q u_t\, dx\,. \tag{8.3.3}
$$

The first integral on the right of (8.3.3) can be transformed by integration by parts (or equivalently, with the use of Green's first formula). We get

$$\int_a^b T_0 u_{xx} u_t \, dx = T_0 u_x u_t \Big|_a^b - \int_a^b T_0 u_x u_{xt} \, dx ,$$

and since

$$T_0 u_x u_{xt} = \frac{\partial}{\partial t}\left(\frac{1}{2} T_0 u_x^2\right),$$

Equation (8.3.3) can be written as

$$\text{(8.3.4)} \qquad \frac{d}{dt}\int_a^b \frac{1}{2}\,[\rho_0 u_t^2 + T_0 u_x^2]\, dx = T_0 u_x u_t \Big|_a^b + \int_a^b \rho_0 q u_t \, dx .$$

The expression

$$\text{(8.3.5)} \qquad \int_a^b \frac{1}{2}\,[\rho_0 u_t^2 + T_0 u_x^2]\, dx$$

is called the **energy integral** of the function $u(x,t)$ on the interval $[a,b]$, and Equation (8.3.4) is called the **energy equation,** for the wave equation in the form (8.3.1). We have shown that every sufficiently smooth solution of the wave equation satisfies the energy equation.

We will examine the physical interpretation of this proposition, which motivates the proposition and justifies the designations energy integral and energy equation, in the case when $u(x,t)$ is the displacement function of a string, as discussed in Sections 1 and 2. In (8.3.4) and (8.3.5) set $a = 0$, $b = L$, so that the energy integral and energy equation become, respectively,

$$\text{(8.3.6)} \qquad E(t) = \int_0^L \left[\frac{1}{2}\rho_0 u_t^2 + \frac{1}{2} T_0 u_x^2\right] dx ,$$

$$\text{(8.3.7)} \qquad \frac{d}{dt}\int_0^L \frac{1}{2}\rho_0 u_t^2\, dx + \frac{d}{dt}\int_0^L \frac{1}{2} T_0 u_x^2\, dx = T_0 u_x u_t \Big|_0^L + \int_0^L \rho_0 q u_t\, dx ,$$

where we have introduced the notation $E(t)$ for the energy integral.

Now, the string has kinetic energy by virtue of its motion. A segment of the string above the small interval $[x_0, x_0 + h]$ has mass $\rho_0 h$ and velocity u_t, so that its kinetic energy is $\frac{1}{2}\rho_0 u_t^2 h$. Subdividing the interval $[0,L]$, summing, and passing to the limit in the usual way, we find that the *kinetic energy*, $K(t)$, of the entire string is given by

$$\text{(8.3.8)} \qquad K(t) = \int_0^L \frac{1}{2}\rho_0 u_t^2\, dx .$$

Thus, the first term on the left of Equation (8.3.7) is the rate of change of the kinetic energy $K(t)$ of the string.

Next, we recall that the work done by a force on a particle moving in a straight line is the product of the component of the force in the direction of the motion and the displacement of the particle. Suppose that the particle moves along the v-axis and its position at time t is $v(t)$, and let $Q(t)$ be the component of the force in the direction of the v-axis. Then the work done in the small time interval $[t,t+h]$ is given approximately by

$$Q(t)[v(t+h) - v(t)]$$

and the rate at which work is done is

$$\lim_{h\to 0} \frac{Q(t)[v(t+h) - v(t)]}{h} = Q(t)v'(t) .$$

Thus, the rate at which work is done by a force is the product of the component of the force in the direction of motion and the velocity of the particle. The rate at which the distributed external force $q(x,t)$ does work on the segment of string above the small interval $[x_0,x_0+h]$ is $\rho_0 q u_t h$, since the segment has mass $\rho_0 h$, velocity u_t, and the force per unit mass q acts in the direction of the u-axis. Thus

$$\int_0^L \rho_0 q u_t \, dx$$

is the rate at which work is done on the entire string by the distributed external force. The external forces acting in the direction of the u-axis at the ends $x = 0$, $x = L$ of the string do work on the string at the rate

$$-T_0 u_x(0,t) u_t(0,t) , \quad T_0 u_x(L,t) u_t(L,t) ,$$

respectively, since these forces are balanced by the u-components of the tension in the string, $T_0 u_x(0,t)$ and $-T_0 u_x(L,t)$ at the respective points. We see therefore that the right side of Equation (8.3.7) represents the rate at which work is done on the string by the *external (boundary and distributed interior) forces* acting on the string.

Finally, to interpret the remaining term of the energy equation, the derivative of

$$V(t) = \int_0^L \frac{1}{2} T_0 u_x^2 \, dx , \tag{8.3.9}$$

we call on the general principle of mechanics which asserts that the rate of change of the kinetic energy of a system of particles is equal to the rate at which work is done on the system by *all* forces acting on the system. It follows from the preceding discussion and Equation (8.3.7) that $-V(t)$ is the work done by the *internal forces* (tension) in the string. We call $V(t)$ the *potential energy* of the entire string.

Thus, observing that

$$E(t) = K(t) + V(t) , \tag{8.3.10}$$

we see that the energy integral can be interpreted as the *total energy* of the entire string, and the energy equation

$$\frac{dE}{dt} = T_0 u_x u_t \Big|_0^L + \int_0^L \rho_0 q u_t \, dx \tag{8.3.11}$$

has the interpretation: *the rate of change of the total energy of the string is equal to the rate at which external forces do work on the string.*

At a fixed end or free end no work is done on the string by the forces on the boundary. If $x = 0$ is fixed then $u(0,t) = 0$, hence $u_t(0,t) = 0$ and the corresponding term in (8.3.11) vanishes. At a free end the factor u_x vanishes. Hence if the string has fixed or free ends and there are no distributed external forces, the total energy is constant for $t > 0$.

The telegraph equation (8.1.17) can always be transformed, as in Section 2.3, into

$$\rho_0 u_{tt} = T_0 u_{xx} - ku + \rho_0 q(x,t) , \tag{8.3.12}$$

where, of course, the value of the constant $k > 0$ will be changed by the transformation. For the telegraph equation in the form (8.3.12) the energy integral of a function $u(x,t)$ is defined by

$$\int_a^b \frac{1}{2} [\rho_0 u_t^2 + T_0 u_x^2 + ku^2] \, dx \tag{8.3.13}$$

and the energy equation, which holds for all sufficiently smooth solutions of (8.3.12), is

$$\frac{d}{dt} \int_a^b \frac{1}{2} [\rho_0 u_t^2 + T_0 u_x^2 + ku^2] \, dx = T_0 u_x u_t \Big|_a^b + \int_a^b \rho_0 q u_t \, dx . \tag{8.3.14}$$

The proof of this proposition differs trivially from that of the corresponding statement for the wave equation, and is left as an exercise.

We will make significant applications of the energy integrals and energy equations for the wave and telegraph equations in Exercises 8b, Problems 1 and 5, and in Sections 5 and 8.

EXERCISES 8a

1. (i) A string of length L, linear density ρ_0, and tension T_0, has the end $x = L$ fixed on the x-axis. The end $x = 0$ is attached to a runner of mass m which slides in a groove on the u-axis under the action of a force $au + bu_t + p(t)$ (a, b constants) in the direction of the positive u-axis. Find the condition satisfied at $x = 0$ by the displacement function $u(x,t)$, for small vibrations. What does this condition become when the mass m is negligibly small, and the runner may be regarded as massless ($m = 0$)?

(ii) State the corresponding conditions if the end $x = 0$ is fixed and the runner slides in a groove on the line $x = L$.

2. A string of length L, linear density ρ_0, and tension T_0, has the end $x = L$ fixed. The end $x = 0$ is attached to a massless runner which slides in a frictionless groove on the u-axis. The runner is connected to a spring which exerts a restoring force toward the origin proportional to the displacement (spring constant κ).

(i) For small vibrations formulate the *linear* boundary condition at $x = 0$.

(ii) Find a formula for the total energy of the system (string + spring) when small vibrations occur. Show that the total energy is constant.

3. A small quantity of powdered iron of neglible mass is imbedded along the part $x = L/4$ to $x = L/2$ of the string in Problem 2. A nearby electromagnet exerts a constant force in the negative u-direction of magnitude Q per unit length of this part of the string. Find the equilibrium displacement.

4. (i) A string of length 1, linear density ρ_0, and tension T_0, with the ends at $x = 0$ and $x = 1$ fixed on the x-axis, is in equilibrium under the action of a constant point force P in the direction of the positive u-axis applied at the point ξ, $0 < \xi < 1$. Find the equilibrium displacement function $u(x)$ of the string, stating your answer in the form

$$u(x) = \frac{P}{T_0} G(x,\xi) .$$

Show that $G(x,\xi) = G(\xi,x)$, and state a physical interpretation of this result.

(ii) The string of part (i) is again in equilibrium, but under the action of a continuous distributed external force $q(x)$ per unit mass, in the positive u-direction, instead of a point force. Show that

$$u(x) = \frac{\rho_0}{T_0} \int_0^1 G(x,\xi) q(\xi)\, d\xi$$

gives the equilibrium displacement of the string in this case, where $G(x,\xi)$ is the function of part (i). State a physical interpretation of this result.

[The function $G(x,\xi)$ is called the "*Green's function*" for the boundary value problem

$$\text{D.E.} \qquad -\frac{d^2u}{dx^2} = f(x)$$

$$\text{B.C.} \qquad u(0) = 0\,, \quad u(1) = 0\,.]$$

5. An endless thread of linear density ρ_0, tension T_0, traveling at constant velocity V, comes out of a small hole in a wall at $x = 0$ and disappears into another small hole in a wall at $x = L$. The visible part of the thread is executing small vibrations.

(i) If no external forces act, find the differential equation satisfied by the displacement function $u(x,t)$, and show that it is of hyperbolic type.

(ii) Find the differential equation if the air exerts a damping force with component in the positive u-direction proportional (constant $r > 0$) to the velocity in the u-direction and oppositively directed.

[Note that for an observer traveling at constant velocity V to the right the motion of the string appears to be a purely transverse vibration, and that the laws of mechanics are the same in a coordinate system moving with constant velocity as they are in a fixed coordinate system.]

6. Consider a perfectly flexible uniform chain of linear density ρ_0 and length L which has one end fixed but is otherwise free to move in a vertical plane under the action of gravity. Choose a coordinate system in the plane with the origin at the fixed point of the chain, the u-axis horizontal and directed to the right, and the x-axis vertical and directed downward.

Assume that the chain oscillates so that its vertical motion is negligible and its horizontal displacement from its equilibrium position along the x-axis is small. Under these assumptions it is approximately correct that the force exerted on the segment of the chain above a point by the segment below the point has magnitude equal to the weight of the segment below the point and is directed along the tangent to the chain at the point.

Let s be the length of chain measured from the origin, and $x = x(s,t)$, $u = u(s,t)$, $0 \leq s \leq L$, $0 \leq t$, be parametric equations of the curve described by the chain at time t. Find the partial differential equation satisfied by $u(s,t)$ under the stated assumptions.

7. Transform the initial-boundary value problem for the telegraph equation

$$\begin{array}{lll} \text{D.E.} & \rho_0 u_{tt} = T_0 u_{xx} - r u_t - k u & 0 < x < L\,, \quad 0 < t \\ \text{B.C.} & u_x(0,t) = 0\,, \quad u(L,t) = 0\,, & \qquad\qquad\quad 0 < t \\ \text{I.C.} & u(x,0) = f(x)\,, \quad u_t(x,0) = g(x) & 0 < x < L \end{array}$$

into an equivalent problem for a function which satisfies the standard form (2.3.3) of the telegraph equation.

8. Show that every solution $u(x,t)$ of the telegraph equation (8.3.12) which is continuous and has continuous first and second derivatives for $a \leq x \leq b$, $0 < t$, satisfies the energy equation (8.3.14). State the physical significance of the term

$$\int_a^b \tfrac{1}{2} k u^2 \, dx$$

in the energy integral (8.3.13).

8.4. *THE SOLUTION OF INITIAL-BOUNDARY VALUE PROBLEMS*

The solution of initial-boundary value problems, homogeneous or inhomogeneous, for the wave equation and related equations can be effected using the methods developed in Chapters 4 and 5 for corresponding problems for the heat equation and related equations. The only novelties are that there are two initial conditions to be satisfied instead of one, and that the ordinary differential equations for the time dependent coefficients are second-order rather than first-order equations with constant coefficients. These purely technical difficulties are easily overcome.

In the discussion of problems for the wave equation and related equations, solutions of homogeneous or inhomogeneous problems are frequently said to describe **free** vibrations or **forced** vibrations, respectively. We will occasionally follow this usage.

The eigenfunction expansions of the solutions of homogeneous initial-boundary value problems, which we have considered so far from a purely mathematical point of view, have interesting and illuminating physical interpretations in the case of problems for the vibrating string. We will discuss these interpretations of free vibrations in connection with the example of a plucked string.

A string of length L, linear density ρ, with ends fixed and under tension T, is held in the position shown in Figure 8.3 by a force applied at x_0. (Here and hereafter we suppress the subscript 0 on the constant density and tension, ρ and T.) At time $t = 0$ the string is released and we wish to find the subsequent motion. Thus, the initial velocity of the string is zero and its initial displacement is

$$f(x) = \begin{cases} \dfrac{hx}{x_0}, & 0 \leq x \leq x_0, \\[2ex] \dfrac{h(L - x)}{L - x_0}, & x_0 \leq x \leq L. \end{cases} \tag{8.4.1}$$

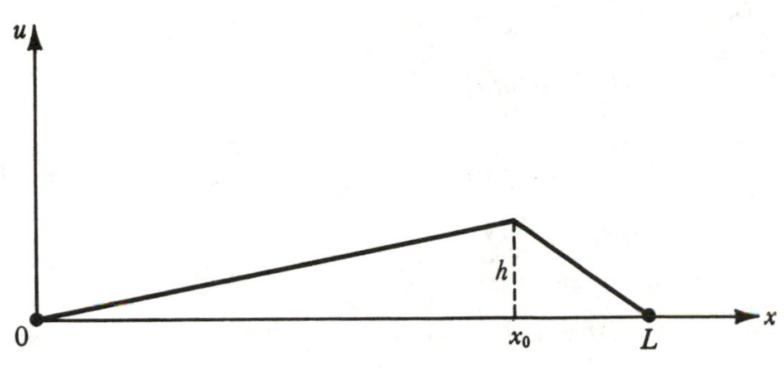

FIG. 8.3.

We therefore have to solve the initial-boundary value problem (8.1.15) where $f(x)$ is given by (8.4.1) and $g(x) = 0$. A function of the form $u(x,t) = \psi(t)\varphi(x)$ which satisfies the D.E. and the B.C. is obtained by solving

$$\text{(8.4.2)} \qquad \begin{aligned} &\varphi'' + \lambda\varphi = 0\,, \qquad\qquad 0 < x < L \\ &\varphi(0) = 0\,, \quad \varphi(L) = 0 \end{aligned}$$

and

$$\text{(8.4.3)} \qquad \psi'' + \lambda c^2 \psi = 0\,, \qquad 0 < t\,.$$

The eigenfunctions and eigenvalues of the familiar eigenvalue problem (8.4.2) are

$$\text{(8.4.4)} \quad \varphi_n(x) = \sin(n\pi x/L)\,, \quad \lambda_n = (n\pi/L)^2\,, \qquad n = 1, 2, 3, \ldots\,.$$

When $\lambda = \lambda_n$ the general solution of (8.4.3) is

$$\text{(8.4.5)} \qquad \psi_n(t) = A_n \cos\left(\frac{n\pi ct}{L}\right) + B_n \frac{\sin(n\pi ct/L)}{(n\pi c/L)}$$

where A_n and B_n are arbitrary constants. Hence the series

$$\text{(8.4.6)} \qquad u(x,t) = \sum_{n=1}^{\infty} \psi_n(t)\varphi_n(x)$$

is a formal solution of the D.E. and the B.C. in (8.1.15). From (8.4.5) we have $\psi_n(0) = A_n$, $\psi'_n(0) = B_n$ and therefore the initial conditions will be satisfied by (8.4.6) provided

$$\text{(8.4.7)} \qquad f(x) = \sum_{n=1}^{\infty} \psi_n(0)\varphi_n(x) = \sum_{n=1}^{\infty} A_n \sin(n\pi x/L)\,,$$

$$\text{(8.4.8)} \qquad 0 = \sum_{n=1}^{\infty} \psi'_n(0)\varphi_n(x) = \sum_{n=1}^{\infty} B_n \sin(n\pi x/L)\,.$$

Using these equations the arbitrary constants can now be determined. We find $B_n = 0$ for every n and

$$A_n = \frac{2}{L}\int_0^L f(x)\sin(n\pi x/L)\,dx$$

$$(8.4.9) \qquad = \frac{2}{L}\left[\int_0^{x_0}\frac{hx}{x_0}\sin(n\pi x/L)\,dx + \int_{x_0}^L \frac{h(L-x)}{L-x_0}\sin(n\pi x/L)\,dx\right]$$

$$= \frac{2h}{x_0(L-x_0)}\cdot\frac{L^2}{n^2\pi^2}\cdot\sin\frac{n\pi x_0}{L}.$$

Hence the series solution is

$$(8.4.10) \qquad u(x,t) = \frac{2L^2h}{\pi^2 x_0(L-x_0)}\sum_{n=1}^{\infty}\frac{1}{n^2}\sin\frac{n\pi x_0}{L}\cos\frac{n\pi ct}{L}\sin\frac{n\pi x}{L},$$

and (8.4.7) can be written as

$$(8.4.11) \qquad f(x) = \frac{2L^2h}{\pi^2 x_0(L-x_0)}\sum_{n=1}^{\infty}\frac{1}{n^2}\sin\frac{n\pi x_0}{L}\sin\frac{n\pi x}{L}.$$

In each of these series the trigonometric factors are bounded, and by comparison with the convergent series $\sum 1/n^2$ we see that both (8.4.11) and (8.4.10) are uniformly convergent.

The displacement function

$$(8.4.12) \qquad u_n(x,t) = A_n\cos\frac{n\pi ct}{L}\sin\frac{n\pi x}{L}$$

describes a possible free vibration of the string, and is called the nth **natural** (or *normal*) **mode of vibration** or the nth **harmonic.** The first harmonic is often called the **fundamental** harmonic, and the subsequent harmonics the first, second, etc. **overtones.**

If, in (8.4.12) we consider the value of x fixed, $u_n(x,t)$ represents the motion of the particle of the string with abscissa x. We see that this is a simple harmonic motion of frequency

$$(8.4.13) \qquad \nu_n = nc/2L,$$

called the nth **natural frequency,** and of **amplitude** $A_n \sin(n\pi x)/L$. Thus, in a natural mode of vibration of the plucked string all points execute simple harmonic motion with a common frequency but with amplitude varying from point to point. The frequency of the first harmonic is, referring to (8.4.13) and (8.1.12),

$$(8.4.14) \qquad \nu_1 = \frac{c}{2L} = \frac{1}{2L}\sqrt{\frac{T}{\rho}},$$

which is called the **fundamental frequency.** The natural frequencies are all integral multiples of the fundamental frequency.

Now, Equation (8.4.10) can be described by the statement that the motion of the plucked string is a superposition of natural modes of vibration, with frequencies which are all integral multiples of the fundamental frequency. The fact that the sound emitted by a plucked string is a *musical* sound can be attributed to this statement. The properties of the musical sound depend on the natural modes (8.4.12). The *pitch* of the sound is given by the fundamental frequency (8.4.14). The *loudness* of the sound, to the extent that it is physically determined, depends on the energy of the vibration, which in turn depends on an infinite linear combination of the squares of the coefficients A_n (see Exercises 8b, Problem 1). Finally, the *timbre* of the sound (the property which distinguishes between the same notes played on different instruments, for example) is determined by the relative values of the coefficients A_n. It is because of this connection with music that the terms harmonic, overtone, etc., are employed in the discussion of vibrations.

Let us return to the consideration of (8.4.12), this time regarding t as fixed. Then $u_n(x,t)$ represents the shape of the string at time t. We see that, for a given natural mode, at different times the shape of the string is altered only by a scale factor. For this reason, the motions described by natural modes are called **standing waves.**

For a standing wave, there are points at which the amplitude vanishes, so that these points are motionless. Such points are called **nodes** of the wave. For the nth harmonic (8.4.12) of the plucked string, the shape of the string is given, except for a time dependent scale factor, by the eigenfunction $\varphi_n(x) = \sin(n\pi x/L)$. The points at which $\varphi_n(x)$ vanishes, namely

$$x = 0\,, \quad \frac{L}{n}\,, \quad \frac{2L}{n}\,, \quad \frac{3L}{n}\,, \ldots, L\,,$$

are the nodes of the nth harmonic. The string displays n *loops* separated by the nodes as in Figure 8.4. Referring again to Equation (8.4.10) which expresses the vibration of the plucked string as a superposition of natural modes of vibration, we see that in this superposition the nth harmonic is absent if the string is plucked at a point x_0 such that $\sin(n\pi x_0/L) = 0$, i.e., at a node of the nth harmonic. The fundamental frequency is always present but, for example, if the string is plucked at the center all the even harmonics

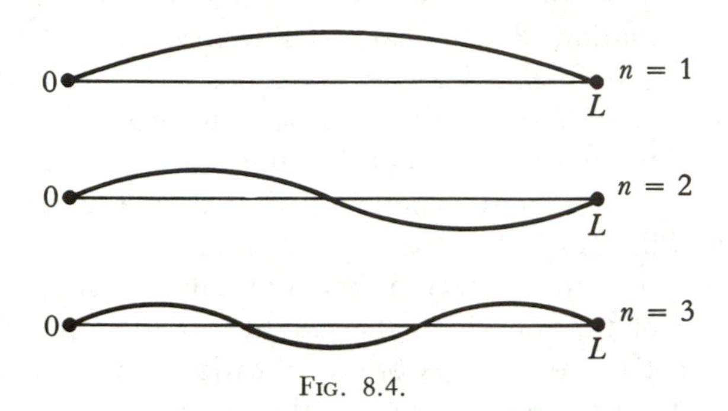

FIG. 8.4.

will be absent. (It may be appropriate to remark at this point, in connection with the notion of timbre referred to above, that the absence of even harmonics is characteristic of the notes produced by those most delightful of musical instruments, the oboe and the bassoon.)

The discussion above extends to more general free vibrations of the string, with some qualifications, as the reader can easily verify. In the case of the general free vibration of the string with either both ends fixed or both ends free, all the statements above hold. For other boundary conditions the same statements hold, but it is not true in general that the nth natural frequency is an integral multiple of the fundamental frequency. In every case, there may occur a phenomenon which we did not encounter in the preceding discussion, namely a difference of phase between different natural modes.

A principal object in the preceding discussion has been to point out that solutions in which the variables are separated, which we have sought on purely formal mathematical grounds, would have been obtained in the case of the vibrating string, if we had asked either for those solutions in which every point executes simple harmonic motion, or for those solutions which represent standing waves.

We shall leave the discussion of forced vibrations for the exercises, only remarking that although the notion of equilibrium solution has the same meaning and employment as in the case of the heat equation, the notion of transient solution has no significance for the wave equation. As a matter of convention, the term "transient" for an inhomogeneous problem for the wave equation is often applied to a solution of the related homogeneous problem, that is, to a free vibration of the system (see Exercises 8b, Problem 6). As in the case of the heat equation, a particular function which satisfies the D.E. and B.C. of an inhomogeneous initial-boundary value problem can be used to solve the problem, but such a function is not properly speaking an asymptotic solution.

EXERCISES 8b

1. Assuming the possibility of term-by-term differentiation, use the energy integral (8.3.6) and Parseval's equations for the Fourier sine and cosine series to calculate the total energy of the plucked string.

2. Discuss the problem of the plucked string for the case when the end $x = 0$ is free and the end $x = L$ is fixed. Carry out the following steps.

(i) Formulate the initial-boundary-value problem, giving an explicit formula for the initial displacement when the string is plucked at x_0.

(ii) Solve for the displacement $u(x,t)$.

(iii) Find the natural frequencies. Which natural frequencies actually are present when the string is plucked at x_0?

3. The string with fixed ends is struck a sudden blow with a hammer at time $t = 0$. As a result the initial displacement and initial velocity are

$$u(x,0) = 0$$

$$u_t(x,0) = g(x) = \begin{cases} 0, & 0 \leq x < \dfrac{L}{4} \\ -V, & \dfrac{L}{4} \leq x \leq \dfrac{L}{2} \\ 0, & \dfrac{L}{2} < x \leq L\,. \end{cases}$$

Find a formula for the resulting motion.

4. Find a complete formula for the solution of problem (8.1.15) for general initial data.

5. (i) Show that the solution (8.4.6) and (8.4.5) of the D.E. and B.C. of problem (8.1.15) can be expressed as a superposition of natural modes $R_n(x) \cos (2\pi\nu_n t + \gamma_n)$, $\nu_n = nc/2L$, expressing $R_n(x)$ and γ_n in terms of A_n and B_n.

(ii) Assuming the possibility of term-by-term differentiation, use the energy integral (8.3.6) and Parseval's equations for the Fourier sine and cosine series to calculate the total energy of the string. Show, in particular, that the contribution of one natural mode to the total energy is independent of the other natural modes.

6. (i) Find the motion of a string with fixed ends under the action of gravity (in the direction of the negative u-axis), and of a *suitably* small force $-2ru_t$ $(0 < r)$ per unit mass due to air resistance, assuming the string initially at rest along the x-axis.

(ii) What is the equilibrium displacement of the string?

(iii) What is the transient motion of the string? In what sense can one speak of a transient motion when $r = 0$?

7. (i) Let Q and ω be constants in

$$\begin{array}{lll} \text{D.E.} & u_{tt} = c^2 u_{xx} + Q \sin \omega t & 0 < x < L\,, \quad 0 < t \\ \text{B.C.} & u(0,t) = 0\,, \quad u(L,t) = 0 & 0 < t\,. \end{array}$$

Show that, excluding certain exceptional values of ω, there is a function of the form $P(x) \sin \omega t$ which satisfies the D.E. and B.C. Find $P(x)$ and the exceptional values of ω.

(ii) Solve the initial-boundary value problem with the D.E. and B.C. of part (i) and the

$$\text{I.C.} \qquad u(x,0) = 0\,, \quad u_t(x,0) = 0 \qquad 0 < x < L\,.$$

8. (i) Use the method of variation of parameters for ordinary differential equations, or Laplace transform methods, to show that if $S(t)$ is piecewise continuous

$$\psi(t) = \int_0^t S(\tau) \frac{\sin \nu(t - \tau)}{\nu} \, d\tau$$

is the solution of the ordinary differential equation

$$\text{D.E.} \qquad \psi'' + \nu^2 \psi = S(t) \qquad 0 < t$$

which satisfies the initial conditions

$$\text{I.C.} \qquad \psi(0) = 0\,, \quad \psi'(0) = 0\,.$$

(ii) Find the solution of the D.E. of part (i) which satisfies the

$$\text{I.C.} \qquad \psi(0) = a\,, \quad \psi'(0) = b\,,$$

where a and b are any constants.

9. Use the method of variation of parameters to find a formula for the solution of the problem of the vibration of a string with fixed ends, initially at rest along the x-axis, under the action of a force $q(x,t)$ per unit mass, that is, the solution of the problem

$$\begin{array}{lll} \text{D.E.} & u_{tt} = c^2 u_{xx} + q(x,t) & 0 < x < L\,, \quad 0 < t \\ \text{B.C.} & u(0,t) = 0\,, \quad u(L,t) = 0 & 0 < t \\ \text{I.C.} & u(x,0) = 0\,, \quad u_t(x,0) = 0 & 0 < x < L\,. \end{array}$$

10. Apply the result of Problem 9 to the case

$$q(x,t) = \begin{cases} Q(x)/\sigma & 0 \le t \le \sigma \\ 0 & \sigma < t\,. \end{cases}$$

Determine the solution in the limit as $\sigma \to 0$. Discuss the connection of this result with Problem 3. (Compare Exercises 5c, Problem 6.)

11. A string of length L under tension T with ends fixed is composed of two materials. The part between $x = 0$ and $x = L/3$ has linear density ρ_1, the remaining part has linear density ρ_2. There are no finite masses or point forces at the junction. Find an equation for the natural frequencies.

12. Find the equilibrium displacement foı the string in Problem 11 when gravity acts in the negative u-direction.

8.5. *EXISTENCE, UNIQUENESS AND REPRESENTATION OF SOLUTIONS*

Uniqueness and representation theorems for the solutions of initial-boundary value problems for the wave equation and related equations can be formulated and proved following the model of the corresponding discussion for the heat equation in Chapter 7. The only new features correspond to those mentioned at the beginning of Section 4 (see Exercises 8c, Problem 1).

A plausible existence statement for the wave equation analogous to that proved for the heat equation in Chapter 7 would assert that problem (8.1.15) has a solution if the initial data $f(x)$ and $g(x)$ are piecewise continuous. In fact, such a statement is not true. The existence statement fails to hold even under the more stringent, and apparently more reasonable, hypothesis that $f(x)$ and $g(x)$ are continuous and piecewise smooth.

We can see this easily by considering the example of the problem of the plucked string discussed in the preceding section. The solution we obtained there, namely (8.4.10), can be written, using the abbreviation b_n for the nth coefficient,

$$u(x,t) = \sum_{n=1}^{\infty} b_n \cos \frac{n\pi ct}{L} \sin \frac{n\pi x}{L} \tag{8.5.1}$$

while the function $f(x)$ defined by (8.4.1) is given by (8.4.11), that is,

$$f(x) = \sum_{n=1}^{\infty} b_n \sin \frac{n\pi x}{L} \qquad 0 < x < L . \tag{8.5.2}$$

Now, from the trigonometric identity

$$\cos \frac{n\pi ct}{L} \sin \frac{n\pi x}{L} = \frac{1}{2} \sin \frac{n\pi}{L}(x + ct) + \frac{1}{2} \sin \frac{n\pi}{L}(x - ct) ,$$

Equation (8.5.1) can be rewritten as

$$u(x,t) = \frac{1}{2} \sum_{n=1}^{\infty} b_n \sin \frac{n\pi}{L}(x + ct) + \frac{1}{2} \sum_{n=1}^{\infty} b_n \sin \frac{n\pi}{L}(x - ct) \tag{8.5.3}$$

or, referring to (8.5.2),

$$u(x,t) = \frac{1}{2}f(x + ct) + \frac{1}{2}f(x - ct) \tag{8.5.4}$$

for values of x and t satisfying the inequalities

$$0 < x + ct < L\,, \quad 0 < x - ct < L\,.$$

Now, from its definition, the derivative of $f(x)$ does not exist at $x = x_0$. It follows from (8.5.4) that the first partial derivatives of $u(x,t)$ do not exist for values of x and t for which one of the equations

$$x + ct = x_0\,, \quad x - ct = x_0 \tag{8.5.5}$$

holds. Since its first partial derivatives do not exist at some points, $u(x,t)$ certainly cannot be a solution of the D.E., and therefore $u(x,t)$ cannot be a solution of the problem of the plucked string. On the other hand, as we indicated above, we can show (see Exercises 8c, Problem 1) that if this problem has a solution it must be the function $u(x,t)$ given by (8.5.1). We are thus forced to conclude that the problem has no solution.

We have here encountered a striking contrast between the heat equation, for which the initial-boundary value problem has a solution with continuous derivatives of all orders when the initial data is merely piecewise continuous, and the wave equation for which the problem fails to have a solution even when the initial data is piecewise smooth.

In fact, an existence theorem for the initial-boundary value problem for the wave equation can be proved under more restrictive hypotheses on the smoothness of the initial data. An example of sufficient, but excessively stringent, conditions is given in Exercises 8c, Problem 2.

This, however, does not alter the feeling that the problem of the plucked string should have a solution, and indeed, the notion of solution of the wave equation and related equations has been generalized in various ways to admit solutions with certain discontinuities in their derivatives. The essence of one such generalization is the physically plausible notion that a function should be considered a generalized solution of a differential equation if it can be approximated uniformly and with arbitrarily small error by actual solutions of the differential equation. We shall discuss this notion more explicitly below.

We first define a **strict solution** of the initial-boundary value problem (8.1.15) to be a function $u(x,t)$ which is continuous together with its first- and second-order partial derivatives and which satisfies the D.E. for $0 \leq x \leq L$, $0 \leq t$, and for which the B.C. and I.C. are satisfied in the sense of equality. A strict solution is certainly a solution in the sense of Exercises 8c, Problem 1 since, for example, the continuity of $u(x,t)$ implies that

$$\lim_{t\to 0} u(x,t) = u(x,0) = f(x)$$

uniformly for $0 \leq x \leq L$, and this in turn implies

$$\lim_{t\to 0} \int_0^L u(x,t)\psi(x)\,dx = \int_0^L f(x)\psi(x)\,dx\,.$$

Thus, for strict solutions the uniqueness and representation theorem of Exercises 8c, Problem 1 holds.

Now [for given $f(x)$, $g(x)$] we define $u(x,t)$ to be a **generalized solution** of the initial-boundary value problem (8.1.15) if there is a sequence of strict solutions $u_N(x,t)$ of the problem with initial data $f_N(x)$, $g_N(x)$, such that

$$\lim_{N\to\infty} f_N(x) = f(x)\,, \quad \lim_{N\to\infty} g_N(x) = g(x) \tag{8.5.6}$$

uniformly for $0 \leq x \leq L$, and

$$\lim_{N\to\infty} u_N(x,t) = u(x,t)\,, \tag{8.5.7}$$

uniformly for $0 \leq x \leq L$, $0 \leq t$.

In this sense, the solution of the problem of the plucked string is a generalized solution. The Nth partial sum

$$u_N(x,t) = \sum_{n=1}^{N} b_n \cos\frac{n\pi ct}{L} \sin\frac{n\pi x}{L}$$

of (8.5.1) is clearly a strict solution of problem (8.1.15) with initial data

$$f_N(x) = \sum_{n=1}^{N} b_n \sin\frac{n\pi x}{L}\,, \quad g_N(x) = 0\,.$$

Equation (8.5.7) holds, by the definition of $u(x,t)$, and Equations (8.5.6) hold, the first by the pointwise convergence theorem for Fourier sine series, and the second trivially. Furthermore, the convergence in (8.5.6) and (8.5.7) is uniform with respect to $0 \leq x \leq L$ and $0 \leq x \leq L$, $0 \leq t$, respectively, as was observed after (8.4.11). This proves our assertion.

The issues raised above will be further illuminated by the discussion in the next section.

To conclude this section we present a method for proving the uniqueness of *strict* solutions of initial-boundary value problems for the wave equation, the **energy integral method,** which is distinguished by its elegant simplicity, and its extensibility to problems in more than one space variable and to other differential equations, even in cases in which a representation of the solution cannot be found. Consider the problem

$$\begin{array}{llll} & \text{D.E.} & u_{tt} = c^2 u_{xx} + q(x,t) & 0 \le x \le L, \quad 0 \le t \\ (8.5.8) & \text{B.C.} & u(0,t) = A(t), \quad u(L,t) = B(t) & 0 \le t \\ & \text{I.C.} & u(x,0) = f(x), \quad u_t(x,0) = g(x) & 0 \le x \le L, \end{array}$$

where we suppose that q, A, B, f, g are such that the problem has a strict solution. We will show that there can be only one such solution by showing that any two solutions are identical.

Let $u_1(x,t)$ and $u_2(x,t)$ be two strict solutions of problem (8.5.8) and set $u(x,t) = u_1(x,t) - u_2(x,t)$. Then, because of linearity, $u(x,t)$ is a strict solution of the problem

$$\begin{array}{llll} & \text{D.E.} & u_{tt} = c^2 u_{xx} & 0 \le x \le L, \quad 0 \le t \\ (8.5.9) & \text{B.C.} & u(0,t) = 0, \quad u(L,t) = 0 & 0 \le t \\ & \text{I.C.} & u(x,0) = 0, \quad u_t(x,0) = 0 & 0 \le x \le L. \end{array}$$

Now consider the energy integral $E(t)$ of u given by (8.3.6) (recall that $c^2 = T_0/\rho_0$). Since u is a strict solution of the D.E. and B.C. of (8.5.9), u satisfies the energy equation (8.3.11), and hence, as in the discussion immediately following (8.3.11), $E(t)$ is constant for $t > 0$. Further, the continuity of u and its derivatives for $0 \le x \le L$, $0 \le t$ implies that $E(t)$ is continuous for $t \ge 0$, so that the constant value of $E(t)$ is $E(0)$; that is, $E(t) = E(0)$ for all $t \ge 0$. But, the I.C. of (8.5.9) give $u_x(x,0) = 0$, $u_t(x,0) = 0$, and substituting in the integrand

$$\frac{1}{2}\rho_0 u_t^2(x,t) + \frac{1}{2} T_0 u_x^2(x,t) \tag{8.5.10}$$

of (8.3.6), we find $E(0) = 0$, so that $E(t) = 0$ for all $t \ge 0$.

Thus, the integral with respect to x of (8.5.10) on the interval $0 \le x \le L$ is 0 for all $t \ge 0$. But (8.5.10) is clearly nonnegative, and the integral over an interval of a continuous nonnegative function of x can be 0 only if the integrand is 0 for all values of x in the interval. Thus (8.5.10), and hence u_x and u_t, are 0 for $0 \le x \le L$, $0 \le t$, so that, for all these values of (x,t), u is constant. Referring again to the I.C. of (8.5.9) and the continuity of u we find that this constant value must be 0.

Finally, from $u(x,t) = 0$ for all $0 \leq x \leq L$, $0 \leq t$ and $u = u_1 - u_2$, we get $u_1(x,t) = u_2(x,t)$ for all $0 \leq x \leq L$, $0 \leq t$. This establishes the statement of uniqueness.

Using the preceding discussion as a model, a uniqueness proof for strict solutions of initial-boundary value problems for the heat equation can be formulated and proved. The details are left as an exercise (Exercises 8c, Problem 4).

EXERCISES 8c

1. Using the discussion in Section 7.4 as a model, prove the following representation and uniqueness theorem for problem (8.1.15).

   Suppose there is a function $u(x,t)$ *such that* (i) u, u_t, u_x, u_{tt}, u_{xx} *are continuous for* $0 < x < L$, $0 < t$ *and satisfy*

$$u_{tt} = c^2 u_{xx} \qquad 0 < x < L\,, \quad 0 < t\,,$$

   and further, for any t_1, t_2, $0 < t_1 < t_2$, u_t *and* u_{tt} *are bounded for* $0 < x < L$, $t_1 < t < t_2$; (ii) $u(0+,t)$, $u(L-,t)$, $u_x(0+,t)$, $u_x(L-,t)$ *exist for all* $t > 0$, *and*

$$u(0+,t) = 0\,, \quad u(L-,t) = 0 \qquad 0 < t\,;$$

   (iii) *for every piecewise continuous function* $\psi(x)$ *on* $[0,L]$

$$\lim_{\substack{t\to 0\\ t>0}} \int_0^L u(x,t)\psi(x)\,dx = \int_0^L f(x)\psi(x)\,dx\,,$$

$$\lim_{\substack{t\to 0\\ t>0}} \int_0^L u_t(x,t)\psi(x)\,dx = \int_0^L g(x)\psi(x)\,dx\,,$$

   where $f(x)$ *and* $g(x)$ *are piecewise continuous on* $[0,L]$. *Then there is only one such function and it is given for* $0 < x < L$, $0 < t$ *by Equations* (8.4.4), (8.4.5), *and* (8.4.6), *with*

$$A_n = \frac{2}{L}\int_0^L f(x)\sin\frac{n\pi x}{L}\,dx\,, \qquad n = 1, 2, 3, \ldots$$

$$B_n = \frac{2}{L}\int_0^L g(x)\sin\frac{n\pi x}{L}\,dx\,.$$

2. Show that the problem (8.1.15) has a solution in the strict sense if $f(x)$ and $g(x)$ each have continuous derivatives of fourth order, and these functions and their first four derivatives vanish at $x = 0$ and $x = L$, as follows:
   (i) Use the mode of estimate of Fourier coefficients employed in Section

6.8 and Exercises 6e, to estimate the coefficients in the formal solution described in Problem 1.

(ii) Using the estimates of part (i), and the theorems on infinite series cited in Section 7.5, show that the formal solution has the required differentiability and continuity properties.

(iii) Using the discussion in Section 7.5 in part as a model, show that the D.E., B.C. and I.C. are satisfied by the formal solution in the required sense.

3. Show that for problem (8.1.15) to have a strict solution it is necessary that $f(x)$, $f'(x)$, $f''(x)$, $g(x)$, $g'(x)$ be continuous for $0 \leq x \leq L$, and that $f(0) = f''(0) = g(0) = 0, f(L) = f''(L) = g(L) = 0$.

4. (i) By multiplying the heat equation

$$u_t = ku_{xx}$$

by u, integrating with respect to x from a to b, and imitating the discussion following (8.3.2), establish the equation

$$\frac{d}{dt}\int_a^b \frac{1}{2}u^2\,dx = kuu_x\Big|_a^b - \int_a^b ku_x^2\,dx\,,$$

which (although without physical significance) is the analogue for solutions of the heat equation of the energy equation for the wave equation.

(ii) Formulate a definition of *strict* solution of the problem

$$\begin{array}{lll} \text{D.E.} & u_t = ku_{xx} + q(x,t) & 0 < x < L\,,\quad 0 < t \\ \text{B.C.} & u(0,t) = A(t)\,,\quad u(L,t) = B(t) & 0 < t \\ \text{I.C.} & u(x,0) = f(x)\,, & 0 < x < L\,, \end{array}$$

and use the result of part (i) to prove that a strict solution of the problem is unique.

8.6. *THE INITIAL VALUE PROBLEM AND RELATED PROBLEMS*

The **initial value problem** for the one-dimensional wave equation is the problem of determining a function $u(x,t)$ which satisfies

$$\text{(8.6.1)}\quad \begin{array}{lll} \text{D.E.} & u_{tt} = c^2u_{xx} & -\infty < x < \infty\,,\quad 0 < t \\ \text{I.C.} & u(x,0) = f(x)\,,\quad u_t(x,0) = g(x) & -\infty < x < \infty\,. \end{array}$$

This problem is an example of a class of problems in which the domain of the space variable is unbounded, and to which the solution techniques developed previously are not applicable. In the next chapter we shall consider general methods for the solution of such problems. For the one-dimensional wave equation, which we consider here, however, special methods are available.

We interpret the solution of problem (8.6.1) physically as describing the transverse vibrations of an infinite string. The notion of an infinite string is, of course, an idealization. Its value lies in the fact that the study of this idealization furnishes extremely useful information about the vibrations of very long strings.

To be more explicit, suppose that we have a very long string that is disturbed from equilibrium only in a small interval near its center, and that we are interested only in the motion of the string near this initial disturbance. Then, as we shall see in the next section, for a very long interval of time the motion in which we are interested will be unaffected by the conditions at the ends of the string. It is this which motivates the study of the initial value problem for the wave equation. What makes this study profitable is that the solution of the problem is particularly simple and illuminating.

We shall make use of a special property of the one-dimensional wave equation: we can determine its general solution. (Except for the transformations of Section 2.3, the one-dimensional wave equation is only nondegenerate second-order equation with constant coefficients which has this property.) Although this general solution could not be used profitably in the discussion of problems on finite intervals, it has great value for the problems we will consider here.

To find the general solution of the equation

$$u_{tt} = c^2 u_{xx} , \tag{8.6.2}$$

where u and its first and second derivatives are assumed to be continuous, we make the change of variables

$$\begin{gathered} \xi = x - ct , \quad \eta = x + ct \\ \omega(\xi,\eta) = u(x,t) . \end{gathered} \tag{8.6.3}$$

Applying the chain rule to (8.6.3) to calculate u_{tt} and u_{xx} in terms of derivatives of ω, and substituting the results in (8.6.2), we find for $\omega(\xi,\eta)$ the equation

$$\omega_{\xi\eta} = 0 . \tag{8.6.4}$$

We easily find the general solution of (8.6.4) to be

$$\omega = F(\xi) + G(\eta) ,$$

where F and G are arbitrary continuous functions with continuous first derivatives. It follows that the general solution of (8.6.2) is

$$u(x,t) = F(x - ct) + G(x + ct)\,, \tag{8.6.5}$$

where F and G have continuous first and second derivatives. Observe that F and G are determined only up to an additive constant, since if K is any constant, (8.6.5) may also be written,

$$u(x,t) = [F(x - ct) + K] + [G(x + ct) - K]\,.$$

It is also sometimes convenient to use the remark that by setting

$$\hat{F}\left(-\frac{\xi}{c}\right) = F(\xi)\,, \quad \hat{G}\left(\frac{\eta}{c}\right) = G(\eta)\,,$$

we may write the general solution (8.6.5) in the alternative form

$$u(x,t) = \hat{F}\left(t - \frac{x}{c}\right) + \hat{G}\left(t + \frac{x}{c}\right). \tag{8.6.6}$$

Now, consider the function $F(x - ct)$, which according to (8.6.5) represents a possible displacement function for an infinite string. The shape of the string is given at time $t = 0$ by $u = F(x)$. At a later fixed time its shape is given by $u = F(x - ct)$ or $u = F(\xi)$, where $\xi = x - ct$ represents a new coordinate obtained by translating the origin a distance ct to the right. Thus the shape of the string remains the same as time changes, but moves to the right with velocity c. To put the matter another way, the displacement at a given point x_0 at time $t = 0$ is the same as that at the points x at times t for which $x - ct = x_0$ or $x = x_0 + ct$. (For pictorial purposes suppose, for example, that x_0 is a point at which F has a maximum or minimum value.) Thus the point x whose displacement is the same as that at x_0 at time $t = 0$ travels to the right according to the equation $x = x_0 + ct$, that is, with velocity

$$\frac{dx}{dt} = \frac{d}{dt}(x_0 + ct) = c\,.$$

We call $F(x - ct)$ a **traveling** or **progressive wave,** traveling to the right with velocity c. Similarly, $G(x + ct)$ is a traveling or progressive wave, traveling to the left with velocity c. The result stated in Equation (8.6.5) can thus be formulated: Every solution $u(x,t)$ of the wave equation defined for $-\infty < x < \infty$ and $-\infty < t < \infty$ is a superposition of two progressing waves, one progressing to the right and one to the left, with the same velocity.

We shall use (8.6.5) to solve the initial value problem (8.6.1),

$$\begin{aligned} &\text{D.E.} \quad u_{tt} = c^2 u_{xx} && -\infty < x < \infty\,, \quad 0 < t \\ &\text{I.C.} \quad u(x,0) = f(x)\,, \quad u_t(x,0) = g(x) && -\infty < x < \infty\,. \end{aligned}$$

Rewriting (8.6.5)

$$u(x,t) = F(x - ct) + G(x + ct) ,$$

and differentiating with respect to t, we get

$$u_t(x,t) = -cF'(x - ct) + cG'(x + ct) . \tag{8.6.7}$$

Substituting (8.6.5) and (8.6.7) in the I.C. of (8.6.1) we obtain the system

$$\begin{cases} F(x) + G(x) = f(x) \\ -cF'(x) + cG'(x) = g(x) \end{cases}$$

or, integrating the second equation of the system,

$$\begin{cases} F(x) + G(x) = f(x) \\ F(x) - G(x) = -\dfrac{1}{c}\displaystyle\int_0^x g(\xi)\, d\xi + K , \end{cases} \tag{8.6.8}$$

where K is a constant. Solving this system algebraically we get

$$\begin{aligned} F(x) &= \frac{1}{2}f(x) - \frac{1}{2c}\int_0^x g(\xi)\, d\xi \\ G(x) &= \frac{1}{2}f(x) + \frac{1}{2c}\int_0^x g(\xi)\, d\xi \end{aligned} \tag{8.6.9}$$

where, utilizing the observation following (8.6.5), we have set the constant K equal to 0. Finally, substituting (8.6.9) in (8.6.5) we have

$$u(x,t) = \frac{f(x + ct) + f(x - ct)}{2} + \frac{1}{2c}\int_{x-ct}^{x+ct} g(\xi)\, d\xi . \tag{8.6.10}$$

Our discussion shows that if the problem (8.6.1) has a solution it must be given by (8.6.10). Conversely, we verify directly that if $f(x)$ has a continuous second derivative and $g(x)$ has a continuous first derivative, then (8.6.10) is a solution of the problem (8.6.1), indeed in the *strict* sense that u and its first and second partial derivatives are continuous for $-\infty < x < \infty$, $0 \leq t$, and the I.C. are satisfied in the sense of equality. Thus we have established the following **existence, uniqueness and representation statement.**

If $f(x)$ has a continuous second derivative and $g(x)$ a continuous first derivative for all x, then the initial value problem (8.6.1) *for the wave equation has one and only one solution, given by* (8.6.10).

We remark that the function (8.6.10) is well-defined and satisfies the I.C. of problem (8.6.1) if we assume only that $f(x)$ is continuous, $f'(x)$ is piecewise smooth, and $g(x)$ is piecewise smooth. It is true that the derivatives of $u(x,t)$ may fail to exist for certain values of (x,t).

However, for values of (x,t) for which the second derivatives of u exist, the D.E. will be satisfied, and it is thus appropriate to call (8.6.10) the **generalized solution** of (8.6.1) in this case.

For the discussion of problem (8.6.1) and its solution (8.6.10) it is useful to remark that we can consider separately the problems in which the I.C. are

$$u(x,0) = f(x)\,, \quad u_t(x,0) = 0 \tag{8.6.11}$$

and

$$u(x,0) = 0\,, \quad u_t(x,0) = g(x)\,. \tag{8.6.12}$$

The solution of the general problem can then be obtained by addition of these separate solutions, as is easily verified.

We consider in detail the solution $u(x,t)$ of the problem (8.6.1) with the special initial conditions (8.6.11), and we suppose that $f(x)$ is a function which vanishes outside a finite interval. Then $u(x,t)$ is given by

$$u(x,t) = \frac{f(x + ct) + f(x - ct)}{2}\,. \tag{8.6.13}$$

From (8.6.13) we see that the motion of the string can be described as follows (see Figures 8.5, 8.6). The initial shape $u = f(x)$ is split into two parts, each of half the original height, and these two parts travel to the left and right with velocity c. For small values of t the nonzero portions of the two traveling waves overlap, and the shape

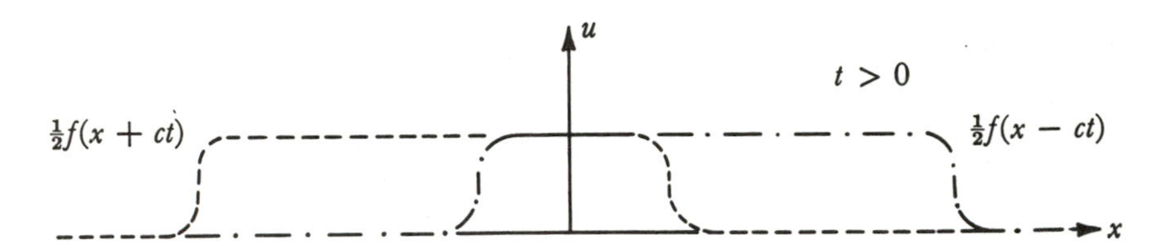

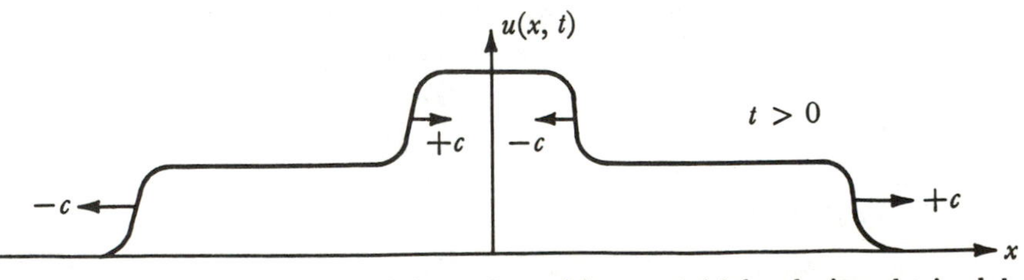

FIG. 8.5. The shape of the infinite string with zero initial velocity obtained by graphical addition for small values of $t > 0$.

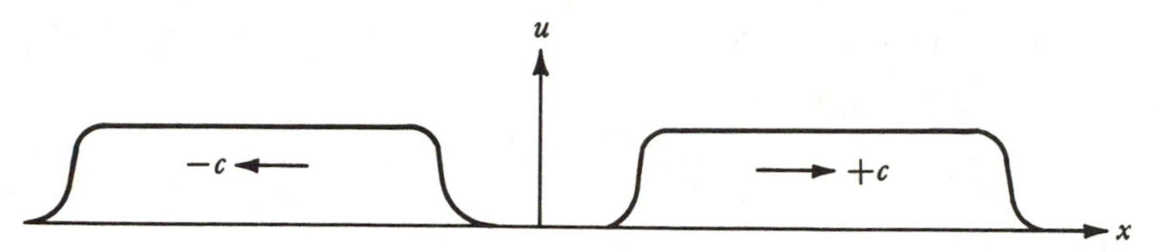

FIG. 8.6. The shape of the infinite string with zero initial velocity for large values of $t > 0$.

of the string is obtained by graphical addition of these portions. For large values of t, the nonzero portions do not overlap, and the motion of the string is described by the statement that there are two identical profiles, each the same as the initial profile but of half the height, separate, and traveling to the left and right with velocity c.

A discussion of the solution of (8.6.1) with the special initial conditions (8.6.12) is left to Exercises 8d.

There are two consequences of the uniqueness of the solution of the general problem (8.6.1) which will be of value in the discussion of later problems, namely if $f(x)$ and $g(x)$ are both odd or are both even, then the solution $u(x,t)$ is correspondingly odd or even. We indicate the proof of the first of these statements. Let $u(x,t)$ be the solution of the problem, and set $v(x,t) = -u(-x,t)$. By differentiating v using the chain rule one verifies directly that v is a solution of the D.E. Further,

$$v(x,0) = -u(-x,0) = -f(-x) = f(x)$$
$$v_t(x,0) = -u_t(-x,0) = -g(-x) = g(x)\,.$$

Thus v also satisfies the I.C., and hence is a solution of the problem. Since the solution is unique, then, we must have $v(x,t) = u(x,t)$, that is, $-u(-x,t) = u(x,t)$, which is the assertion to be proved. A consequence of the statements above and the differentiability of the solution $u(x,t)$ is that for odd initial data $u(0,t) = 0$, while for even initial data $u_x(0,t) = 0$, for all $t \geq 0$.

We return to the consideration of a very long string, initially disturbed from equilibrium over a small interval, but suppose now that the interval of initial disturbance, and of subsequent interest, is near an end of the string. For definiteness, assume that the left end of the string is fixed and that we are interested in the motion near this end. Then, for reasons similar to those presented in the discussion of the initial value problem (8.6.1), it is appropriate to study the problem

$$
\begin{array}{llll}
 & \text{D.E.} & u_{tt} = c^2 u_{xx} & 0 < x < \infty\,,\ 0 < t \\
(8.6.14) & \text{B.C.} & u(0,t) = 0 & 0 < t \\
 & \text{I.C.} & u(x,0) = f(x)\,,\quad u_t(x,0) = g(x) & 0 < x < \infty
\end{array}
$$

which is called an (homogeneous) *initial-boundary value problem for a semi-infinite interval* for the wave equation.

We can solve this problem by solving the initial value problem for the wave equation with $u(x,0)$ and $u_t(x,0)$ set equal to the odd extensions of $f(x)$ and $g(x)$, respectively. This is so because, first, the solution $u(x,t)$ of the initial value problem satisfies the D.E. Second, we have $u(x,0) = f(x)$ and $u_t(x,0) = g(x)$ for $0 < x$, since for $0 < x$, $f(x)$ and $g(x)$ are equal to their odd extensions. Finally, by the result stated immediately above, $u(0,t) = 0$ for all $t \geq 0$. Thus the solution of the initial value problem satisfies all the conditions of the initial-boundary value problem.

This procedure of solution has an illuminating physical interpretation. We consider the case in which $g(x) = 0$ and $f(x)$ vanishes outside a finite interval. Then our procedure amounts to the extension of the semi-infinite string to an infinite string by the addition of a fictional semi-infinite string on the left. The initial displacement in the infinite string consists of the given displacement on the right, and its reflection in the x-axis (its negative) on the left. After a period of time sufficient for the original displacement to split into separate traveling waves, the description of the infinite string in the neighborhood of $x = 0$ can be given as follows. On the right, traveling toward $x = 0$ is a profile while on the left traveling toward $x = 0$ is a profile which is the reflection in the x-axis of the profile on the right. As time passes they meet and cross the origin (see Figure 8.7). During this period the shape of the string is obtained by graphical

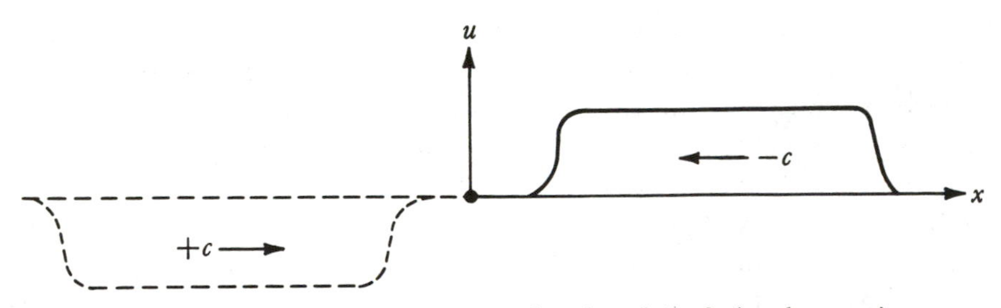

FIG. 8.7. The semi-infinite string with fixed end and its fictional extension to an infinite string; just before the beginning of reflection at the fixed end.

addition of the two traveling waves. This period ends and the two profiles continue their independent motions.

This description of the behavior in the fictional infinite string yields the following description of the motion of the real semi-infinite string. A profile travels toward the fixed end $x = 0$ unchanged in shape. When it reaches the end its shape changes, but after a period it again travels with unchanged shape, but now away from the fixed end of the string, and with its shape the reflection in the x-axis of the shape it had as it approached the end $x = 0$.

The solution of the initial value problem for the wave equation can also be employed for the solution of the initial-boundary value problem for a semi-infinite interval when the boundary condition is of the second kind and homogeneous. For other boundary conditions, including inhomogeneous conditions of the first and second kind and conditions of the third kind, a more direct approach must be used.

To describe this procedure we shall solve the initial-boundary value problem

$$\text{(8.6.15)}\quad \begin{array}{lll} \text{D.E.} & u_{tt} = c^2 u_{xx} & 0 < x < \infty\,,\ 0 < t \\ \text{B.C.} & u(0,t) = \Phi(t) & 0 < t \\ \text{I.C.} & u(x,0) = f(x)\,,\quad u_t(x,0) = g(x) & 0 < x < \infty\,, \end{array}$$

and find the motion of an infinite string, with a given initial displacement and initial velocity, when its left end is moved in a prescribed way. We seek a solution of the form (8.6.5), observing that since $x \geq 0$, $t \geq 0$, and $c \geq 0$, we have $x + ct \geq 0$, so that G must be defined and its values found only for nonnegative values of its argument, but we have $x - ct \geq 0$ for $x \geq ct$ while $x - ct \leq 0$ for $0 \leq x \leq ct$, so that F must be defined and its values found for all real values of its argument.

Substituting (8.6.5) in the I.C. of (8.6.15), we obtain equations (8.6.9), just as in the discussion of the initial value problem, but now these equations hold only for $x > 0$. We have, writing ζ in place of x in (8.6.9),

$$\text{(8.6.16)}\quad \begin{aligned} F(\zeta) &= \frac{1}{2}f(\zeta) - \frac{1}{2c}\int_0^{\zeta} g(\xi)\,d\xi \\ G(\zeta) &= \frac{1}{2}f(\zeta) + \frac{1}{2c}\int_0^{\zeta} g(\xi)\,d\xi \end{aligned} \qquad \zeta > 0\,.$$

Thus the values of F and G are determined for positive values of their arguments. To determine the values of F for negative values of its argument, we substitute (8.6.5) in the B.C. of (8.6.15). We get

$$F(-ct) + G(ct) = \Phi(t) \qquad 0 < t$$

or, setting $\zeta = -ct$ and noting that $\zeta < 0$ for $0 < t$,

$$F(\zeta) + G(-\zeta) = \Phi\left(-\frac{\zeta}{c}\right) \qquad \zeta < 0\,. \tag{8.6.17}$$

But, $\zeta < 0$ implies $-\zeta > 0$, so that $G(-\zeta)$ for $\zeta < 0$ can be found by replacing ζ by $-\zeta$ in the second of Equations (8.6.16). Substituting this determination of $G(-\zeta)$ in (8.6.17) we obtain

$$F(\zeta) = \Phi\left(-\frac{\zeta}{c}\right) - \frac{1}{2}f(-\zeta) - \frac{1}{2c}\int_0^{-\zeta} g(\xi)\,d\xi \qquad \zeta < 0\,, \tag{8.6.18}$$

which furnishes the value of F for negative values of its argument.

Now, using (8.6.5), (8.6.16), and (8.6.18) we find the formal solution $u(x,t)$ of problem (8.6.15),

$$u(x,t) = \begin{cases} \Phi\left(t - \dfrac{x}{c}\right) + \dfrac{f(ct + x) - f(ct - x)}{2} \\ \qquad\qquad + \dfrac{1}{2c}\displaystyle\int_{ct-x}^{ct+x} g(\xi)\,d\xi & 0 \le x < ct \\ \dfrac{f(x + ct) + f(x - ct)}{2} \\ \qquad\qquad + \dfrac{1}{2c}\displaystyle\int_{x-ct}^{x+ct} g(\xi)\,d\xi & ct < x\,. \end{cases} \tag{8.6.19}$$

Note that $u(x,t)$ is not defined for $x = ct$, $0 \le t$. It is clear that for $u(x,t)$ to be defined for all $0 \le x$, $0 \le t$, and to be a strict solution of the problem, $\Phi(t)$ and its first and second derivatives must be defined and continuous for $0 \le t$, $f(x)$ and its first and second derivatives, and $g(x)$ and its first derivative must be defined and continuous for $0 \le x$. The reader can verify that $u(x,t)$ will be defined for all $x \ge 0$, $t \ge 0$, and will be a strict solution of the problem if, in addition to the statements above, we require that the I.C., B.C., and D.E. continue to hold when $x = 0$ or $t = 0$, and are compatible when $x = t = 0$, so that $\Phi(0) = f(0)$, $\Phi'(0) = g(0)$, and $\Phi''(0) = c^2 f''(0)$.

EXERCISES 8d

1. Discuss the generalized solution (8.6.10) of the initial value problem (8.6.1) in the case $f(x) = 0$, $-\infty < x < \infty$,

$$g(x) = \begin{cases} 1 & |x| < L \\ 0 & |x| > L\,. \end{cases}$$

Include graphs of the displacement function $u(x,t)$ for $t = L/2c$, $t = L/c$, and $t = 2L/c$, and referring to these graphs, describe the graphs of $u(x,t)$ for $0 < t < L/c$ and $L/c < t$.

2. Verify that if $f(x)$ and its first and second derivatives, and $g(x)$ and its first derivative are continuous for $-\infty < x < \infty$, then (8.6.10) furnishes a strict solution of the initial value problem (8.6.1).

3. (i) Use the solution (8.6.10) of the initial value problem and the discussion in Section 6 of the initial-boundary value problem (8.6.14) to obtain an explicit formula for the solution of (8.6.14).

(ii) Verify that the result of part (i) is a strict solution if $f(x)$ and its first and second derivatives, and $g(x)$ and its first derivative are continuous for $0 \leq x \leq \infty$, and satisfy $f(0) = f''(0) = g(0) = 0$.

4. (i) Use the solution (8.6.10) of the initial value problem (8.6.1) to solve the initial-boundary value problem

$$\begin{array}{lll} \text{D.E.} & u_{tt} = c^2 u_{xx} & 0 < x < \infty\,, \quad 0 < t \\ \text{B.C.} & u_x(0,t) = 0 & 0 < t \\ \text{I.C.} & u(x,0) = f(x)\,, \quad u_t(x,0) = g(x) & 0 < x < \infty\,. \end{array}$$

(ii) Discuss, including appropriate graphs, the solution of (i) when $g(x) = 0$, $0 < x < \infty$, and $f(x)$ is positive inside a finite interval and equal to 0 outside that interval.

(iii) Solve the problem of (i) using the general solution (8.6.5) directly.

5. (i) Solve the initial-boundary value problem

$$\begin{array}{lll} \text{D.E.} & u_{tt} = c^2 u_{xx} & 0 < x < \infty\,, \quad 0 < t \\ \text{B.C.} & u_x(0,t) = \Phi(t) & 0 < t \\ \text{I.C.} & u(x,0) = 0\,, \quad u_t(x,0) = 0 & 0 < x < \infty\,. \end{array}$$

State the physical interpretation of the problem.

(ii) State the conditions that Φ must satisfy for the solution of (i) to be a strict solution, and verify your statement.

6. Solve the initial-boundary value problem

$$\begin{array}{lll} \text{D.E.} & u_{tt} = c^2 u_{xx} & 0 < x < \infty\,, \quad 0 < t \\ \text{B.C.} & u_x(0,t) = hu(0,t) & 0 < t \\ \text{I.C.} & u(x,0) = f(x)\,, \quad u_t(x,0) = 0 & 0 < x < \infty\,. \end{array}$$

State the physical interpretation of the problem.

7. An infinite string undergoing small transverse displacement in the (x,u)-plane is composed of two different uniform strings joined at $x = 0$, the first of density ρ_1 extending from $-\infty$ to 0, the second of density ρ_2 extending from 0 to ∞. The equilibrium tension in the string is T_0, and we set

$c_1^2 = T_0/\rho_1$, $c_2^2 = T_0/\rho_2$. There is no finite mass or point force acting at the junction, and no external forces act on the string.

For some time before the instant $t = 0$, a wave traveling to the right with velocity c_1, $\Phi(t - x/c_1)$, where Φ is a function which vanishes outside a finite interval, has been moving along the portion $x < 0$ of the string. At time $t = 0$ the wave has not reached $x = 0$. From the observation of the string at time $t = 0$ we wish to find the transverse displacement $u(x,t)$ of the string for all times $t \geq 0$.

The conditions satisfied by the function $u(x,t)$ are

$$\text{D.E.} \qquad u_{tt} = c_1^2 u_{xx} \qquad x < 0\,, \quad t > 0$$

$$u_{tt} = c_2^2 u_{xx} \qquad x > 0\,, \quad t > 0$$

$$\text{J.C.} \qquad u(0-,t) = u(0+,t)$$

$$u_x(0-,t) = u_x(0+,t)$$

$$\text{I.C.} \qquad u(x,0) = \begin{cases} \Phi\left(-\dfrac{x}{c_1}\right) & x < 0 \\ 0 & 0 \leq x \end{cases}$$

$$u_t(x,0) = \begin{cases} \Phi'\left(-\dfrac{x}{c_1}\right) & x < 0 \\ 0 & 0 \leq x\,. \end{cases}$$

Use the representation

$$u(x,t) = \begin{cases} F_1(x - c_1t) + G_1(x + c_1t) & x < 0 \\ F_2(x - c_2t) + G_2(x + c_2t) & 0 < x\,, \end{cases}$$

which ensures the satisfaction of the D.E., to express $u(x,t)$ explicitly in terms of Φ in the intervals $x \leq -c_1t$, $-c_1t \leq x \leq 0$, $0 \leq x \leq c_2t$, $c_2t \leq x$.

8. Show that if $f(x)$ and $g(x)$ are periodic with period p and satisfy the hypothesis of the existence, uniqueness, and representation statement for the initial value problem (8.6.1), then the solution $u(x,t)$ of this problem is also periodic with period p.

9. Prove the existence of a strict solution of the initial-boundary value problem (8.1.15) for a finite interval, assuming that the initial data satisfy the conditions of Exercises 8c, Problem 3, by applying the existence theorem of the preceding section and the result of Problem 8 above to the periodic extensions with period $2L$ of the odd extensions of the initial data.

8.7. *PROPAGATION SPEED AND RELATED NOTIONS*

The explicit solution

$$u(x,t) = \frac{f(x + ct) + f(x - ct)}{2} + \frac{1}{2c}\int_{x-ct}^{x+ct} g(\xi)\, d\xi \tag{8.7.1}$$

of the initial value problem for the wave equation enables us to discover a fundamental mathematical property of solutions of the wave equation which corresponds to a distinguishing feature of the physical phenomena described by the wave equation and other hyperbolic equations.

Suppose that $f(x)$ and $g(x)$ vanish outside an interval $|x| \leq a$. We interpret $f(x)$ and $g(x)$ as a disturbance from equilibrium in the interval $|x| \leq a$ of an infinite string and inquire into the effect of this disturbance outside the interval.

Consider (8.7.1), and suppose that either $x \geq a + ct$ or $x \leq -a - ct$. In the first case we have, for $t > 0$, $x - ct \geq a$ and, a fortiori, $x + ct \geq a$. It follows that $f(x + ct) = f(x - ct) = 0$ since $f(\xi) = 0$ for $\xi \geq a$, and also, since $g(\xi) = 0$ for $\xi \geq a$, that

$$\int_{x-ct}^{x+ct} g(\xi)\, d\xi = 0 .$$

Thus, if $x \geq a + ct$, we have $u(x,t) = 0$. Similarly $u(x,t) = 0$ if $x \leq -a - ct$. We can combine these statements into one: $u(x,t) = 0$ outside the interval $-a - ct \leq x \leq a + ct$. The ends of this interval outside of which the string remains in equilibrium travel with velocities $\pm c$. Consequently we say that for the wave equation *a disturbance is propagated with finite speed.* This is the distinguishing property to which we referred in the first paragraph of this section.

The preceding discussion can be illuminated by diagrams in the (x,t) or space-time plane. Let (x_0,t_0) be a point in the plane. According to (8.7.1) the value $u(x_0,t_0)$ depends only on the values of f and g in the interval $[x_0 - ct_0, x_0 + ct_0]$. This interval is called the **domain of dependence** of the point (x_0,t_0). In particular, if f and g vanish there, then $u(x_0,t_0) = 0$. In Figure 8.8 the domain of dependence is seen to be the base of the triangle in the (x,t)-plane formed by the x-axis and the straight lines of slope $\pm 1/c$ through (x_0,t_0). These lines are called the **characteristic lines** (or, briefly, **characteristics**) of the wave equation through the point.

In Figure 8.9 we have drawn the characteristics (of slope $\pm 1/c$) through the endpoints $\pm a$ of the interval $[-a,a]$. The region of the (x,t)-plane bounded by these characteristics and the interval is called the **region of influence** of the interval. The statement we have proved above is that the segment of the string which is disturbed at time $t = t_0$ in consequence of an initial disturbance confined to the interval $[-a,a]$ must lie on the segment of the line $t = t_0$ included in the region of influence of the interval $[-a,a]$. This is now immediately

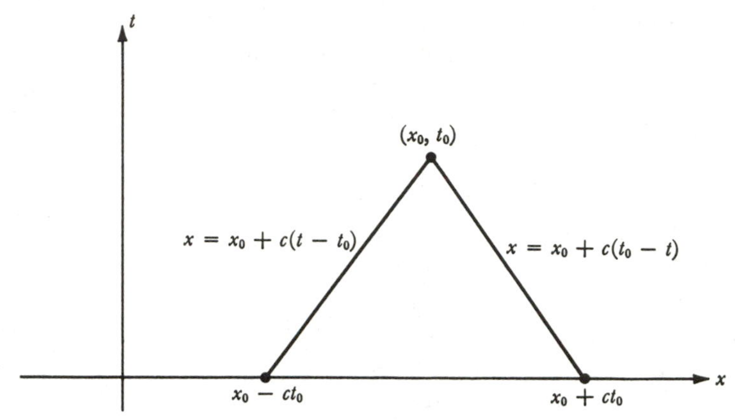

FIG. 8.8. Domain of dependence of (x_0,t_0).

clear if one considers the point (x_0,t_0) moving along the line $t = t_0$, using the fact that u must vanish at all such points whose domain of dependence does not intersect the segment $[-a,a]$.

It should be observed that the only information from formula (8.7.1) that we have employed is the following. If the initial data $u(x,0)$, $u_t(x,0)$ vanish in the domain of dependence $[x_0 - ct_0, x_0 + ct_0]$ of a point (x_0,t_0), then $u(x_0,t_0) = 0$. From this statement follows the statement that disturbances are propagated with finite speed. There is an alternative way—using the energy integral—of proving the first and hence the second of these statements. This method has the advantage of being applicable, with appropriate modification, to other hyperbolic differential equations, and also to problems in finite and

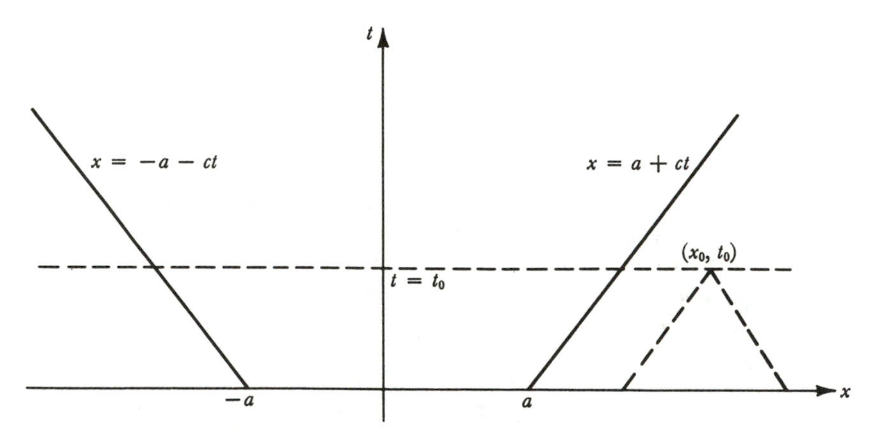

FIG. 8.9. Region of influence of $[-a,a]$.

semi-infinite intervals (with obvious limitation on the interval of time considered). Here we will apply the **energy inequality method** to the homogeneous telegraph equation.

We write the homogeneous telegraph equation, (8.3.12) with $q(x,t) = 0$, in the form

$$u_{tt} = c^2 u_{xx} - hu\,, \tag{8.7.2}$$

with $c^2 = T_0/\rho_0$ and $h = k/\rho_0 > 0$. The energy integral (8.3.13) over an interval $[a,b]$ becomes

$$\frac{\rho_0}{2}\int_a^b (u_t^2 + c^2u_x^2 + hu^2)\,dx\,.$$

Let $u(x,t)$ be a strict solution of (8.7.2), in the sense that u and its first and second derivatives exist and are continuous, and (8.7.2) is satisfied for all values of (x,t) considered. Our object is to show that if u and u_t are both 0 on the base of the triangle of Figure 8.7, then u is 0 everywhere in this triangle, in particular, at the vertex (x_0,t_0). To this end we set, for $0 \le t < t_0$,

$$W(t) = \int_{x_0+c(t-t_0)}^{x_0+c(t_0-t)} (u_t^2 + c^2u_x^2 + hu^2)\,dx \tag{8.7.3}$$

which is, except for the inessential positive factor $\rho_0/2$, the energy integral of u at time t over the x-interval $[x_0 + c(t - t_0), x_0 + c(t_0 - t)]$. The significance of the interval of integration is evident from Figure 8.8.

Since the integrand in (8.7.3) is nonnegative, we have

$$W(t) \ge 0\,, \qquad 0 \le t < t_0\,, \tag{8.7.4}$$

and $W(t) = 0$ for some t if and only if the integrand, and hence in particular $u(x,t)$, is identically 0 in the interval of integration. Our hypothesis is that $u(x,0) = 0$ and $u_t(x,0) = 0$ for $x_0 - ct_0 \le x \le x_0 + ct_0$. Since the first of these conditions implies that $u_x(x,0) = 0$ in the same interval, we have

$$W(0) = 0\,. \tag{8.7.5}$$

The crucial step in the argument is the proof of the inequality

$$\frac{dW(t)}{dt} \le 0 \qquad 0 \le t < t_0 \tag{8.7.6}$$

that is, the statement that $W(t)$ is a decreasing function of t. This statement with (8.7.5) implies

$$W(t) \le 0 \qquad 0 \le t < t_0\,. \tag{8.7.7}$$

But (8.7.4) and (8.7.7) can both hold only if

$$W(t) = 0 \qquad 0 \le t < t_0 \, . \tag{8.7.8}$$

As already observed, it follows that $u(x,t) = 0$ for $0 \le t < t_0$, $x_0 + c(t - t_0) \le x \le x + c(t_0 - t)$, that is, at all points of the triangle of Figure 8.7, except the vertex (x_0,t_0). That $u(x_0,t_0) = 0$ is then a consequence of the continuity of u.

To prove (8.7.6) and thus complete our argument we need an extension of the theorem stated in Section 7.4 on the differentiation of an integral depending on a parameter. The statement of the extended theorem which follows can be established using the theorem cited in Section 7.4, the Fundamental Theorem of the calculus and the chain rule for functions of several variables.

If $\varphi(x,t)$ and $\varphi_t(x,t)$ are continuous for $x_1 \le x \le x_2$, $t_1 \le t \le t_2$, and $\alpha(t)$, $\beta(t)$, $\alpha'(t)$, $\beta'(t)$ are continuous and satisfy $x_1 \le \alpha \le x_2$, $x_1 \le \beta \le x_2$ for $t_1 \le t \le t_2$, then

$$\Phi(t) = \int_{\alpha(t)}^{\beta(t)} \varphi(x,t)\, dx$$

has a continuous derivative $\Phi'(t)$ for $t_1 \le t \le t_2$, given by

$$\Phi'(t) = \int_{\alpha(t)}^{\beta(t)} \varphi_t(x,t)\, dx + \varphi[\beta(t),t]\beta'(t) - \varphi[\alpha(t),t]\alpha'(t) \, .$$

We apply this theorem to (8.7.3), obtaining (with obvious notation)

$$\begin{aligned} W'(t) = {} & 2 \int_{x_0+c(t-t_0)}^{x_0+c(t_0-t)} (u_t u_{tt} + c^2 u_x u_{xt} + h u u_t)\, dx \\ & + [u_t^2 + c^2 u_x^2 + h u^2]_{x = x_0 + c(t_0 - t)}(-c) \\ & - [u_t^2 + c^2 u_x^2 + h u^2]_{x = x_0 + c(t - t_0)}(c) \, . \end{aligned} \tag{8.7.9}$$

If we transform the second term in the integral of Equation (8.7.9) by integration by parts (compare Section 3), the equation becomes

$$\begin{aligned} W'(t) = {} & 2 \int_{x_0+c(t-t_0)}^{x_0+c(t_0-t)} u_t(u_{tt} - c^2 u_{xx} + hu)\, dx \\ & - c[u_t^2 - 2cu_x u_t + c^2 u_x^2 + hu^2]_{x = x_0 + c(t_0 - t)} \\ & - c[u_t^2 + 2cu_x u_t + c^2 u_x^2 + hu^2]_{x = x_0 + c(t - t_0)} \, . \end{aligned}$$

Finally, substituting (8.7.2) in the preceding integral, and algebraically transforming the terms evaluated at the boundary, we get

$$\begin{aligned} W'(t) = {} & -c[(u_t - cu_x)^2 + hu^2]_{x = x_0 + c(t_0 - t)} \\ & -c[(u_t + cu_x)^2 + hu^2]_{x = x_0 + c(t - t_0)} \, . \end{aligned} \tag{8.7.10}$$

From (8.7.10) it is evident that (8.7.6) holds, and our argument is thereby completed.

We have shown that for all homogeneous equations of *hyperbolic* type given in the list of standard forms in Chapter 2, it is true that disturbances are propagated with finite speed. We shall see later that no other equations in the list have this property, so that the property is characteristic of the hyperbolic type.

EXERCISES 8e

1. Show that the interval $[x_0 - ct_0, x_0 + ct_0]$ is the domain of dependence of the point (x_0,t_0) for the telegraph equation (8.7.2) in the following sense: if $v(x,t)$ and $w(x,t)$ are two strict solutions of (8.7.2) and $v(x,0) = w(x,0)$, $v_t(x,0) = w_t(x,0)$ for $x_0 - ct_0 \leq x \leq x_0 + ct_0$, then $v(x_0,t_0) = w(x_0,t_0)$.

2. By setting $h = 0$ in the discussion of the telegraph equation in Section 7, establish the properties of the solutions of the wave equation stated in that section, without reference to (8.7.1).

3. Show that the initial value problem for the telegraph equation

D.E. $\quad u_{tt} = c^2u_{xx} - hu \qquad -\infty < x < \infty\,, \quad 0 \leq t$

I.C. $\quad u(x,0) = f(x)\,, \quad u_t(x,0) = g(x) \qquad -\infty < x < \infty$

has a unique strict solution.

9

Problems on infinite and semi-infinite intervals

9.1. *INITIAL-BOUNDARY VALUE PROBLEMS AND INITIAL VALUE PROBLEMS*

In this chapter we shall consider generally one-dimensional problems in which the domain of the space variable is unbounded. We have already discussed such problems in Section 8.6, for the very special case of the one-dimensional homogeneous wave equation. There are many other problems in infinite domains, for the solution of which a general procedure is needed. As typical examples, we have the following problems,

$$(9.1.1)\quad \begin{array}{lll} \text{D.E.} & u_t = ku_{xx} & 0 < x < \infty\,, \quad 0 < t \\ \text{B.C.} & u(0,t) = 0 & \qquad\qquad\qquad\ \ 0 < t \\ \text{I.C.} & u(x,0) = f(x) & 0 < x < \infty \end{array}$$

and

$$(9.1.2)\quad \begin{array}{lll} \text{D.E.} & u_t = ku_{xx} & -\infty < x < \infty\,, \quad 0 < t \\ \text{I.C.} & u(x,0) = f(x) & -\infty < x < \infty\,. \end{array}$$

These problems describe, respectively, the conduction of heat in an infinite rod whose left end is maintained at temperature zero, and heat conduction in an infinite rod with, so to speak, no boundaries. We call the first an **initial-boundary value problem for a semi-infinite interval,** and the second the **initial value problem,** for the heat equation.

The notion of an infinite rod is, of course, an idealization, and it is natural to ask when and why do we consider such problems.

Suppose, for example, we are studying heat conduction in a very long bar whose left end is maintained at zero temperature, and that we are interested in the temperature distribution just near the left end of the rod. Then (9.1.1) would seem an appropriate model to consider, if there were anything to be gained. There is, as we shall see, a considerable gain, in that the solution of (9.1.1) can be given in a relatively simple form, whose analytical manipulation is far easier than that of the infinite series solution for a finite rod. Similarly (9.1.2) would be an appropriate model to study if we were interested in the temperature distribution near the center of a very long rod. As we have seen, similar statements hold for corresponding problems for the wave equation.

The preceding remarks suggest that, as a general procedure, we undertake to find the solutions to problems such as (9.1.1) and (9.1.2) by a passage to the limit as $L \to \infty$ in the corresponding problems for a finite interval of length L. Here "corresponding" is not quite defined, since for the finite interval boundary conditions must be specified at the ends which tend to $\pm\infty$. This is only an apparent difficulty since, as a little reflection will indicate, the nature of these boundary conditions should not affect the result, if the result is to have the significance we desire.

A rigorous discussion of the limiting procedure we have proposed turns out to be very difficult. However a formal discussion of such a limiting process is a valuable heuristic procedure. It becomes simpler and more valuable if we recall, from the discussion in Chapters 5 and 7, that the solution of initial-boundary value problems (whether homogeneous or inhomogeneous) can be effected by the method of variation of parameters once we have determined the appropriate eigenfunction expansion. Since the same eigenfunction expansion is associated with a number of different problems, we achieve greater generality, and hence economy, by applying the limiting process to the eigenfunction expansion rather than to the particular problem.

We thus arrive at the following **heuristic principle:** For the solution of a problem in an infinite or semi-infinite domain, consider the eigenfunction expansion associated with a corresponding problem for a finite interval of length L, and make a formal passage to the limit as $L \to \infty$.

If the resulting representation statement can be proved, then we can use this and the method of variation of parameters to solve the problem.

9.2. *FOURIER INTEGRALS*

Following the heuristic principle of the preceding section, we will obtain the representation appropriate to the initial value problem for the heat equation and the corresponding problems for the wave equation and for related equations. Let $f(x)$ be defined and be piecewise smooth for $-\infty < x < \infty$. Then for each value of L we may represent $f(x)$ in the interval $-L < x < L$ by its Fourier series (in the narrow sense),

$$f(x) = \frac{a_0}{2} + \sum_{n=1}^{\infty} \left(a_n \cos \frac{n\pi x}{L} + b_n \sin \frac{n\pi x}{L} \right) \tag{9.2.1}$$

$$\begin{aligned} a_n &= \frac{1}{L} \int_{-L}^{L} f(\xi) \cos \frac{n\pi\xi}{L}\, d\xi \qquad n = 0, 1, 2, \ldots \\ b_n &= \frac{1}{L} \int_{-L}^{L} f(\xi) \sin \frac{n\pi\xi}{L}\, d\xi \qquad n = 1, 2, \ldots, \end{aligned} \tag{9.2.2}$$

where the left side of Equation (9.2.1) must be replaced by $[f(x+) + f(x-)]/2$ at all points x at which $f(x)$ is discontinuous. Substituting (9.2.2) in (9.2.1) we obtain, using the addition theorem for the cosine,

$$f(x) = \frac{1}{2L} \int_{-L}^{L} f(\xi)\, d\xi + \sum_{n=1}^{\infty} \frac{1}{L} \int_{-L}^{L} f(\xi) \cos \frac{n\pi}{L} (x - \xi)\, d\xi\,. \tag{9.2.3}$$

Now holding x fixed, we let $L \to \infty$ and attempt formally to find the limit of the right side of (9.2.3). Observe that if the first integral on the right side is bounded, then the first term converges to 0, and this will certainly be true if f is absolutely integrable on $(-\infty, \infty)$; that is, if

$$\int_{-\infty}^{\infty} |f(\xi)|\, d\xi < \infty \tag{9.2.4}$$

exists. We make this hypothesis so that (9.2.3) becomes

$$f(x) = \lim_{L\to\infty} \sum_{n=1}^{\infty} \frac{1}{L} \int_{-L}^{L} f(\xi) \cos \frac{n\pi}{L} (x - \xi)\, d\xi\,. \tag{9.2.5}$$

Let us set

$$\lambda_n = \frac{n\pi}{L}, \quad \Delta\lambda = \lambda_{n+1} - \lambda_n = \frac{\pi}{L}$$

(λ_n is not here intended to denote an eigenvalue), and

$$I(\lambda; L) = \int_{-L}^{L} f(\xi) \cos \lambda(x - \xi)\, d\xi\,.$$

Then (9.2.5) can be written as

$$f(x) = \lim_{L\to\infty} \frac{1}{\pi} \sum_{n=1}^{\infty} I(\lambda_n;L)\Delta\lambda .$$

Now $\Delta\lambda \to 0$ when $L \to \infty$, and the sum on the right side of the last equation resembles the sum appearing in the definition of a definite integral. So apparently,

$$\begin{aligned} f(x) &= \lim_{L\to\infty} \frac{1}{\pi} \int_0^{\infty} I(\lambda;L)\, d\lambda \\ &= \lim_{L\to\infty} \frac{1}{\pi} \int_0^{\infty} d\lambda \int_{-L}^{L} f(\xi) \cos \lambda(x - \xi)\, d\xi \\ &= \frac{1}{\pi} \int_0^{\infty} d\lambda \int_{-\infty}^{\infty} f(\xi) \cos \lambda(x - \xi)\, d\xi . \end{aligned}$$

However, in reaching this conclusion we assumed that $\Delta\lambda \to 0$ independent of L, which is not so, and we ignored the fact that an integral over an infinite interval is defined as an improper integral and not directly as the limit of a sum. Thus the preceding discussion serves merely to lead us to the following conjecture.

If $f(x)$ is defined for $-\infty < x < \infty$, is piecewise smooth in every finite interval, and is absolutely integrable on $(-\infty,\infty)$, then

$$\text{(9.2.6)} \qquad \frac{f(x+) + f(x-)}{2} = \frac{1}{\pi} \int_0^{\infty} d\lambda \int_{-\infty}^{\infty} f(\xi) \cos \lambda(x - \xi)\, d\xi .$$

Formula (9.2.6) is called the **Fourier integral formula.** To prove the above statement, we do not attempt to justify the preceding formal calculation. Instead we will start afresh and present a proof of (9.2.6) analogous to the convergence proof for Fourier series given in Section 6.7. Since the two proofs are so similar, a brief sketch will be sufficient.

The essential theorem concerning limits in Section 6.7 was Riemann's lemma (6.7.14). For the proof of the Fourier integral formula, it is necessary to extend this lemma to the case in which the limits of integration are $\pm\infty$. This extension is easily made under the hypothesis that f is absolutely integrable on $(-\infty,\infty)$, and the task is left as an exercise. We also leave as an easy exercise the proof of the formulas

$$\text{(9.2.7)} \qquad \int_0^{\infty} \frac{\sin \Lambda t}{t}\, dt = \int_{-\infty}^{0} \frac{\sin \Lambda t}{t}\, dt = \frac{\pi}{2}, \qquad \Lambda > 0 .$$

Now consider the integral on the right of (9.2.6). We have

$$\begin{aligned}
&\frac{1}{\pi}\int_0^\infty d\lambda \int_{-\infty}^\infty f(\xi)\cos\lambda(x-\xi)\,d\xi \\
&\quad = \frac{1}{\pi}\lim_{\Lambda\to\infty}\int_0^\Lambda d\lambda \int_{-\infty}^\infty f(\xi)\cos\lambda(x-\xi)\,d\xi \\
&\quad = \frac{1}{\pi}\lim_{\Lambda\to\infty}\int_{-\infty}^\infty \int_0^\Lambda f(\xi)\cos\lambda(x-\xi)\,d\lambda\,d\xi \\
&\quad = \frac{1}{\pi}\lim_{\Lambda\to\infty}\int_{-\infty}^\infty f(\xi)\,\frac{\sin\Lambda(x-\xi)}{(x-\xi)}\,d\xi
\end{aligned}$$

where the interchange of integrations above is justified by the absolute integrability of the integrand, which is a consequence of the absolute integrability of $f(\xi)$ on $(-\infty,\infty)$ and the inequality

$$|f(\xi)\cos\lambda(x-\xi)| \le |f(\xi)|\,.$$

Making the change of variable $\xi = x + t$ in the last integral above, we obtain

$$\text{(9.2.8)}\qquad \begin{aligned}
&\frac{1}{\pi}\int_0^\infty d\lambda \int_{-\infty}^\infty f(\xi)\cos\lambda(x-\xi)\,d\xi \\
&\quad = \lim_{\Lambda\to\infty}\frac{1}{\pi}\int_{-\infty}^\infty f(x+t)\,\frac{\sin\Lambda t}{t}\,dt\,.
\end{aligned}$$

Using (9.2.8) and (9.2.7), we see that (9.2.6) is equivalent to the statement

$$\text{(9.2.9)}\qquad \begin{aligned}
\lim_{\Lambda\to\infty}\Bigg\{&\frac{1}{\pi}\int_{-\infty}^0 [f(x+t)-f(x-)]\,\frac{\sin\Lambda t}{t}\,dt \\
&+ \frac{1}{\pi}\int_0^\infty [f(x+t)-f(x+)]\,\frac{\sin\Lambda t}{t}\,dt\Bigg\} = 0\,.
\end{aligned}$$

The second integral in (9.2.9) can be written as

$$\frac{1}{\pi}\int_0^\infty [f(x+t)-f(x+)]\,\frac{\sin\Lambda t}{t}\,dt = I_1 + I_2 - I_3$$

where

$$\begin{aligned}
I_1 &= \frac{1}{\pi}\int_0^1 [f(x+t)-f(x+)]\,\frac{\sin\Lambda t}{t}\,dt\,, \\
I_2 &= \frac{1}{\pi}\int_1^\infty \frac{f(x+t)}{t}\,\sin\Lambda t\,dt\,, \\
I_3 &= \frac{f(x+)}{\pi}\int_1^\infty \frac{\sin\Lambda t}{t}\,dt\,.
\end{aligned}$$

By the substitution $\tau = \Lambda t$ we obtain

$$\int_1^\infty \frac{\sin\Lambda t}{t}\,dt = \int_\Lambda^\infty \frac{\sin\tau}{\tau}\,d\tau\,,$$

and the integral on the right converges to zero as $\Lambda \to \infty$. Hence $I_3 \to 0$ when $\Lambda \to \infty$. Moreover, because $g(t) = f(x + t)/t$ is absolutely integrable on the interval $1 \leq t < \infty$, it follows from Riemann's lemma that $I_2 \to 0$ as $\Lambda \to \infty$. Finally, it can be shown that $I_1 \to 0$ as $\Lambda \to \infty$. We omit the proof, which is similar to the proof that the second integral in (6.7.8) approaches zero when $n \to \infty$. Thus

$$\lim_{\Lambda \to \infty} (I_1 + I_2 - I_3) = 0$$

and the second integral in (9.2.9) has the limit zero when $\Lambda \to \infty$. By a similar discussion, the first integral in (9.2.9) is shown to have the limit zero, and (9.2.9) is established. This completes the proof of Fourier's integral formula.

By applying the addition theorem for the cosine to formula (9.2.6), we can write the formula in the equivalent form,

$$\begin{aligned} f(x) &= \frac{1}{\pi} \int_0^\infty [a(\lambda) \cos \lambda x + b(\lambda) \sin \lambda x] \, d\lambda \,, \\ a(\lambda) &= \int_{-\infty}^{\infty} f(\xi) \cos \lambda\xi \, d\xi \,, \\ b(\lambda) &= \int_{-\infty}^{\infty} f(\xi) \sin \lambda\xi \, d\xi \end{aligned} \tag{9.2.10}$$

which exhibits clearly the analogy between the Fourier integral formula and the Fourier series formula. (Here and hereafter we write $f(x)$ instead of $[f(x+) + f(x-)]/2$, with the understanding that the equation with the former holds at points of continuity, and the latter must be employed at points of discontinuity.)

We can also, by a simple transformation, write the Fourier integral formula in complex form. In the integral of (9.2.6) we replace $\cos \lambda(x - \xi)$ by its expression in terms of exponentials, obtaining

$$\begin{aligned} \int_0^\infty d\lambda \int_{-\infty}^{\infty} & f(\xi) \cos \lambda(x - \xi) \, d\xi \\ &= \int_0^\infty d\lambda \int_{-\infty}^{\infty} f(\xi) \frac{e^{i\lambda(x-\xi)} + e^{-i\lambda(x-\xi)}}{2} \, d\xi \\ &= \frac{1}{2} \int_0^\infty d\lambda \int_{-\infty}^{\infty} f(\xi) e^{i\lambda(x-\xi)} \, d\xi + \frac{1}{2} \int_0^\infty d\lambda \int_{-\infty}^{\infty} f(\xi) e^{-i\lambda(x-\xi)} \, d\xi \,. \end{aligned}$$

In the first integral we make the change of variable $\lambda = -\sigma$, to get

$$\int_0^\infty d\lambda \int_{-\infty}^{\infty} f(\xi) e^{i\lambda(x-\xi)} \, d\xi = \int_{-\infty}^{0} d\sigma \int_{-\infty}^{\infty} f(\xi) e^{-i\sigma(x-\xi)} \, d\xi \,.$$

Substitution of this last equation in the preceding, with the observation that σ is a dummy variable, yields

$$\int_0^\infty d\lambda \int_{-\infty}^\infty f(\xi) \cos \lambda(x - \xi)\, d\xi = \tfrac{1}{2} \int_{-\infty}^\infty d\lambda \int_{-\infty}^\infty f(\xi) e^{-i\lambda(x-\xi)}\, d\xi \,.$$

Thus (9.2.6) becomes

$$f(x) = \frac{1}{2\pi} \int_{-\infty}^\infty \int_{-\infty}^\infty f(\xi) e^{-i\lambda(x-\xi)}\, d\lambda\, d\xi \tag{9.2.11}$$

which is the **complex form of the Fourier integral formula.**

Formula (9.2.11) can also be written in a form which exhibits an analogy with the complex form of Fourier series. If we set

$$F(\lambda) = \int_{-\infty}^\infty f(\xi) e^{i\lambda\xi}\, d\xi \,, \tag{9.2.12}$$

then from (9.2.11)

$$f(x) = \frac{1}{2\pi} \int_{-\infty}^\infty F(\lambda) e^{-i\lambda x}\, d\lambda \,. \tag{9.2.13}$$

The function $F(\lambda)$ is called the **Fourier transform** of $f(x)$, and (9.2.13) is the **inversion formula for the Fourier transform.** (Instead of including the factor $1/2\pi$ in (9.2.13) we could have included it in (9.2.12) or included $\sqrt{1/2\pi}$ in both (9.2.12) and (9.2.13), and the resulting formulas are often used.) Formulas (9.2.12) and (9.2.13) together are equivalent to the Fourier integral formula (9.2.6). The formulation (9.2.12)–(9.2.13) is the most useful form of the Fourier integral formula in applications.

To find the representations appropriate for initial-boundary value problems such as (9.1.1) which involve a boundary condition of the first kind, we could apply our heuristic principle to the representation appropriate to the corresponding problem for a finite interval, namely the Fourier sine series. We could also seek to find the representation appropriate to similar problems involving a boundary condition of the second kind by first considering the Fourier cosine series.

Instead, although we keep the heuristic principle in mind in applications, we recall that the Fourier sine and cosine series formulas were derived from the Fourier series formula by applying the last to, respectively, the odd and even extensions of a function. In exactly the same way, employing (9.2.10) we can prove the following statements.

If $f(x)$ is absolutely integrable on $(0,\infty)$ and is piecewise smooth on every finite subinterval, then

$$f(x) = \frac{2}{\pi}\int_0^\infty b(\lambda) \sin \lambda x \, d\lambda \tag{9.2.14}$$

where

$$b(\lambda) = \int_0^\infty f(\xi) \sin \lambda\xi \, d\xi \tag{9.2.15}$$

at every point of continuity of $f(x)$, *and* (9.2.14) *holds also at points of discontinuity of* f *if* $f(x)$ *is replaced by* $[f(x+) + f(x-)]/2$.

With the same qualifications

$$f(x) = \frac{2}{\pi}\int^\infty a(\lambda) \cos \lambda x \, dx \tag{9.2.16}$$

where

$$a(\lambda) = \int_0^\infty f(\xi) \cos \lambda\xi \, d\xi \, . \tag{9.2.17}$$

The functions (9.2.15) and (9.2.17), respectively, are called the **Fourier sine transform** and the **Fourier cosine transform** of $f(x)$. The pairs (9.2.14)–(9.2.15) and (9.2.16)–(9.2.17) (or the equivalent statements obtained by substituting the second equation of a pair in the first), respectively, are called the **Fourier sine integral formula** and the **Fourier cosine integral formula.**

Many problems in infinite domains can be effectively solved by finding the Fourier transform, or the Fourier sine or cosine transform of the unknown function. The heuristic principle of Section 1 is the most generally useful guide for choosing the kind of transform to be used in a particular problem. When the transform of the unknown function has been found, the function itself is given by the inversion formula, which we regard as an integral representation analogous to an eigenfunction series representation. In simple cases it may be possible to evaluate in closed form the integral in the inversion formula. To facilitate the evaluation of such integrals, extensive tables of Fourier transforms and of Fourier sine and cosine transforms have been compiled and published. Although such simple cases are of rare occurrence in our problems, it is nevertheless useful to have at hand a short table of transforms, such as will be found in Exercises 9a.

For Fourier transforms there is an analogue of Parseval's equation for Fourier series. The **Parseval equation for Fourier transforms** can be found by expressing the integral

$$J = \int_{-\infty}^\infty f(x)\overline{g(x)} \, dx \tag{9.2.18}$$

in terms of the Fourier transforms $F(\lambda)$ and $G(\lambda)$. We will assume that $f(x)$ and $g(x)$ are continuous, piecewise smooth, and absolutely integrable, and that $G(\lambda)$ is absolutely integrable. This implies that $g(x)$ is bounded, because by the convergence theorem for Fourier integrals we have

$$g(x) = \frac{1}{2\pi}\int_{-\infty}^{\infty} G(\lambda)e^{-i\lambda x}\, d\lambda \tag{9.2.19}$$

and hence

$$|g(x)| \leq \frac{1}{2\pi}\int_{-\infty}^{\infty} |G(\lambda)|\, d\lambda = C < \infty\ .$$

Consequently (9.2.18) is convergent. We substitute (9.2.19) in (9.2.18) to get

$$J = \frac{1}{2\pi}\int_{-\infty}^{\infty} f(x)\left[\int_{-\infty}^{\infty} \overline{G(\lambda)}e^{i\lambda x}\, d\lambda\right] dx\ . \tag{9.2.20}$$

The double integral corresponding to the iterated integral (9.2.20) is absolutely convergent because both $f(x)$ and $\overline{G(\lambda)}$ are absolutely integrable. Therefore the order of integration in (9.2.20) can be inverted,

$$\begin{aligned} J &= \frac{1}{2\pi}\int_{-\infty}^{\infty}\left[\int_{-\infty}^{\infty} f(x)e^{i\lambda x}\, dx\right]\overline{G(\lambda)}\, d\lambda \\ &= \frac{1}{2\pi}\int_{-\infty}^{\infty} F(\lambda)\overline{G(\lambda)}\, d\lambda\ . \end{aligned}$$

Thus we have the following form of *Parseval's equation*,

$$\int_{-\infty}^{\infty} f(x)\overline{g(x)}\, dx = \frac{1}{2\pi}\int_{-\infty}^{\infty} F(\lambda)\overline{G(\lambda)}\, d\lambda\ . \tag{9.2.21}$$

By choosing $f(x) = g(x)$ we obtain the alternative form,

$$\int_{-\infty}^{\infty} |g(x)|^2\, dx = \frac{1}{2\pi}\int_{-\infty}^{\infty} |G(\lambda)|^2\, d\lambda\ . \tag{9.2.22}$$

The assumptions made in proving these formulas are unnecessarily restrictive, but since we will make no use of the formulas we will not pursue the matter.

We note that the Fourier transform is *linear;* that is, if $f_1(x)$ and $f_2(x)$ have transforms $F_1(\lambda)$ and $F_2(\lambda)$, then $c_1f_1(x) + c_2f_2(x)$ has transform $c_1F_1(\lambda) + c_2F_2(\lambda)$. Some other general properties of Fourier transforms will be found in Exercises 9a below.

EXERCISES 9a

1. Use the statement of Riemann's lemma in Section 6.7 to prove the following extended statement of that lemma: Let $g(x)$ be absolutely

integrable on $(-\infty,\infty)$ and be piecewise smooth on every finite interval. Then

$$\lim_{\lambda\to\infty} \int_{-\infty}^{\infty} g(t) \sin \lambda t \, dt = 0 .$$

Show also that

$$\lim_{\lambda\to\infty} \int_{-\infty}^{\infty} g(t) \cos \lambda t \, dt = 0 ,$$

and that if $G(\lambda)$ is the Fourier transform of $g(x)$ then $G(\lambda) \to 0$ when $\lambda \to +\infty$ and when $\lambda \to -\infty$.

2. Use the convergence theorem of Section 6.7 and Equation (6.7.7), applied to the function $f(t) = (\sin t/2)/t$ and the value $x = 0$ to show that

$$\int_{-\infty}^{\infty} \frac{\sin t}{t} \, dt = \pi ,$$

and with this establish (9.2.7).

3. Prove the Fourier sine and cosine formulas stated in Section 2 in the manner suggested there.

4. (i) By a heuristic calculation show that the limiting form, when $L \to \infty$, of the expansion of a function $f(x)$ in terms of the eigenfunctions of the problem

$$\begin{array}{lll} \text{D.E.} & \varphi'' + \lambda\varphi = 0 , & 0 < x < L , \\ \text{B.C.} & \varphi(0) = 0 , \quad \varphi(L) = 0 , & \end{array}$$

is the Fourier sine transform

$$f(x) = \frac{2}{\pi} \int_0^{\infty} F(s) \sin sx \, ds$$

where

$$F(s) = \int_0^{\infty} f(x) \sin sx \, dx .$$

(ii) Repeat the heuristic calculation for the problem

$$\begin{array}{lll} \text{D.E.} & \varphi'' + \lambda\varphi = 0 , & 0 < x < L , \\ \text{B.C.} & \varphi(0) = 0 , \quad \varphi'(L) = 0 . & \end{array}$$

5. Complete the following table: [Number (ix) will be discussed in Section 4.]

	$f(x)$	$F(\lambda) = \int_{-\infty}^{\infty} f(x)e^{i\lambda x}\,dx$
(i)		$F(a\lambda) \qquad a > 0$
(ii)		$e^{i\lambda b}F(\lambda)$
(iii)	$if'(x)$	
(iv)		$F'(\lambda)$
(v)	$f(x) = \begin{cases} 1, & \lvert x\rvert < a \\ 0, & \lvert x\rvert > a \end{cases}$	$a > 0$
(vi)	$f(x) = \begin{cases} 1 - \dfrac{\lvert x\rvert}{a}, & \lvert x\rvert \le a \\ 0, & \lvert x\rvert > a \end{cases}$	$a > 0$
(vii)		$e^{-a\lvert\lambda\rvert} \qquad a > 0$
(viii)		$\dfrac{a}{\lambda^2 + a^2} \qquad a > 0$
(ix)	$\dfrac{1}{\sqrt{4\pi a}} e^{-x^2/4a}$	$e^{-a\lambda^2} \qquad a > 0$

6. Use the preceding table to find the functions whose Fourier transforms are

(i) $(\cos \lambda b)F(\lambda)$
(ii) $\lambda^2 F(\lambda)$
(iii) $\lambda^4 F(\lambda)$
(iv) $\lambda/(\lambda^2 + a^2)$.

7. For each of the following functions find the Fourier sine transform, the Fourier cosine transform, and the representations of the function given by the corresponding inversion formulas:

(i) $f(x) = e^{-ax}$,
(ii) $f(x) = e^{-ax}\cos bx$,
(iii) $f(x) = \begin{cases} 1, & 0 < x < a, \\ 0, & x \ge a, \end{cases}$

where $a > 0$ in each case.

8. Let $f(x)$ be absolutely integrable on $-\infty < x < \infty$, and have Fourier transform $F(\lambda)$. Show that

(i) If $a > 0$, the transform of $f(ax)$ is $(1/a)F(\lambda/a)$.
(ii) For any real b, the transform of $f(x - b)$ is $e^{i\lambda b}F(\lambda)$.
(iii) The transform of $F(x)$ is $2\pi f(-\lambda)$.

(iv) The transform of $\overline{f(x)}$ is $\overline{F(-\lambda)}$.

(v) If both $f(x)$ and $xf(x)$ are piecewise continuous and absolutely integrable, the transform of $xf(x)$ is $-iF'(\lambda)$. What is the transform of $x^n f(x)$?

(vi) If $f(x)$ is continuous and differentiable and both $f(x)$ and $f'(x)$ are absolutely integrable, then $f(x) \to 0$ as $x \to +\infty$ or $x \to -\infty$. Deduce that the transform of $f'(x)$ is $-i\lambda F(\lambda)$. What is the transform of $d^n/dx^n\, f(x)$?

9. Find, by a heuristic calculation, the limiting form $(L \to \infty)$ of the expansion of a function $f(x)$ in terms of the eigenfunctions of the problem

$$\begin{aligned} &\text{D.E.} \quad \varphi'' + \lambda\varphi = 0\,, && 0 < x < L\,, \\ &\text{B.C.} \quad \varphi'(0) - h\varphi(0) = 0\,, \quad \varphi(L) = 0\,. \end{aligned}$$

Show that if $h > 0$ the result is

$$f(x) = \frac{2}{\pi}\int_0^\infty F(s)\left(\cos sx + \frac{h \sin sx}{s}\right)\frac{s^2}{h^2 + s^2}\, ds$$

where

$$F(s) = \int_0^\infty f(x)\left(\cos sx + \frac{h \sin sx}{s}\right) dx\,,$$

but that if $h < 0$ the result is

$$f(x) = c_0 e^{hx} + \frac{2}{\pi}\int_0^\infty f(s)\left(\cos sx + \frac{h \sin sx}{s}\right)\frac{s^2}{h^2 + s^2}\, ds\,,$$

where $F(s)$ is as above and

$$c_0 = \frac{-1}{2h}\int_0^\infty f(x) e^{hx}\, dx\,.$$

10. Give an example of a function of x which is piecewise constant and absolutely integrable on $(0,\infty)$ but whose limit as $x \to \infty$ does not exist.

9.3. *THE CONVOLUTION THEOREM*

In the next section and in later sections we shall use Fourier transforms to solve initial value problems in a manner which we now outline briefly. We first find the Fourier transform $U(\lambda,t)$ of the unknown function $u(x,t)$. When $U(\lambda,t)$ has been found, the inversion formula gives an integral representation for the solution:

$$u(x,t) = \frac{1}{2\pi}\int_{-\infty}^{\infty} U(\lambda,t) e^{-i\lambda x}\, d\lambda\,. \tag{9.3.1}$$

This integral representation can be regarded as a kind of eigenfunction expansion of $u(x,t)$. Occasionally, when the function $U(\lambda,t)$ is particularly simple, the integral (9.3.1) can be evaluated in a closed form, for example, when $U(\lambda,t)$ can be found in a table of Fourier transforms.

In a number of important problems, the integral representation of the solution cannot be evaluated in closed form, but can be transformed into a different and frequently more useful representation. This situation arises when $U(\lambda,t)$ is a product of two functions of λ, each of which is the Fourier transform of a known function. We need the answer to the following question: If $f(x)$ and $g(x)$ are known functions, with Fourier transforms $F(\lambda)$ and $G(\lambda)$, what is the function $h(x)$ whose Fourier transform is $H(\lambda) = F(\lambda)G(\lambda)$? The answer is given by the following theorem.

Let $f(x)$, $g(x)$, $h(x)$ be piecewise smooth and absolutely integrable on $(-\infty,\infty)$ and let their Fourier transforms be $F(\lambda)$, $G(\lambda)$, $H(\lambda)$, respectively. Suppose that $G(\lambda)$ is absolutely integrable on $(-\infty,\infty)$. If

$$H(\lambda) = F(\lambda)G(\lambda) \tag{9.3.2}$$

then

$$h(x) = \int_{-\infty}^{\infty} f(\xi)g(x - \xi)\, d\xi\,. \tag{9.3.3}$$

The function $h(x)$ defined by (9.3.3) is called the **convolution** of $f(x)$ and $g(x)$. This is sometimes written $h = f*g$, with an asterisk to distinguish convolution from ordinary multiplication of functions. If, in place of ξ in (9.3.2), we introduce a new variable of integration η by the formula $\xi = x - \eta$, we obtain

$$h(x) = \int_{-\infty}^{\infty} g(\eta)f(x - \eta)\, d\eta\,.$$

In other words $h = f*g = g*f$, or convolution is commutative.

To prove the convolution theorem stated above, we first use the Fourier inversion formula for $h(x)$,

$$\begin{aligned} h(x) &= \frac{1}{2\pi}\int_{-\infty}^{\infty} H(\lambda)e^{-i\lambda x}\, d\lambda \\ &= \frac{1}{2\pi}\int_{-\infty}^{\infty} F(\lambda)G(\lambda)e^{-i\lambda x}\, d\lambda\,. \end{aligned} \tag{9.3.4}$$

Substituting the definition of $F(\lambda)$ in (9.3.4), we get

$$h(x) = \frac{1}{2\pi}\int_{-\infty}^{\infty}\left(\int_{-\infty}^{\infty} f(\xi)e^{i\lambda\xi}\, d\xi\right)G(\lambda)e^{-i\lambda x}\, d\lambda\,. \tag{9.3.5}$$

Now, since $f(\xi)$ is absolutely integrable on $-\infty < \xi < \infty$ and $G(\lambda)$ is absolutely integrable on $-\infty < \lambda < \infty$, we can interchange the order of integration and obtain

$$(9.3.6) \qquad h(x) = \int_{-\infty}^{\infty} f(\xi)\left(\frac{1}{2\pi}\int_{-\infty}^{\infty} G(\lambda)e^{-i\lambda(x-\xi)}\,d\lambda\right)d\xi\,.$$

We have

$$(9.3.7) \qquad g(x) = \frac{1}{2\pi}\int_{-\infty}^{\infty} G(\lambda)e^{-i\lambda x}\,d\lambda\,,$$

and by replacing x by $x - \xi$ in (9.3.7) we see that the inner integral in (9.3.6) is $g(x - \xi)$. Thus (9.3.3) holds, and the proof is complete.

The hypotheses of the convolution theorem stated above are excessively restrictive. In particular, the condition that $G(\lambda)$ be absolutely integrable on $(-\infty,\infty)$ is unnecessary. However, this condition will be satisfied in all of our applications.

EXERCISES 9b

1. (a) Compute the convolution $h(x)$ of $f(x)$ and $g(x)$ when

$$f(x) = g(x) = \begin{cases} 1 & \text{if } -a \le x \le a \\ 0 & \text{if } |x| > a\,. \end{cases}$$

(b) Verify the formula $H(\lambda) = F(\lambda)G(\lambda)$ for the example of part (a).

(c) Use your results in (a) and (b) to evaluate

$$\int_{-\infty}^{\infty}\left(\frac{\sin\lambda}{\lambda}\right)^2 d\lambda\,.$$

(d) Use your result in (c) to verify Parseval's equation for $f(x)$.

2. Let

$$f(x) = \sum_{n=-\infty}^{\infty} a_n e^{inx}\,, \quad g(x) = \sum_{n=-\infty}^{\infty} b_n e^{inx}\,.$$

(a) Find a convolution type of formula for

$$h(x) = \sum_{n=-\infty}^{\infty} a_n b_n e^{inx}\,.$$

(b) Express the coefficients $\{c_n\}$ of the series

$$f(x)g(x) = \sum_{n=-\infty}^{\infty} c_n e^{inx}$$

in terms of the coefficients $\{a_n\}$ and $\{b_n\}$.

3. Let $f(x)$ and $g(x)$ be continuous, even, absolutely integrable functions on $(-\infty,\infty)$. Let $F(\lambda)$, $G(\lambda)$ be their Fourier *cosine* transforms. By a *formal* calculation show that the function $h(x)$ whose Fourier *cosine* transform is $H(\lambda) = F(\lambda)G(\lambda)$ is given by

$$h(x) = \int_0^{\infty} f(\xi)\left[\frac{g(x+\xi) + g(x-\xi)}{2}\right]d\xi\,.$$

9.4. *SOLUTION OF PROBLEMS FOR THE HEAT EQUATION*

We consider first the *initial value problem*

$$(9.4.1)\qquad \begin{array}{lll} \text{D.E.} & u_t = ku_{xx} & -\infty < x < \infty\,, \quad 0 < t \\ \text{I.C.} & u(x,0) = f(x) & -\infty < x < \infty\,. \end{array}$$

The appropriate representation is given by the Fourier integral formula, which we shall use in its complex formulation (9.2.12)–(9.2.13). We require accordingly that $f(x)$ be piecewise smooth on every finite interval and be absolutely integrable on $(-\infty,\infty)$,

$$(9.4.2)\qquad \int_{-\infty}^{\infty} |f(x)|\, dx < \infty\,.$$

Now suppose that the problem (9.4.1) has a solution $u(x,t)$. For our discussion we shall have to require that $u(x,t)$ have several properties not included in the statement of the problem. We shall state these properties as the need for them arises, and assume that they hold. The question as to whether or not these assumptions are valid we postpone for discussion in Sections 6 and 7. Similar statements hold for the other problems considered in this and the next section.

Assume first that

$$(9.4.3)\qquad \int_{-\infty}^{\infty} |u(x,t)|\, dx < \infty \qquad t > 0\,.$$

Then, since for each $t > 0$, $u(x,t)$ is a smooth function of x, we have, using the Fourier integral formula,

$$(9.4.4)\qquad u(x,t) = \frac{1}{2\pi}\int_{-\infty}^{\infty} U(\lambda,t)e^{-i\lambda x}\, d\lambda\,,$$

where

$$(9.4.5)\qquad U(\lambda,t) = \int_{-\infty}^{\infty} u(x,t)e^{i\lambda x}\, dx\,.$$

We differentiate (9.4.5) with respect to t, assuming that the differentiation may be carried out under the integral sign. We thus obtain, referring to the D.E. of (9.4.1),

$$(9.4.6)\qquad \frac{\partial U}{\partial t} = \int_{-\infty}^{\infty} u_t(x,t)e^{i\lambda x}\, dx = \int_{-\infty}^{\infty} ku_{xx}(x,t)e^{i\lambda x}\, dx\,.$$

Using Green's formula (with the understanding that all statements involving values at $\pm\infty$ are statements concerning limits), we get

$$(9.4.7)\qquad \frac{\partial U}{\partial t} = k[u_x e^{i\lambda x} - i\lambda u e^{i\lambda x}]_{-\infty}^{\infty} - \lambda^2 k\int_{-\infty}^{\infty} u e^{i\lambda x}\, dx\,.$$

Assuming that

(9.4.8) $$u_x(\pm\infty,t) = 0\,, \quad u(\pm\infty,t) = 0$$

and referring to (9.4.5), we obtain from (9.4.7)

(9.4.9) $$\frac{\partial U}{\partial t} = -\lambda^2 k U \qquad 0 < t\,.$$

Letting $t \to 0$ in (9.4.5), we have by the I.C. of (9.4.1),

(9.4.10) $$U(\lambda,0) = \int_{-\infty}^{\infty} f(x)e^{i\lambda x}\,dx = F(\lambda)\,,$$

where $F(\lambda)$ is the Fourier transform of $f(x)$.

Equation (9.4.9) is an ordinary differential equation for $U(\lambda,t)$ as a function of t, and Equation (9.4.10) is an initial condition for this differential equation. The solution of (9.4.9) that satisfies (9.4.10) is

(9.4.11) $$U(\lambda,t) = F(\lambda)e^{-\lambda^2 kt}\,,$$

and hence the solution of (9.4.1) is

(9.4.12) $$\begin{aligned} u(x,t) &= \frac{1}{2\pi}\int_{-\infty}^{\infty} U(\lambda,t)e^{-i\lambda x}\,d\lambda \\ &= \frac{1}{2\pi}\int_{-\infty}^{\infty} e^{-\lambda^2 kt}F(\lambda)e^{-i\lambda x}\,d\lambda\,. \end{aligned}$$

This problem provides the first occasion for the application of the convolution theorem. If we can find a function $w(x,t)$ whose Fourier transform, with respect to x, is

(9.4.13) $$W(\lambda,t) = e^{-\lambda^2 kt}\,,$$

then we can write (9.4.11) in the form

(9.4.14) $$U(\lambda,t) = F(\lambda)W(\lambda,t)\,,$$

and conclude from the convolution theorem that

(9.4.15) $$u(x,t) = \int_{-\infty}^{\infty} f(\xi)w(x-\xi,t)\,d\xi\,.$$

We will find $w(x,t)$ by evaluating the integral

(9.4.16) $$\begin{aligned} w(x,t) &= \frac{1}{2\pi}\int_{-\infty}^{\infty} W(\lambda,t)e^{-i\lambda x}\,d\lambda \\ &= \frac{1}{2\pi}\int_{-\infty}^{\infty} e^{-\lambda^2 kt}e^{-i\lambda x}\,d\lambda \\ &= \frac{1}{2\pi}\int_{-\infty}^{\infty} e^{-\lambda^2 kt}(\cos\lambda x + i\sin\lambda x)\,d\lambda\,. \end{aligned}$$

Since $e^{-\lambda^2 kt} \sin \lambda x$ is an odd function of λ,

$$\int_{-\infty}^{\infty} e^{-\lambda^2 kt} \sin \lambda x \, d\lambda = 0$$

and hence, using the fact that $e^{-\lambda^2 kt} \cos \lambda x$ is an even function of λ,

$$\begin{aligned} w(x,t) &= \frac{1}{2\pi} \int_{-\infty}^{\infty} e^{-\lambda^2 kt} \cos \lambda x \, d\lambda \\ &= \frac{1}{\pi} \int_0^{\infty} e^{-\lambda^2 kt} \cos \lambda x \, d\lambda \, . \end{aligned}$$

We introduce a new variable of integration by the substitution $\lambda = z/\sqrt{kt}$, and obtain

$$\begin{aligned} w(x,t) &= \frac{1}{\pi\sqrt{kt}} \int_0^{\infty} e^{-z^2} \cos \frac{xz}{\sqrt{kt}} \, dz \\ &= \frac{1}{\pi\sqrt{kt}} I(\mu) \end{aligned} \tag{9.4.17}$$

where $\mu = x/\sqrt{kt}$ and

$$I(\mu) = \int_0^{\infty} e^{-z^2} \cos \mu z \, dz \, . \tag{9.4.18}$$

The integral $I(\mu)$ is an elementary function which we can determine explicitly. We have

$$\begin{aligned} \frac{dI}{d\mu} &= \frac{d}{d\mu} \int_0^{\infty} e^{-z^2} \cos \mu z \, dz \\ &= \int_0^{\infty} - z e^{-z^2} \sin \mu z \, dz \\ &= -\left[-\frac{e^{-z^2}}{2} \sin \mu z \right]_0^{\infty} + \int_0^{\infty} \frac{e^{-z^2}}{2} \mu \cos \mu z \, dz \Bigg] \\ &= -\frac{\mu}{2} I(\mu) \, . \end{aligned}$$

Solving this ordinary differential equation, we get

$$I(\mu) = Ce^{-\mu^2/4} \, ,$$

and we evaluate the constant C by setting $\mu = 0$ and using the known value

$$I(0) = \int_0^{\infty} e^{-z^2} \, dz = \frac{\sqrt{\pi}}{2} \, .$$

Thus

$$I(\mu) = \frac{\sqrt{\pi}}{2} e^{-\mu^2/4} \, ,$$

and

$$w(x,t) = \frac{1}{\sqrt{4\pi kt}} e^{-x^2/4kt} . \tag{9.4.19}$$

For every fixed $t > 0$ this function is smooth and absolutely integrable on $-\infty < x < \infty$. Consequently, if we assume that the function $f(x)$ in (9.4.1) is piecewise smooth and absolutely integrable on $(-\infty,\infty)$, then the convolution theorem can be applied to deduce (9.4.15) from (9.4.14).

We now combine (9.4.15) and (9.4.19) to obtain

$$u(x,t) = \frac{1}{\sqrt{4\pi kt}} \int_{-\infty}^{\infty} f(\xi) e^{-(x-\xi)^2/4kt}\, d\xi . \tag{9.4.20}$$

Formulas (9.4.12) and (9.4.20) are equivalent representations of the solution of problem (9.4.1).

Next we consider the initial-boundary value problem for the semi-infinite interval

$$\begin{array}{llll} \text{D.E.} & u_t = ku_{xx} & 0 < x < \infty\,, & 0 < t \\ \text{B.C.} & u(0,t) = 0 & & 0 < t \\ \text{I.C.} & u(x,0) = f(x) & 0 < x < \infty\,. & \end{array} \tag{9.4.21}$$

Our heuristic principle indicates that the appropriate representation for the solution of this problem is given by the Fourier sine integral formula. Accordingly we make the hypothesis that $f(x)$ is piecewise smooth on every finite subinterval of $[0,\infty)$ and absolutely integrable on that interval, and suppose that the problem has a solution $u(x,t)$ which satisfies

$$\int_0^{\infty} |u(x,t)|\, dx < \infty \qquad t > 0 . \tag{9.4.22}$$

Then, from formulas (9.2.14)–(9.2.15)

$$u(x,t) = \frac{2}{\pi} \int_0^{\infty} U(\lambda,t) \sin \lambda x \, d\lambda \tag{9.4.23}$$

where

$$U(\lambda,t) = \int_0^{\infty} u(x,t) \sin \lambda x \, dx . \tag{9.4.24}$$

Differentiating (9.4.24) with respect to t, assuming the possibility of differentiating under the integral sign, we obtain, referring to the D.E. of (9.4.21)

$$\frac{\partial U}{\partial t} = \int_0^{\infty} u_t \sin \lambda x \, dx = \int_0^{\infty} ku_{xx} \sin \lambda x \, dx . \tag{9.4.25}$$

Applying Green's formula to the last integral of (9.4.25), we get

$$\frac{\partial U}{\partial t} = [ku_x \sin \lambda x - k\lambda u \cos \lambda x]_0^\infty - k\lambda^2 \int_0^\infty u \sin \lambda x \, dx \,, \tag{9.4.26}$$

so that, assuming

$$u(\infty,t) = u_x(\infty,t) = 0 \,, \tag{9.4.27}$$

and referring to (9.4.24) and the I.C. of (9.4.21), we have

$$\frac{\partial U}{\partial t} = -\lambda^2 k U \,, \tag{9.4.28}$$

$$U(\lambda,0) = \int_0^\infty f(x) \sin \lambda x \, dx = F(\lambda) \,, \tag{9.4.29}$$

where $F(\lambda)$ is the Fourier sine transform of $f(x)$. Solving the problem (9.4.28)–(9.4.29) and substituting in (9.4.23), we obtain the solution

$$u(x,t) = \frac{2}{\pi} \int_0^\infty e^{-\lambda^2 kt} F(\lambda) \sin \lambda x \, d\lambda \,. \tag{9.4.30}$$

There is an alternative approach to problem (9.4.21). Namely, it is easily verified that if the initial function in problem (9.4.1) is odd then the solution is also odd, and since it is continuous, must vanish for $x = 0$. This suggests that given the initial function $f(x)$ of our initial-boundary value problem, we solve the initial value problem (9.4.1) for the odd extension of $f(x)$, and thereby obtain the solution of the initial-boundary value problem. This device is, of course, successful only for the homogeneous initial-boundary value problem, while our first procedure applies to inhomogeneous as well as homogeneous problems.

We have seen that, for a problem on a finite interval, if the boundary conditions are inhomogeneous it is often advantageous to first transform the problem, if possible, into a problem with homogeneous boundary conditions. This procedure may be inadvisable, however, when the interval is infinite and Fourier transform methods are to be used. Consider, for example, the problem

$$\begin{array}{llll} \text{D.E.} & v_t = kv_{xx} \,, & 0 < x < \infty \,, & 0 < t \,, \\ \text{B.C.} & v(0,t) = A \,, & & 0 < t \,, \\ \text{I.C.} & v(x,0) = 0 \,, & 0 < x < \infty \,, & \end{array} \tag{9.4.31}$$

whose solution can be interpreted as the temperature in a semi-infinite rod, initially at temperature zero, when one end is kept at constant temperature A. The temperature at each point of the rod will approach the equilibrium value A; that is,

$$\lim_{t\to\infty} v(x,t) = A$$

for each x. However, at any fixed time, the temperature will be nearly zero at points sufficiently far from the end $x = 0$, and it is reasonable to assume that

$$\int_0^\infty |v(x,t)|\, dt < \infty .$$

Hence we would try to solve (9.4.31) by finding the Fourier sine transform of $v(x,t)$. If, on the other hand, we let $u = v - A$, then u will satisfy problem (9.4.21) with $f(x) = -A$. Since $f(x)$, in this case, is not absolutely integrable, $f(x)$ does not have a Fourier sine transform, and the method of Fourier transforms could not be used to find the solution.

The discussion of initial-boundary value problems with a boundary condition of the second kind parallels the discussion above and is left to the exercises.

To treat the initial-boundary value problem

$$\text{(9.4.32)}\qquad \begin{array}{lll} \text{D.E.} & u_t = ku_{xx} & 0 < x < \infty, \quad 0 < t \\ \text{B.C.} & u_x(0,t) - u(0,t) = 0 & 0 < t \\ \text{I.C.} & u(x,0) = f(x) & 0 < x < \infty \end{array}$$

involving a boundary condition of the third kind, we need the appropriate representation theorem. This was conjectured in Exercises 9a, Problem 9, but has not been established. Accordingly, we present an alternative discussion.

Our discussion is based on the following proposition, which is often useful in the discussion of equations with constant coefficients, and is easily verified. *Every partial derivative of a sufficiently differentiable solution of a homogeneous linear partial differential equation with constant coefficients is a solution of the same equation.* For example, if $u(x,t)$ is a sufficiently differentiable solution of the heat equation $u_t = ku_{xx}$, then $u_t, u_x, u_{xx}, u_{xt}, \ldots$ are also solutions of the heat equation. It follows from this, and the principle of superposition, that any linear combination of a solution of such an equation and its derivatives is also a solution, provided it is sufficiently differentiable.

We apply this proposition to the problem (9.4.32), assuming that $f(x)$ is a continuous, absolutely integrable function with a piecewise smooth, absolutely integrable derivative, and that the partial derivative of u with respect to x has continuous partial derivatives of second order.

Let u be the solution of (9.4.32), and set

$$v(x,t) = u_x(x,t) - u(x,t) \,. \tag{9.4.33}$$

Then v, according to the proposition above and direct calculation, is a solution of the problem

$$\begin{array}{llll} \text{D.E.} & v_t = kv_{xx} & 0 < x < \infty\,, & 0 < t \\ \text{B.C.} & v(0,t) = 0 & & 0 < t \\ \text{I.C.} & v(x,0) = f'(x) - f(x) & 0 < x < \infty\,. & \end{array} \tag{9.4.34}$$

This is a problem with a boundary condition of the first kind, which we have already solved under hypotheses that are satisfied by v. Having determined v, we can determine u from (9.4.33), which can be solved as a first-order, linear, essentially ordinary differential equation. Multiplying (9.4.33) by the integrating factor e^{-x} and integrating from ∞ to x, we obtain

$$\int_\infty^x \frac{\partial}{\partial x} e^{-x}u(x,t)\,dx = \int_\infty^x e^{-x}v(x,t)\,dx$$

which becomes, on the assumption that $e^{-x}u(x,t) \to 0$ as $x \to \infty$,

$$u(x,t) = e^x \int_\infty^x e^{-\xi}v(\xi,t)\,d\xi \,. \tag{9.4.35}$$

The plausibility of the assumption above is the motive for the choice of ∞ as a limit of integration in (9.4.35).

EXERCISES 9c

1. Solve problem (9.4.21) using the alternative method described in Section 4 and (9.4.20).

2. Solve in two ways, and verify that the answers are the same:

$$\begin{array}{llll} \text{D.E.} & u_t = ku_{xx} & 0 < x < \infty\,, & 0 < t \\ \text{B.C.} & u_x(0,t) = 0 & & 0 < t \\ \text{I.C.} & u(x,0) = e^{-\alpha x} & 0 < x < \infty\,, & \end{array}$$

where $\alpha > 0$.

3. Solve

$$\begin{array}{llll} \text{D.E.} & u_t = ku_{xx} + q(x,t) & -\infty < x < \infty\,, & 0 < t \\ \text{I.C.} & u(x,0) = 0 & -\infty < x < \infty\,. & \end{array}$$

4. (i) Find an asymptotic solution $v(x,t)$ for the problem

$$\begin{array}{llll} \text{D.E.} & u_t = ku_{xx} & 0 < x < \infty\,, & 0 < t \\ \text{B.C.} & u(0,t) = A \sin \omega t & & 0 < t \\ \text{I.C.} & u(x,0) = f(x)\,, & & \end{array}$$

such that $\int_0^\infty |v(x,t)|\, dx < \infty$, by first finding solutions of the heat equation of the form

$$Ae^{i\omega t+\alpha x}.$$

(ii) The temperature fluctuations in the earth due to the daily variation of heating at the surface are barely detectable in average rock at a depth of 4 feet. Estimate the depth at which fluctuations due to the annual variation of surface heating can be detected, if the amplitude A at the surface is the same for the two cases.

5. Use (9.4.20) to find the solution of (9.4.1) when

$$f(x) = \begin{cases} \dfrac{1}{2\delta}, & |x| \leq \delta \\ 0, & |x| > \delta \end{cases}$$

and find the limit of $u(x,t)$ when $\delta \to 0$. On the basis of your result, give a physical interpretation of the function (9.4.19) and of formula (9.4.20).

6. The function $(2/\sqrt{\pi}) \int_0^x e^{-\xi^2}\, d\xi$ is called the **error function** and is denoted by erf x. Find a series representation of erf x, and use this to obtain a simple approximation to erf x for small values of x.

7. Let $u(x,t)$ be the solution of problem (9.4.1) with

$$f(x) = \begin{cases} u_0 & |x| \leq L \\ 0 & |x| > L\,. \end{cases}$$

Find $u(x,t)$ and express this solution in terms of the error function defined in Problem 6. Find the value of $u(x,t)$ approximately for a fixed value of x and for large values of t by using the result of Problem 6.

8. Writing $\int_x^\infty e^{-\xi^2}\, d\xi = \int_x^\infty e^{-\xi^2}\xi/\xi\, d\xi$ and integrating by parts obtain

$$\int_x^\infty e^{-\xi^2}\, d\xi = \frac{e^{-x^2}}{2x} - \frac{1}{2}\int_x^\infty \frac{e^{-\xi^2}}{\xi^2}\, d\xi$$

and conclude that the integral on the left is given approximately by $e^{-x^2}/2x$ with a relative error

$$\frac{1}{2}\left(\frac{e^{-x^2}}{2x}\right)^{-1} \int_x^\infty \frac{e^{-\xi^2}}{\xi^2}\, d\xi\,.$$

Treating the integral immediately above in a fashion similar to that employed to transform the first integral, show that

$$\int_x^\infty \frac{e^{-\xi^2}}{\xi^2}\, d\xi \leq \frac{e^{-x^2}}{2x^3}$$

and thus infer that the relative approximation error above converges to 0 as x becomes infinite.

Use this result and the fact that $\operatorname{erf} \infty = 1$ to obtain a similar kind of approximation for $\operatorname{erf} x$ for large values of x.
Apply your result to determine the approximate value of the solution of Problem 7 for a fixed value of x and small values of t.

9. (i) Verify that if the function $f(x)$ in problem (9.4.1) is odd or even, respectively, then the solution found in Section 4 is correspondingly odd or even.
(ii) Show that this statement could be established without knowing the solution of the problem if it were known that the solution was unique.

10. Write the solution of problem (9.4.21) for the function

$$f(x) = \begin{cases} u_0 & 0 < x < L \\ 0 & L < x < \infty \end{cases}$$

in terms of the error function.

11. Solve the inhomogeneous problem

$$\begin{array}{llll} \text{D.E.} & u_t = ku_{xx} & 0 < x < \infty\,, & 0 < t \\ \text{B.C.} & u(0,t) = U_0 & & 0 < t \\ \text{I.C.} & u(x,0) = 0 & 0 < x < \infty\,, & \end{array}$$

where U_0 is constant, and show that the answer can be transformed into

$$u(x,t) = U_0\left(1 - \operatorname{erf} \frac{x}{\sqrt{4kt}}\right).$$

12. Solve the inhomogeneous problem

$$\begin{array}{llll} \text{D.E.} & u_t = ku_{xx} & 0 < x < \infty\,, & 0 < t \\ \text{B.C.} & u(0,t) = U_0(t) & & 0 < t \\ \text{I.C.} & u(x,0) = 0 & 0 < x < \infty & \end{array}$$

and show that the answer can be written as

$$u(x,t) = \int_0^t \frac{\partial}{\partial t} F(x,\tau,t - \tau)\, d\tau\,,$$

where $F(x,\tau,t)$ is the solution of the problem with *constant* boundary condition

$$\begin{array}{llll} \text{D.E.} & F_t = kF_{xx} & 0 < x < \infty\,, & 0 < t \\ \text{B.C.} & F(0,t) = U_0(\tau) & & 0 < t \\ \text{I.C.} & F(x,0) = 0 & 0 < x < \infty\,. & \end{array}$$

13. Combine the results of Problems 11 and 12, and simplify by a change of variables.

14. Solve the initial value problem

$$\begin{array}{llll} \text{D.E.} & u_t = -u_{xxxx} & t > 0\,, & -\infty < x < \infty \\ \text{I.C.} & u(x,0) = f(x) & & -\infty < x < \infty\,. \end{array}$$

9.5. *SOLUTION OF PROBLEMS FOR THE WAVE EQUATION AND RELATED EQUATIONS*

For the solution of initial value problems and initial-boundary value problems on semi-infinite intervals for the wave equation and related equations, we can employ the techniques based on an appropriate representation theorem which we used in the preceding section. The only new feature that occurs in the formal calculation of the solutions of these problems is that the ordinary differential equations satisfied by the transforms of the solutions are of the second instead of the first order. Accordingly, we shall leave most of these calculations as exercises for the reader (Exercises 9d).

In the case of the telegraph equation and the inhomogeneous wave equation, the solutions obtained using the methods of this chapter are new results. However, in Section 8.6 we found solutions of a number of problems for the homogeneous wave equation. Much of this section will be devoted to the demonstration of the identity of the solutions given in Section 8.6 with those obtained using the methods of this chapter.

Consider first the initial value problem for the wave equation

$$\text{(9.5.1)}\quad \begin{array}{lll} \text{D.E.} & u_{tt} = c^2 u_{xx} & -\infty < x < \infty\,, \quad 0 < t \\ \text{I.C.} & u(x,0) = f(x)\,, & \\ & u_t(x,0) = g(x) & -\infty < x < \infty\,. \end{array}$$

This problem can be solved (Exercises 9d, Problem 1) by imitating the method employed in solving the initial value problem (9.4.1) for the heat equation. We state the result. Let $F(\lambda)$ and $G(\lambda)$ be, respectively, the Fourier transforms of $f(x)$ and $g(x)$, so that

$$\text{(9.5.2)}\qquad f(x) = \frac{1}{2\pi}\int_{-\infty}^{\infty} F(\lambda)e^{-i\lambda x}\,d\lambda\,,$$

$$\text{(9.5.3)}\qquad g(x) = \frac{1}{2\pi}\int_{-\infty}^{\infty} G(\lambda)e^{-i\lambda x}\,d\lambda\,.$$

Then an integral representation of the solution $u(x,t)$ of (9.5.1) is given by

$$\text{(9.5.4)}\quad u(x,t) = \frac{1}{2\pi}\int_{-\infty}^{\infty}\left[F(\lambda)\cos c\lambda t + G(\lambda)\frac{\sin c\lambda t}{c\lambda}\right]e^{-i\lambda x}\,d\lambda\,.$$

If we express the sine and cosine in the integral of (9.5.4) in terms of exponentials, then (9.5.4) can be rewritten as

$$(9.5.5)\quad u(x,t) = \frac{1}{2\pi}\int_{-\infty}^{\infty} F(\lambda)\,\frac{e^{-i\lambda(x-ct)} + e^{-i\lambda(x+ct)}}{2}\,d\lambda$$

$$+ \frac{1}{2\pi}\int_{-\infty}^{\infty} G(\lambda)\,\frac{e^{-i\lambda(x-ct)} - e^{-i\lambda(x+ct)}}{2ci\lambda}\,d\lambda\,.$$

From the linearity of the Fourier transform and Exercises 9a, Problem 5(ii) we see immediately that the first integral of (9.5.5) is equal to

$$\frac{f(x-ct) + f(x+ct)}{2}\,.$$

To simplify the second integral of (9.5.5) we first integrate (9.5.3) with respect to x from $x = a$ to $x = b$, and formally change the order of integration. We get

$$(9.5.6)\qquad \int_a^b g(x)\,dx = \frac{1}{2\pi}\int_{-\infty}^{\infty} G(\lambda)\,\frac{e^{-i\lambda a} - e^{-i\lambda b}}{i\lambda}\,d\lambda\,.$$

Introducing the dummy variable of integration ξ in place of x, and setting $b = x + ct$, $a = x - ct$ in (9.5.6), we see that the second integral on the right of (9.5.5) is equal to

$$\frac{1}{2c}\int_{x-ct}^{x+ct} g(\xi)\,d\xi\,.$$

Thus the formula (9.5.5) is seen to be equivalent to the formula (8.6.10) for the solution of the initial value problem for the wave equation.

We shall apply the transform method, presenting all details, to one more problem discussed in Section 8.6, and compare the results. Consider the inhomogeneous initial-boundary value problem

$$(9.5.7)\qquad \begin{array}{lll} \text{D.E.} & u_{tt} = u_{xx} & 0 < x < \infty\,,\quad 0 < t \\ \text{B.C.} & u(0,t) = \Phi(t) & \qquad\qquad\qquad\quad 0 < t \\ \text{I.C.} & u(x,0) = 0\,, & \\ & u_t(x,0) = 0 & 0 < x < \infty\,, \end{array}$$

which is problem (8.6.15) with the simplifying assumptions $c = 1$, $f = g = 0$.

The appropriate representation statement for this problem is the Fourier sine integral formula. Thus, we assume

$$(9.5.8)\qquad \int_0^\infty |u(x,t)|\,dx < \infty$$

and we have then

$$(9.5.9)\qquad u(x,t) = \frac{2}{\pi}\int_0^\infty U(\lambda,t)\sin\lambda x\,d\lambda\,,$$

where

$$U(\lambda,t) = \int_0^\infty u(x,t) \sin \lambda x \, dx . \tag{9.5.10}$$

Differentiating (9.5.10) twice with respect to t, assuming the possibility of differentiating under the integral sign, and using the D.E. of problem (9.5.7) we obtain

$$\frac{\partial^2 U}{\partial t^2} = \int_0^\infty u_{tt} \sin \lambda x \, dx = \int_0^\infty u_{xx} \sin \lambda x \, dx . \tag{9.5.11}$$

Transforming the last integral of (9.5.11) using Green's formula and Equation (9.5.10), we have

$$\frac{\partial^2 U}{\partial t^2} = [u_x \sin \lambda x - \lambda u \cos \lambda x]_0^\infty - \lambda^2 U . \tag{9.5.12}$$

Assuming that

$$u_x(\infty,t) = u(\infty,t) = 0 \tag{9.5.13}$$

and using the B.C. of (9.5.7), Equation (9.5.12) becomes

$$\frac{\partial^2 U}{\partial t^2} = \lambda\Phi(t) - \lambda^2 U , \tag{9.5.14}$$

to which we adjoin the conditions

$$U(\lambda,0) = 0 , \quad U_t(\lambda,0) = 0 , \tag{9.5.15}$$

obtained from (9.5.10) and the I.C. of (9.5.7). The problem (9.5.14) and (9.5.15) can be solved using the result of Exercises 8b, Problem 8, yielding

$$U(\lambda,t) = \int_0^t \Phi(\tau) \sin \lambda(t - \tau) \, d\tau . \tag{9.5.16}$$

Substituting (9.5.16) in (9.5.9) we have

$$u(x,t) = \frac{2}{\pi} \int_0^\infty \int_0^t \Phi(\tau) \sin \lambda(t - \tau) \sin \lambda x \, d\tau \, d\lambda , \tag{9.5.17}$$

which is an integral representation of the solution of (9.5.7).

Now set $\xi = t - \tau$ in (9.5.17) to get

$$u(x,t) = \frac{2}{\pi} \int_0^\infty \int_0^t \Phi(t - \xi) \sin \lambda\xi \sin \lambda x \, d\xi \, d\lambda . \tag{9.5.18}$$

If we regard t as fixed, and define

$$\Psi(\xi) = \begin{cases} \Phi(t - \xi) & 0 < \xi < t \\ 0 & t < \xi \end{cases} , \tag{9.5.19}$$

then the integral on the right of (9.5.18) can be written as

$$(9.5.20) \qquad \frac{2}{\pi}\int_0^\infty \int_0^\infty \Psi(\xi) \sin \lambda\xi \sin \lambda x \, d\xi \, d\lambda .$$

But by the Fourier sine integral formula, (9.5.20) is equal to $\Psi(x)$. Thus (9.5.18) is equivalent to

$$(9.5.21) \qquad u(x,t) = \begin{cases} \Phi(t - x) & 0 < x < t \\ 0 & t < x \end{cases},$$

which gives the solution of the problem (9.5.7). This is identical to the solution (8.6.19) of problem (8.6.15) for the special case $c = 1$, $f = g = 0$.

EXERCISES 9d

1. Show how the integral representation (9.5.4) of problem (9.5.1) is determined.
2. Find an integral representation of the solution of the initial value problem

$$\begin{array}{lll} \text{D.E.} & u_{tt} = c^2 u_{xx} - ku & -\infty < x < \infty , \quad 0 < t \\ \text{I.C.} & u(x,0) = f(x) , \quad u_t(x,0) = g(x) & -\infty < x < \infty . \end{array}$$

3. (i) Use a transform method to solve the initial-boundary value problem

$$\begin{array}{lll} \text{D.E.} & u_{tt} = c^2 u_{xx} & 0 < x < \infty , \quad 0 < t \\ \text{B.C.} & u_x(0,t) = 0 & 0 < t \\ \text{I.C.} & u(x,0) = f(x) , \quad u_t(x,0) = 0 & 0 < x < \infty . \end{array}$$

(ii) Show that the solution obtained in (i) can be written in the form

$$u(x,t) = \begin{cases} \dfrac{f(x + ct) + f(ct - x)}{2} & 0 < x < ct \\ \dfrac{f(x + ct) + f(x - ct)}{2} & ct < x . \end{cases}$$

4. (i) Solve the initial value problem for the inhomogeneous wave equation,

$$\begin{array}{lll} \text{D.E.} & u_{tt} = c^2 u_{xx} + q(x,t) & -\infty < x < \infty , \quad 0 < t \\ \text{I.C.} & u(0,t) = 0 , \quad u_t(x,0) = 0 & -\infty < x < \infty . \end{array}$$

(ii) Show that the solution obtained in (i) can be written in the form

$$u(x,t) = \frac{1}{2c}\int_0^t \int_{x-c(t-\tau)}^{x+c(t-\tau)} q(\xi,\tau) \, d\xi \, d\tau .$$

5. (i) Use a transform method to find an integral representation of the solution of the inhomogeneous initial-boundary value problem

D.E.	$u_{tt} = u_{xx}$	$0 < x < \infty\,, \quad 0 < t$
B.C.	$u_x(0,t) = \Phi(t)$	$0 < t$
I.C.	$u(x,0) = 0\,, \quad u_t(x,0) = 0$	$0 < x < \infty\,.$

(ii) Solve the problem of (i) using the fact that if u is the solution of the problem then $v = u_x$ satisfies the same D.E.

(iii) Verify the identity of the solutions obtained in parts (i) and (ii).

6. Find an integral representation of the solution of

D.E.	$u_{tt} = u_{xx}$	$0 < x < \infty\,, \quad 0 < t$
B.C.	$u(0,t) - u_x(0,t) = \Phi(t)$	$0 < t$
I.C.	$u(x,0) = 0\,, \quad u_t(x,0) = 0$	$0 < x < \infty\,.$

7. Complex-valued solutions of the wave equation of the form $h(x)e^{2\pi i\nu t}$, and the real and imaginary parts of such solutions, are called *monochromatic* waves of frequency ν.

(i) Show that all complex-valued *monochromatic traveling* waves are given by

$$Ae^{2\pi i\nu(t+x/c)}\,, \qquad Be^{2\pi i\nu(t-x/c)}$$

where A and B are constants. The period of these functions as functions of x is called the *wavelength* of the wave, and is equal to c/ν.

(ii) Use the result of (i) and the principle of superposition in the form stated on p. 9 to give a physical interpretation of the integral representation of the solution of (9.5.1).

8. Solve

D.E.	$u_{tt} = -u_{xxxx}$	$t > 0\,, \quad -\infty < x < \infty$
I.C.	$u(x,0) = f(x)\,, \quad u_t(x,0) = g(x)$	$-\infty < x < \infty\,.$

9.6. *EXISTENCE, UNIQUENESS, AND REPRESENTATION QUESTIONS*

In this section we will consider questions of existence, uniqueness and representation for the initial value problems for the wave equation and the heat equation. Since the solution of homogeneous initial-boundary value problems on semi-infinite intervals for these equations can be related to the solution of the corresponding initial value problems, as in the preceding sections, our results have immediate implications for a wider class of problems.

A complete statement concerning existence, uniqueness, and representation for the initial value problem for the wave equation was made and proved in Section 8.6. We refer to the matter again here to point out that when we approach the problem via the transform method, our argument requires assumptions about the initial data and the solution which do not appear in the final statement concerning the solution. To some extent this is true also for the heat equation.

We shall prove the following *existence and representation* statement for the initial value problem for the heat equation.

Let $f(x)$ be a function which is piecewise continuous on every finite interval, and bounded on $(-\infty,\infty)$. Then the function

$$u(x,t) = \frac{1}{\sqrt{4\pi kt}} \int_{-\infty}^{\infty} f(\xi) e^{-(x-\xi)^2/4kt}\, d\xi \tag{9.6.1}$$

is continuous together with its first- and second-order partial derivatives for $-\infty < x < \infty$, $0 < t$, satisfies the

D.E. $\quad u_t = k u_{xx} \qquad -\infty < x < \infty\,, \quad 0 < t$

and, at every point x of continuity of f, the

I.C. $\quad \lim_{t \to 0^+} u(x,t) = f(x).$

The function

$$(4\pi kt)^{-1/2} e^{-(x-\xi)^2/4kt} = w(x-\xi,t) \tag{9.6.2}$$

is easily verified to be a solution of the heat equation for each ξ. Hence, if we can differentiate (9.6.1) under the integral sign, we obtain

$$\begin{aligned} u_t - k u_{xx} &= \int_{-\infty}^{\infty} f(\xi) \left[\frac{\partial}{\partial t} w(x-\xi,t) - k \frac{\partial^2}{\partial x^2} w(x-\xi,t) \right] d\xi \\ &= 0\,. \end{aligned} \tag{9.6.3}$$

The possibility of differentiating an improper integral depending on a parameter under the integral sign is given by standard theorems which are analogous to the theorems on term-by-term differentiation of infinite series stated in Section 7.5. It is easy to apply these theorems to show that (9.6.3) is justified. Thus it can be shown that (9.6.1) satisfies the D.E. (Exercises 9e, Problems 1 and 2).

It remains to justify the statement concerning the initial condition. We will base this on the following theorem due to Weierstrass.

Let $\psi(x)$ be a positive, continuous, and integrable function on the interval $(-\infty,\infty)$, and suppose that

$$\int_{-\infty}^{\infty} \psi(x)\, dx = 1\,. \tag{9.6.4}$$

If $f(x)$ is piecewise continuous on every finite interval, and bounded on $(-\infty,\infty)$, then

$$\lim_{\delta \to 0} \frac{1}{\delta} \int_{-\infty}^{\infty} \psi\left(\frac{\xi - x}{\delta}\right) f(\xi)\, d\xi = f(x) \tag{9.6.5}$$

at each point x of continuity of f.

To prove this theorem we first make the change of variable $t = (\xi - x)/\delta$ in the integral on the left of (9.6.5), obtaining

$$\frac{1}{\delta} \int_{-\infty}^{\infty} \psi\left(\frac{\xi - x}{\delta}\right) f(\xi)\, d\xi = \int_{-\infty}^{\infty} \psi(t) f(x + \delta t)\, dt\,. \tag{9.6.6}$$

Using (9.6.4) and (9.6.6) we see that (9.6.5) is equivalent to

$$\lim_{\delta \to 0} \int_{-\infty}^{\infty} [f(x + \delta t) - f(x)]\psi(t)\, dt = 0\,. \tag{9.6.7}$$

We denote the integral in (9.6.7) by I, and write it as a sum of integrals, as follows:

$$I = \int_{-M}^{M} [f(x + \delta t) - f(x)]\psi(t)\, dt + \int_{-\infty}^{-M} [\quad]\psi\, dt + \int_{M}^{\infty} [\quad]\psi\, dt\,. \tag{9.6.8}$$

To the first integral on the right we apply the mean-value theorem, and use (9.6.4) and the fact that ψ is positive, to get

$$\begin{aligned} \left|\int_{-M}^{M} [f(x + \delta t) - f(x)]\psi(t)\, dt\right| &= \left|[f(x + \delta \bar{t}) - f(x)] \int_{-M}^{M} \psi(t)\, dt\right| \\ &\leq |f(x + \delta \bar{t}) - f(x)| \int_{-\infty}^{\infty} \psi(t)\, dt \\ &\leq |f(x + \delta \bar{t}) - f(x)|\,, \end{aligned} \tag{9.6.9}$$

where $\bar{t}$ is a value such that $-M \leq \bar{t} \leq M$.

Now, by hypothesis there is a constant C such that

$$|f(x)| \leq C \tag{9.6.10}$$

for all x. Applying (9.6.10) and the fact that ψ is positive we obtain from (9.6.8) and (9.6.9), the estimate

$$|I| \leq |f(x + \delta \bar{t}) - f(x)| + 2C\left[\int_{-\infty}^{-M} \psi\, dt + \int_{M}^{\infty} \psi\, dt\right]\cdot \tag{9.6.11}$$

Since ψ is positive and integrable we can choose M so large that each of the integrals in (9.6.11) is $\leq \epsilon/8C$, where ϵ is any given positive number. With this choice of M, (9.6.11) becomes

$$|I| \leq |f(x + \delta \bar{t}) - f(x)| + \frac{\epsilon}{2}, \tag{9.6.12}$$

where $-M \leq \bar{t} \leq M$, with M now fixed, so that

$$-\delta M \leq (x + \delta \bar{t}) - x \leq \delta M . \tag{9.6.13}$$

Inequality (9.6.13) implies $x + \delta \bar{t} \to x$ as $\delta \to 0$, so that if f is continuous at x, we have for δ sufficiently small

$$|f(x + \delta \bar{t}) - f(x)| \leq \frac{\epsilon}{2} \cdot \tag{9.6.14}$$

Substituting (9.6.14) in (9.6.12) we get

$$|I| \leq \frac{\epsilon}{2} + \frac{\epsilon}{2} = \epsilon \tag{9.6.15}$$

for sufficiently small positive δ, and this completes the proof of the theorem.

Now we apply the theorem just proved to (9.6.1) by setting

$$\psi(x) = \frac{1}{\sqrt{\pi}} e^{-x^2}, \quad \delta = \sqrt{4kt}$$

where we use the fact that

$$\int_{-\infty}^{\infty} e^{-x^2}\, dx = \sqrt{\pi} .$$

The statement of the I.C. follows immediately, and the proof of the existence and representation theorem is concluded.

It would seem desirable now to complete our discussion of the initial value problem for the heat equation by showing that the solution (9.6.1) is unique. In fact the solution is not unique. The function

$$v(x,t) = \frac{\partial}{\partial x} w(x,t) = \frac{1}{4\sqrt{\pi}} x(kt)^{-3/2} e^{-[x^2/4kt]} \tag{9.6.16}$$

is easily seen to be a solution of the heat equation, and to satisfy

$$\lim_{t \to 0} v(x,t) = 0 \qquad -\infty < x < \infty . \tag{9.6.17}$$

Thus the initial value problem for the heat equation with the initial data $f(x)$ has at least two solutions: $u(x,t)$, given by (9.6.1), and $u(x,t) + v(x,t)$ with $v(x,t)$ given by (9.6.16). The reason a uniqueness statement does not hold is that the formulation

$$\lim_{t \to 0^+} u(x,t) = f(x) \tag{9.6.18}$$

of the I.C. is not sufficiently stringent. This is partly the reason for our different formulation of the I.C. in Chapter 7. Further discussion of the uniqueness question here, however, is beyond our scope.

9.7. *PROPAGATION SPEED*

In Section 8.7 we showed that if $u(x,t)$ is the solution of the initial value problem for the wave equation or telegraph equation with initial data which are zero outside a finite interval $[-a,a]$, then, for given $t > 0$, $u(x,t)$ is zero outside the interval $[-a - ct, a + ct]$. We interpreted this result by the statement that for equations of hyperbolic type, disturbances are propagated with finite speed.

Now consider the solution (9.6.1) of the initial value problem for the heat equation with initial data f. Suppose that $f(x)$ vanishes outside an interval $|x| \leq a$ and suppose further that $f(x)$ is positive in this interval. We interpret $f(x)$ as a disturbance from equilibrium in the interval $|x| \leq a$, and inquire into the effect of this disturbance outside the interval. Since the integrand in (9.6.1) is nonnegative for all ξ and positive for $|\xi| \leq a$, it follows that for *every* value of x and $t > 0$ we have $u(x,t) > 0$. Thus for the heat equation, a disturbance from equilibrium limited to a finite interval results in a disturbance from equilibrium in the whole rod at every instant $t > 0$, no matter how small. Hence we say that for parabolic equations *disturbances are propagated with infinite speed.*

The assumptions (9.4.3) and (9.4.8) made in finding the solution of the initial value problem for the heat equation thus are open to question. In the corresponding case of the wave equation it follows from the discussion in Sections 8.6 and 8.7, that although these conditions are not satisfied in general, they are satisfied if the initial data vanish outside a finite interval. The same statement holds for the heat equation, although it is not, as in the case of the wave equation, associated with a finite speed of propagation of disturbances.

To demonstrate the statement, assume that in (9.6.1) the function $f(\xi)$ vanishes for $|\xi| > a$, so that (9.6.1) becomes

$$u(x,t) = \frac{1}{\sqrt{4\pi kt}} \int_{-a}^{a} f(\xi) e^{-(x-\xi)^2/4kt} \, d\xi \,. \tag{9.7.1}$$

Then we easily obtain the estimates

$$|u(x,t)| \leq \begin{cases} \dfrac{1}{\sqrt{4\pi kt}} e^{-(x-a)^2/4kt} \displaystyle\int_{-a}^{a} |f(\xi)| \, d\xi \,, & x > a \\ \dfrac{1}{\sqrt{4\pi kt}} e^{-(x+a)^2/4kt} \displaystyle\int_{-a}^{a} |f(\xi)| \, d\xi \,, & x < -a \end{cases} \tag{9.7.2}$$

and

$$(9.7.3)\quad |u_x(x,t)| \le \begin{cases} \dfrac{x+a}{4\pi^{1/2}(kt)^{3/2}} e^{-(x-a)^2/4kt} \displaystyle\int_{-a}^{a} |f(\xi)|\, d\xi\,, & x > a \\ \dfrac{a-x}{4\pi^{1/2}(kt)^{3/2}} e^{-(x+a)^2/4kt} \displaystyle\int_{-a}^{a} |f(\xi)|\, d\xi\,, & x < -a\,, \end{cases}$$

from which (9.4.3) and (9.4.8) follow immediately.

EXERCISES 9e

1. Verify that the function $w(x-\xi,t)$ defined by (9.6.2) is a solution of the heat equation for each fixed ξ.

2. A sufficient condition that a function

$$G(y) = \int_{-\infty}^{\infty} g(\eta,y)\, d\eta \qquad y_1 \le y \le y_2$$

defined by a convergent improper integral have a derivative given by

$$\frac{dG}{dy} = \int_{-\infty}^{\infty} \frac{\partial g}{\partial y}(\eta,y)\, d\eta\,,$$

is that $\partial g/\partial y$ be continuous for $-\infty < \eta < \infty, y_1 < y < y_2$, and that there exist a function $H(\eta)$ such that

$$\left|\frac{\partial g}{\partial y}(\eta,y)\right| \le H(\eta) \qquad -\infty < \eta < \infty\,, \qquad y_1 \le y \le y_2$$

and

$$\int_{-\infty}^{\infty} H(\eta)\, d\eta < \infty\,.$$

Use this statement to verify (9.6.3).

3. Show that if $f(x)$ is absolutely integrable on $(-\infty,\infty)$, and $u(x,t)$ is given in terms of $f(x)$ by (9.6.1) then, for all $t > 0$, $\int_{-\infty}^{\infty} u(x,t)\, dx$ exists and

$$\int_{-\infty}^{\infty} u(x,t)\, dx = \int_{-\infty}^{\infty} f(x)\, dx\,.$$

State a physical interpretation of this result.

4. Show that if $u(x,t)$ is given in terms of $f(x)$ by (9.6.1), and if m and M are constants such that $m \le f(x) \le M$ for $-\infty < x < \infty$, then $m \le u(x,t) \le M$ for $-\infty < x < \infty$, $0 < t$. State a physical interpretation of this result.

5. Let u be given in terms of f by (9.6.1). Show that $u(x_0,t_0)$ depends on the values of f on all of $(-\infty,\infty)$ and not just on the values in some interval $[a,b]$. Explicitly, show that for any interval $[a,b]$ there are functions f, g and corresponding solutions u, v such that $f(x) = g(x)$ for $a \le x \le b$, but $u(x_0,t_0) \ne v(x_0,t_0)$.

6. Let $f(x)$ be continuous and absolutely integrable on $(-\infty,\infty)$ with Fourier transform $F(\lambda)$ and assume that the series

$$\text{(A)} \qquad g(x) = \sum_{m=-\infty}^{\infty} f(x + 2m\pi)$$

converges for all x. Then $g(x)$ is a periodic function of period 2π.

(i) Show that if the series (A) converges uniformly on the interval $-\pi \leq x \leq \pi$ then the Fourier coefficients of $g(x)$,

$$g_n = \frac{1}{2\pi} \int_{-\pi}^{\pi} g(x) e^{inx}\, dx$$

can be expressed in terms of the Fourier transform $F(\lambda)$ of $f(x)$.

(ii) Deduce that if $g(x)$ is represented by its Fourier series then

$$\frac{1}{2\pi} \sum_{n=-\infty}^{\infty} F(n) e^{-inx} = \sum_{m=-\infty}^{\infty} f(x + 2m\pi) .$$

This is called the *Poisson summation formula.*

(iii) The most famous special case is obtained when

$$f(x) = \frac{1}{\sqrt{4\pi t}} e^{-x^2/4t}, \qquad t > 0 .$$

Verify that all the assumptions are valid in this case, and hence that for $t > 0$ and all real x

$$\text{(B)} \quad \frac{1}{\pi}\left[\frac{1}{2} + \sum_{n=1}^{\infty} e^{-n^2 t} \cos nx\right] = \sum_{m=-\infty}^{\infty} \frac{1}{\sqrt{4\pi t}} \exp\left[-\frac{(x + 2m\pi)^2}{4t}\right].$$

(iv) Show that the solution of (4.5.1)–(4.5.4), when $k = 1$, $L = \pi$, can be written in the form

$$u(x,t) = \int_{-\pi}^{\pi} k(x - \xi, t) f(\xi)\, d\xi$$

and use (B) to show that if the initial temperature $f(x)$ is everywhere non-negative, then the temperature at all subsequent times is everywhere non-negative.

(v) Show that the solution of (7.1.1) can be transformed to

$$u(x,t) = \int_0^{\pi} k(x,\xi;t) f(\xi)\, d\xi ,$$

where

$$k(x,\xi;t) = \int_{x-\xi}^{x+\xi} -\frac{d}{dy} \frac{1}{\pi} \left\{\frac{1}{2} + \sum_{n=1}^{\infty} e^{-n^2 t} \cos ny\right\} dy$$

and then use (B) to show that if the initial temperature is nonnegative for $0 < x < \pi$, then the temperature at all subsequent times is nonnegative for $0 < x < \pi$.

(vi) Show that under suitable conditions

$$\frac{1}{2L}\sum_{n=-\infty}^{\infty} F\left(\frac{n\pi}{L}\right)\exp\left[-i\frac{n\pi x}{L}\right] = \sum_{m=-\infty}^{\infty} f(x+2mL)\,.$$

10

Initial-boundary value problems with two or more space variables

10.1. *HEAT FLOW IN TWO AND THREE DIMENSIONS*

In this and the next chapter we turn our attention to simple problems of heat conduction and of vibration in two and three dimensions. This chapter will be concerned primarily with time dependent problems, while the next chapter will deal with equilibrium problems.

In a discussion of heat flow in more than one dimension it is necessary to remark that some heat conductors have preferred directions in which heat flows with less resistance than in other directions. This is the case, for example, with many crystals. A medium which does not exhibit such preferred directions is called *isotropic*. In our discussion it will always be assumed that the medium is isotropic.

Consider a body composed of an isotropic heat-conducting medium. We refer to a rectangular cartesian coordinate system, and suppose that the body fills out a domain D in the coordinate space. The *temperature* at a point (x,y,z) at time t will be a function $u(x,y,z,t)$. If ρ, c and κ are, respectively, the *density*, *specific heat per unit mass*, and *conductivity* of the medium then, in the absence of sources, the temperature in the body will satisfy the three-dimensional heat equation

$$\frac{\partial u}{\partial t} = \frac{1}{\sigma}\left[\frac{\partial}{\partial x}\left(\kappa\frac{\partial u}{\partial x}\right) + \frac{\partial}{\partial y}\left(\kappa\frac{\partial u}{\partial y}\right) + \frac{\partial}{\partial z}\left(\kappa\frac{\partial u}{\partial z}\right)\right] \tag{10.1.1}$$

where $\sigma = c\rho$ is the *specific heat per unit volume*. If there are sources, so that heat is generated at a rate $q_1 = q_1(x,y,z,t,u,u_x,u_y,u_z)$ per unit volume, then the equation satisfied by the temperature is

$$\frac{\partial u}{\partial t} = \frac{1}{\sigma}\left[\frac{\partial}{\partial x}\left(\kappa\frac{\partial u}{\partial x}\right) + \frac{\partial}{\partial y}\left(\kappa\frac{\partial u}{\partial y}\right) + \frac{\partial}{\partial z}\left(\kappa\frac{\partial u}{\partial z}\right)\right] + q \tag{10.1.2}$$

where $q = q(x,y,z,t,u,u_x,u_y,u_z) = q_1/\sigma$. These two equations may be accepted as plausible on the basis of our earlier discussion, and we state them without derivation.

In the case of a uniform medium σ and κ are constants and Equation (10.1.2) reduces to

$$\frac{\partial u}{\partial t} = k\left[\frac{\partial^2 u}{\partial x^2} + \frac{\partial^2 u}{\partial y^2} + \frac{\partial^2 u}{\partial z^2}\right] + q\,, \tag{10.1.3}$$

where $k = \kappa/\sigma$ is the *diffusivity*.

Sometimes it is clear from physical considerations that the temperature is, and remains, independent of one of the coordinates, say the z-coordinate. Equation (10.1.3) can then be replaced by

$$\frac{\partial u}{\partial t} = k\left[\frac{\partial^2 u}{\partial x^2} + \frac{\partial^2 u}{\partial y^2}\right] + q\,, \tag{10.1.4}$$

For example, this equation is satisfied by the temperature in a uniform thin plate whose plane faces are insulated, if the strength of the sources and the initial temperature are independent of the z-coordinate. This will be the situation envisaged whenever, in the sequel, we speak of a temperature distribution in a plate.

The operators $\partial^2/\partial x^2 + \partial^2/\partial y^2$ and $\partial^2/\partial x^2 + \partial^2/\partial y^2 + \partial^2/\partial z^2$ which occur in the right members of (10.1.4) and (10.1.3) are called the two- and three-dimensional **Laplace operators.** They occur also in the differential equations satisfied by equilibrium temperature distributions, since such a temperature distribution u does not depend on the time, so that $u_t = 0$. The two-dimensional Laplace equation

$$\frac{\partial^2 u}{\partial x^2} + \frac{\partial^2 u}{\partial y^2} = 0 \tag{10.1.5}$$

is satisfied by any source-free equilibrium temperature distribution in a uniform plate. Similarly, the three-dimensional Laplace equation

$$\frac{\partial^2 u}{\partial x^2} + \frac{\partial^2 u}{\partial y^2} + \frac{\partial^2 u}{\partial z^2} = 0 \tag{10.1.6}$$

is satisfied by any source-free equilibrium temperature distribution in a uniform three-dimensional medium. We shall use the symbol $\boldsymbol{\Delta}$ to denote either the two-dimensional Laplace operator

$$\Delta = \frac{\partial^2}{\partial x^2} + \frac{\partial^2}{\partial y^2} \tag{10.1.7}$$

or the three-dimensional Laplace operator

$$\Delta = \frac{\partial^2}{\partial x^2} + \frac{\partial^2}{\partial y^2} + \frac{\partial^2}{\partial z^2}. \tag{10.1.8}$$

It is also customary to call

$$\Delta = \frac{d^2}{dx^2} \tag{10.1.9}$$

the one-dimensional Laplace operator.

The equation

$$\frac{\partial u}{\partial t} = k\,\Delta u \tag{10.1.10}$$

is called the **heat equation.** It will be called the one-, two- or three-dimensional heat equation according to whether Δ is given by (10.1.9), (10.1.7) or (10.1.8).

Consider heat flowing in a body bounded by a surface S. In general there will be a flow of heat across the boundary surface S. The *flux* F is defined to be the local rate of escape of heat, per unit surface area per unit time. We take F to be positive at points on the surface where heat is leaving the body, and negative at points where heat is entering the body.

Our discussion of one-dimensional heat flow was based on the statement that the flux is proportional to the derivative of the temperature in the direction perpendicular to the end cross section. Analogously, we assume in the present case that the flux is proportional to the derivative of the temperature in the direction normal to the surface. To make this explicit, we define the outward normal derivative of a function. Let P (Fig. 10.1) be a point on the surface

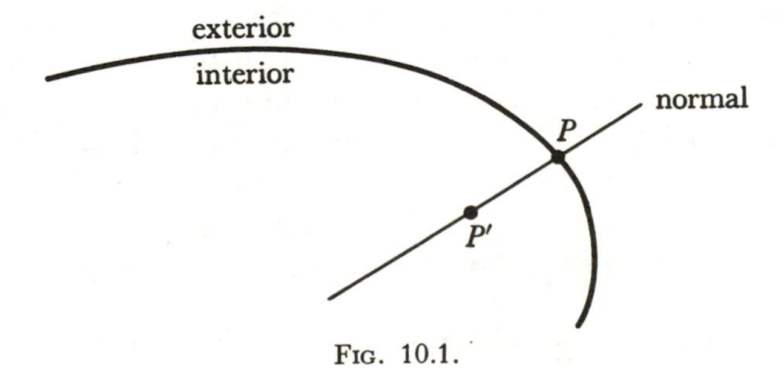

Fig. 10.1.

S and let P' be a nearby point inside the body and such that the line $P'P$ is normal to the surface at P. Then the *outward normal derivative* of the function u at P, denoted by $\partial u/\partial \nu$, is

$$\frac{\partial u}{\partial \nu} = \lim_{P' \to P} \frac{u(P) - u(P')}{|P'P|}, \tag{10.1.11}$$

where $|P'P|$ is the distance between P' and P. For example, suppose that the body is a rectangular block $a_1 < x < a_2$, $b_1 < y < b_2$, $c_1 < z < c_2$. Then, on the part of the surface where $z = c_2$, the outward normal points in the positive z-direction and $\partial u/\partial \nu = \partial u/\partial z$; but, on the part of the surface where $z = c_1$, the outward normal points in the negative z-direction and $\partial u/\partial \nu = -\partial u/\partial z$.

Since the outward normal derivative is the rate of increase of the temperature along the outward normal, our assumption, stated explicitly, is

$$F = -\kappa \frac{\partial u}{\partial \nu}. \tag{10.1.12}$$

The minus sign in (10.1.12) expresses the fact that heat flows from hot regions to cold regions.

A formula similar to (10.1.12) applies in the consideration of heat conduction in a thin plate. We assume that the plate lies on the (x, y)-plane, covering a domain D which is bounded by a curve C. The flux F is the rate of escape of heat energy per unit length and unit height and per unit time across the edges of the plate, and is given at points on C by

$$F = -\kappa \frac{\partial u}{\partial \nu},$$

where now $\partial u/\partial \nu$ denotes the derivative of u in the direction of the outward normal of the curve C.

In three dimensions, a typical heat conduction problem is the following problem of cooling. A body occupies a domain D bounded by a surface S. At time $t = 0$ the temperature distribution throughout the body is known, and at all subsequent times the entire surface S is kept at temperature zero. The problem is to find the temperature distribution throughout the body at all later times. From the physical point of view this problem seems well determined; that is, it should have one and only one solution. The initial-boundary value problem whose solution is the required temperature distribution is given by

$$\text{D.E.} \qquad \frac{\partial u}{\partial t} = k\,\Delta u \qquad \text{in } D \text{ for } t > 0,$$

(10.1.13) B.C. $u = 0$ on S for $t > 0$,
I.C. $u = f$ in D when $t = 0$,

where f is a known function defined in D.

The formal problem (10.1.13) can be interpreted either as a one-, two- or three-dimensional problem. The one-dimensional version is the problem of cooling of a rod which we have discussed extensively. The two-dimensional version is the problem of cooling of a plate whose plane faces are insulated. In this case, the domain D is the plane domain covered by the plate, and S is the boundary curve of D.

If the surface of the body is insulated at all times $t > 0$, then the B.C. in (10.1.13) must be replaced by

(10.1.14) B.C. $$\frac{\partial u}{\partial \nu} = 0 \qquad \text{on } S \text{ for } t > 0 .$$

When problem (10.1.13) is modified in this way we still expect, on the basis of physical intuition, that there is one and only one solution.

If the body exchanges heat through a thin surface film with exterior surroundings that are at temperature zero, then considerations similar to those leading to Equation (3.6.6) give the boundary condition

(10.1.15) B.C. $$\frac{\partial u}{\partial \nu} + hu = 0 \qquad \text{on } S \text{ for } t > 0 ,$$

where h is a known function, possibly constant, defined on the boundary S.

Inhomogeneous forms of the boundary conditions can also occur, just as in one-dimensional problems. The inhomogeneous version of the B.C. in (10.1.13) is

(10.1.16) B.C. $u = g$ on S for $t > 0$,

where g is a known function which is defined on S and which may also depend on t.

In treating heat conduction problems in one dimension, we always stated the differential equation in a form such as

$$\frac{\partial u}{\partial t} = k \frac{\partial^2 u}{\partial x^2}, \qquad 0 < x < L, \quad 0 < t .$$

The point we wish to make refers to the requirement that the differential equation be satisfied for $0 < x < L$ rather than $0 \leq x \leq L$. We require that u satisfy the differential equation at *interior* points of the rod but not necessarily at the end points. Indeed, there are natural problems in which the value of u at the end $x = 0$ is a prescribed function $u(0,t) = A(t)$ which may even have discontinuities, so that $u_t(0,t)$ fails to exist. Thus the D.E. may not even be meaningful at the boundary point $x = 0$. For similar reasons we require only that the D.E. in (10.1.13) should be satisfied at interior points of the body.

As a consequence it is useful to have a conventional name for the set of all interior points of the body. An **interior point** of a set of points is a point such that some sphere (in the two-dimensional case, some disc) centered at the point lies entirely within the set. A set D of points is called a **domain** if it consists entirely of interior points, and if it is connected in the sense that any two points of D can be joined by a curve lying entirely in D. For example, the set of all points in the (x,y)-plane such that $0 < x < a$, $0 < y < b$ is a rectangular domain. The boundary of this domain consists of four straight-line segments. The points of the boundary are not interior points and they do not belong to the domain. It is to be understood that, when D is a domain, a statement such as

$$\Delta u = 0 \qquad \text{in } D$$

means that u satisfies the differential equation at every interior point but not necessarily at the boundary points.

10.2. *THE VIBRATING MEMBRANE*

The two-dimensional **wave equation**

$$u_{tt} = c^2(u_{xx} + u_{yy}) = c^2\,\Delta u \tag{10.2.1}$$

occurs in a great variety of important physical applications. Because the vibrating string is easy to visualize, we chose the string for the interpretation of problems involving the one-dimensional wave equation, in spite of the complexity of the physical assumptions needed to derive the equation. To illustrate the two-dimensional wave equation we choose the vibrating membrane, again because of its pictorial appeal, and despite the even more complex physical background.

If a thin sheet of rubber, or some other flexible elastic material which we call a membrane, is stretched over a frame to form a drumhead, it will vibrate when disturbed from its equilibrium position. The ideal membrane is a uniform sheet of material which is perfectly flexible, that is, does not resist bending, but which resists stretching. We consider a membrane stretched on a supporting frame which forms a plane curve C bounding a domain D in the plane. When the membrane is at rest, and no external forces are acting, the membrane lies in this plane, which we take to be the (x,y)-plane.

The stretching forces, which are applied at the boundary, are transmitted throughout the membrane. It is assumed that the membrane is stretched uniformly at each point and in all directions. This means that the mutual force between the material on two sides of a line segment has a constant magnitude T_0 per unit length. The constant T_0 is called the *tension* in the membrane.

The *surface density* of the membrane is the mass per unit area (when the membrane is at rest), and will be denoted by ρ_0.

We will consider transverse vibrations in which each material point moves perpendicular to the (x,y)-plane. We choose the u-axis perpendicular to the (x,y)-plane. The position of the membrane at any fixed time t then coincides with the surface described by an equation $u = u(x,y,t)$. The function $u(x,y,t)$ is called the *displacement function* of the membrane.

For *small* transverse vibrations, in the absence of external forces, the displacement function of the membrane satisfies the two-dimensional wave equation

$$\rho_0 \frac{\partial^2 u}{\partial t^2} = T_0 \left(\frac{\partial^2 u}{\partial x^2} + \frac{\partial^2 u}{\partial y^2} \right) \tag{10.2.2}$$

or

$$u_{tt} = c^2 \Delta u \,, \tag{10.2.3}$$

where $c^2 = T_0/\rho_0$. If there is an external force of magnitude $q_1 = q_1(x,y,t,u,u_x,u_y,u_t)$ per unit area, acting on the membrane, the equation of motion assumes the form

$$\rho_0 u_{tt} = T_0 \Delta u + q_1 \tag{10.2.4}$$

or

$$u_{tt} = c^2 \Delta u + q \tag{10.2.5}$$

where $q = q_1/\rho_0$ is the force per unit mass. These equations are plausible because of the analogy between the vibrating membrane and the vibrating string, and we present them without derivation.

For a membrane which is rigidly attached to its supporting frame, the boundary condition is of the first kind,

$$u = 0 \qquad \text{on } C \,. \tag{10.2.6}$$

Problems with boundary conditions of the second or third kind can also be formulated, and will be found among the exercises. Although the interpretation of these problems as membrane problems is somewhat forced, they do occur as natural problems in connection with other applications of the wave equation.

The vibrating membrane is a two-dimensional analogue of the vibrating string. Consequently, it can be expected that, in an initial-boundary value problem, both the initial displacement $u(x,y,0)$ and the initial velocity $u_t(x,y,0)$ must be specified. Thus the displacement function of a membrane rigidly attached to its supporting frame, with no external forces acting, and with prescribed initial displacement $f(x,y)$ and initial velocity $g(x,y)$, will be the solution of the problem

$$
\text{(10.2.7)}\quad
\begin{array}{lll}
\text{D.E.} & u_{tt} = c^2\,\Delta u & \text{in } D\,, \quad t > 0\,, \\
\text{B.C.} & u = 0 & \text{on } C\,, \quad t > 0\,, \\
\text{I.C.} & u(x,y,0) = f(x,y)\,, & \\
 & u_t(x,y,0) = g(x,y) & \text{in } D\,.
\end{array}
$$

10.3. *THE RECTANGULAR MEMBRANE*

We shall solve problem (10.2.7) for the case of a rectangular membrane. If the sides of the membrane are of lengths a and b we can choose the coordinate system so that the domain D is the set of all points (x,y) such that $0 < x < a$, $0 < y < b$. The boundary C of D consists of segments of the four lines $x = 0$, $x = a$, $y = 0$, $y = b$. The initial-boundary value problem satisfied by the displacement function is, in detailed form,

$$
\text{(10.3.1)}\quad
\begin{array}{llllll}
\text{D.E.} & u_{tt} = c^2(u_{xx} + u_{yy})\,, & 0 < x < a\,, & 0 < y < b\,, & t > 0\,, \\
\text{B.C.} & u(0,y,t) = 0\,, & & 0 < y < b\,, & t > 0\,, \\
 & u(a,y,t) = 0\,, & & 0 < y < b\,, & t > 0\,, \\
 & u(x,0,t) = 0\,, & 0 < x < a\,, & & t > 0\,, \\
 & u(x,b,t) = 0\,, & 0 < x < a\,, & & t > 0\,, \\
\text{I.C.} & u(x,y,0) = f(x,y)\,, & 0 < x < a\,, & 0 < y < b\,, & \\
 & u_t(x,y,0) = g(x,y)\,, & 0 < x < a\,, & 0 < y < b\,. &
\end{array}
$$

Our solution procedure will be analogous to that employed in the discussion of one-dimensional problems, and begins with the determination of particular solutions of the D.E. and B.C. by separation of variables. The separation of variables will be carried out in two stages. We first seek functions of the form

$$
\text{(10.3.2)}\qquad u(x,y,t) = \psi(t)\varphi(x,y)
$$

which satisfy the D.E. and B.C. in (10.3.1). In order that this function be nontrivial, the factors must be nontrivial solutions of

$$\frac{d^2\psi}{dt^2} + \lambda c^2 \psi = 0\,, \qquad t > 0\,, \tag{10.3.3}$$

and

$$\begin{array}{lll} \text{D.E.} & \varphi_{xx} + \varphi_{yy} + \lambda\varphi = 0\,, & 0 < x < a\,,\; 0 < y < b\,, \\ \text{B.C.} & \varphi(0,y) = 0\,, & 0 < y < b\,, \\ & \varphi(a,y) = 0\,, & 0 < y < b\,, \\ & \varphi(x,0) = 0\,, & 0 < x < a\,, \\ & \varphi(x,b) = 0\,, & 0 < x < a\,, \end{array} \tag{10.3.4}$$

where λ is a constant. Equation (10.3.3) is a familiar ordinary differential equation whose solution we know. On the other hand, the system (10.3.4) for the determination of φ is a boundary value problem involving a partial differential equation. The system (10.3.4) is a *two-dimensional eigenvalue problem*, which could be written in the more compact form

$$\begin{array}{lll} \text{D.E.} & \Delta\varphi + \lambda\varphi = 0 & \text{in } D\,, \\ \text{B.C.} & \varphi = 0 & \text{on } C\,. \end{array} \tag{10.3.5}$$

We need to determine the eigenvalues, that is, the values of λ for which (10.3.5) has a nontrivial solution, and the corresponding nontrivial solutions, the eigenfunctions.

We undertake to solve this problem by further separation of variables; that is, we seek eigenfunctions of the form

$$\varphi(x,y) = X(x)Y(y)\,. \tag{10.3.6}$$

The factors $X(x)$, $Y(y)$ must satisfy

$$\begin{array}{rl} X'' + (\lambda - \mu)X = 0 & 0 < x < a, \\ X(0) = 0 & \\ X(a) = 0\,, & \end{array} \tag{10.3.7}$$

$$\begin{array}{rl} Y'' + \mu Y = 0 & 0 < y < b, \\ Y(0) = 0 & \\ Y(b) = 0\,, & \end{array} \tag{10.3.8}$$

where μ is a constant. In order that (10.3.8) have a nontrivial solution μ must have one of the values

$$\mu_n = \frac{n^2\pi^2}{b^2}\,, \qquad n = 1, 2, 3, \ldots$$

and the corresponding nontrivial solutions are constant multiples of

$$Y_n(y) = \sin\frac{n\pi y}{b}\,.$$

Then, with $\mu = \mu_n$ in (10.3.7), the values of λ which permit a nontrivial solution of (10.3.7) are given by

$$\lambda_{m,n} - \mu_n = \frac{m^2\pi^2}{a^2}, \qquad m = 1, 2, 3, \ldots$$

and the corresponding solutions are constant multiples of

$$X_m(x) = \sin \frac{m\pi x}{a}.$$

Thus we have found eigenvalues $\lambda_{m,n} = \mu_n + (m^2\pi^2/a^2)$ or

$$\lambda_{m,n} = \left(\frac{m^2}{a^2} + \frac{n^2}{b^2}\right)\pi^2, \qquad \begin{matrix} m = 1, 2, 3, \ldots, \\ n = 1, 2, 3, \ldots \end{matrix} \tag{10.3.9}$$

and corresponding eigenfunctions

$$\varphi_{m,n}(x,y) = X_m(x)Y_n(y) = \sin \frac{m\pi x}{a} \sin \frac{n\pi y}{b}. \tag{10.3.10}$$

Equation (10.3.9) gives the only eigenvalues to which there correspond eigenfunctions of the separated form (10.3.6), but it is not immediately clear that these are the only eigenvalues of the problem. Conceivably there are other eigenvalues with corresponding eigenfunctions which are not of the separated form. We shall show in Section 5 that, in fact, (10.3.9) does give all the eigenvalues of the problem and that the corresponding eigenfunctions are the functions given by (10.3.10), or are finite linear combinations of these functions.

The eigenvalues may be arranged in a double sequence

$$\begin{matrix} \lambda_{1,1} & \lambda_{1,2} & \cdots & \lambda_{1,n} & \cdots \\ \lambda_{2,1} & \lambda_{2,2} & \cdots & \lambda_{2,n} & \cdots \\ \vdots & \vdots & & \vdots & \\ \lambda_{m,1} & \lambda_{m,2} & \cdots & \lambda_{m,n} & \cdots \\ \vdots & \vdots & & \vdots & \end{matrix} \tag{10.3.11}$$

and there is similarly a double sequence of eigenfunctions

$$\begin{matrix} \varphi_{1,1} & \varphi_{1,2} & \cdots & \varphi_{1,n} & \cdots \\ \varphi_{2,1} & \varphi_{2,2} & \cdots & \varphi_{2,n} & \cdots \\ \vdots & \vdots & & \vdots & \\ \varphi_{m,1} & \varphi_{m,2} & \cdots & \varphi_{m,n} & \\ \vdots & \vdots & & \vdots & \end{matrix} \tag{10.3.12}$$

The eigenfunctions (10.3.12) form an *orthogonal system* of functions on the rectangle in the sense that

$$\iint_D \varphi_{m,n}(x,y)\varphi_{k,l}(x,y)\,dx\,dy = \begin{cases} 0 & \text{if } m \neq k \text{ or } n \neq l\,, \\ \dfrac{ab}{4} & \text{if } m = k \text{ and } n = l\,. \end{cases} \tag{10.3.13}$$

This is easily verified by expressing the double integral as an iterated integral and using (10.3.10). We get

$$\iint_D \varphi_{m,n}\varphi_{k,l}\,dx\,dy = \left(\int_0^a \sin\frac{m\pi x}{a}\sin\frac{k\pi x}{a}\,dx\right)\left(\int_0^b \sin\frac{n\pi y}{b}\sin\frac{l\pi y}{b}\,dy\right),$$

and since

$$\int_0^a \sin\frac{m\pi x}{a}\sin\frac{k\pi x}{a}\,dx = \begin{cases} 0 & \text{if } m \neq k\,, \\ \dfrac{a}{2} & \text{if } m = k\,, \end{cases}$$

$$\int_0^b \sin\frac{n\pi y}{b}\sin\frac{l\pi y}{b}\,dy = \begin{cases} 0 & \text{if } n \neq l\,, \\ \dfrac{b}{2} & \text{if } n = l\,, \end{cases}$$

the result follows.

Now the solution of Equation (10.3.3) with $\lambda = \lambda_{m,n}$ is

$$\psi_{m,n}(t) = A_{m,n}\cos ct\sqrt{\lambda_{m,n}} + B_{m,n}\frac{\sin ct\sqrt{\lambda_{m,n}}}{c\sqrt{\lambda_{m,n}}}\,. \tag{10.3.14}$$

Therefore, the series

(10.3.15)

$$\begin{aligned} u(x,y,t) &= \sum_{n=1}^{\infty}\sum_{m=1}^{\infty} \psi_{m,n}(t)\varphi_{m,n}(x,y) \\ &= \sum_{n=1}^{\infty}\sum_{m=1}^{\infty}\left[A_{m,n}\cos ct\sqrt{\lambda_{m,n}} + B_{m,n}\frac{\sin ct\sqrt{\lambda_{m,n}}}{c\sqrt{\lambda_{m,n}}}\right]\varphi_{m,n}(x,y) \end{aligned}$$

will be a formal solution of problem (10.3.1) provided the constants $A_{m,n}$, $B_{m,n}$ can be chosen so that the I.C. is satisfied. For $t = 0$, Equation (10.3.15) gives

$$f(x,y) = \sum_{n=1}^{\infty}\sum_{m=1}^{\infty} A_{m,n}\varphi_{m,n}(x,y)\,, \tag{10.3.16}$$

and the derivative of (10.3.15) with respect to t, at $t = 0$, gives

$$g(x,y) = \sum_{n=1}^{\infty}\sum_{m=1}^{\infty} B_{m,n}\varphi_{m,n}(x,y)\,. \tag{10.3.17}$$

We use the orthogonality relation (10.3.13) in the usual way to determine the constants $A_{m,n}$ and $B_{m,n}$ in (10.3.16) and (10.3.17), and find

$$A_{m,n} = \frac{\iint_D f(x,y)\varphi_{m,n}(x,y)\,dx\,dy}{\iint_D [\varphi_{m,n}(x,y)]^2\,dx\,dy} = \frac{4}{ab}\iint_D f(x,y)\varphi_{m,n}(x,y)\,dx\,dy\,, \tag{10.3.18}$$

$$B_{m,n} = \frac{\iint_D g(x,y)\varphi_{m,n}(x,y)\,dx\,dy}{\iint_D [\varphi_{m,n}(x,y)]^2\,dx\,dy} = \frac{4}{ab}\iint_D g(x,y)\varphi_{m,n}(x,y)\,dx\,dy\,. \tag{10.3.19}$$

With this determination of the coefficients, the formal solution of the vibrating membrane problem is complete.

Like the vibrating string, the membrane can execute *free* simple harmonic vibrations, that is, free vibrations with displacement functions of the form

$$u(x,y,t) = R(x,y)\cos(2\pi\nu t + \alpha)\,. \tag{10.3.20}$$

These motions are called the *natural modes* of the membrane, and their frequencies are called the *natural frequencies*. Since (10.3.20) must satisfy the wave equation and the boundary conditions we find that

$$\Delta R + \frac{4\pi^2\nu^2}{c^2} R = 0 \qquad \text{in } D\,,$$
$$R = 0 \qquad \text{on } C\,.$$

Thus $(4\pi^2\nu^2)/c^2$ is an eigenvalue of problem (10.3.5) and R is an eigenfunction. The natural frequencies of the rectangular membrane are therefore

$$\nu_{m,n} = \frac{c}{2\pi}\sqrt{\lambda_{m,n}} = \frac{c}{2}\sqrt{\frac{m^2}{a^2} + \frac{n^2}{b^2}}\,, \qquad m,n = 1, 2, \ldots\,. \tag{10.3.21}$$

Each of the displacement functions,

$$u_{m,n}(x,y,t) = \psi_{m,n}(t)\varphi_{m,n}(x,y)\,, \tag{10.3.22}$$

where $\psi_{m,n}$ and $\varphi_{m,n}$ are defined by (10.3.14) and (10.3.10), represents a natural mode, and thus we see that the general free vibration (10.3.15) is a superposition of natural modes of vibration.

It is familiar that the sound of a freely vibrating membrane is not in general a pure musical note like the sound of a string. This may be attributed to the fact that the natural frequencies of a membrane are not all integral multiples of the lowest natural frequency, nor are they integral multiples of any single fixed frequency. We will verify this for rectangular membranes.

Consider, for example, the natural frequencies

$$\nu_{m,n} = \frac{c}{2a}\sqrt{m^2 + n^2}$$

of a square membrane with side a. If all of these frequencies were integral multiples of some fixed frequency ν_0, then the ratio of any two of them would be of the form $p\nu_0/q\nu_0 = p/q$ where p and q are integers; that is, the ratio would be a rational number. But this is not the case since, for example, $\nu_{3,1}/\nu_{1,1} = \sqrt{5}$, which is not rational. Thus for the square membrane there is no "fundamental" frequency ν_0 such that all the natural frequencies are integral multiples of ν_0. (See also Exercises 10a, Problem 3.)

The analogues of the nodes of a vibrating string are the nodal curves of the membrane. If the membrane vibrates in such a way that all the points on a curve remain at rest throughout the motion, then the curve is called a **nodal curve.** The nodal curves of the natural mode (10.3.22) are the curves along which the function $\varphi_{m,n}(x,y)$ is zero. These curves are the straight lines

$$x = 0, \frac{a}{m}, \frac{2a}{m}, \frac{3a}{m}, \ldots, a\,,$$

and the straight lines

$$y = 0, \frac{b}{n}, \frac{2b}{n}, \frac{3b}{n}, \ldots, b\,.$$

Although the nodal curves of the natural modes given by (10.3.22) are exclusively straight lines, there may be other natural modes of vibration with nodal curves which are not straight lines. When there are repeated eigenvalues in the sequence (10.3.11) we can find such modes as follows. Suppose, for example, $\lambda_{m,n} = \lambda_{p,q}$. Then the corresponding natural frequencies are also equal, $\nu_{m,n} = \nu_{p,q}$, and hence for any constants α,β

$$u(x,y,t) = [\alpha\varphi_{m,n}(x,y) + \beta\varphi_{p,q}(x,y)]\cos 2\pi\nu_{m,n}t$$

is a natural mode of vibration. The nodal curves for this motion are the curves along which $\alpha\varphi_{m,n} + \beta\varphi_{p,q}$ is zero. These curves depend on the ratio β/α, and are not in general straight lines.

A specific example with multiple eigenvalues is provided by the square membrane, for if $b = a$ it is clear from (10.3.9) that $\lambda_{m,n} = \lambda_{n,m}$. Hence the zero curves of $\alpha\varphi_{m,n} + \beta\varphi_{n,m}$ are nodal curves of natural modes. A few examples are illustrated in Figure 10.2.

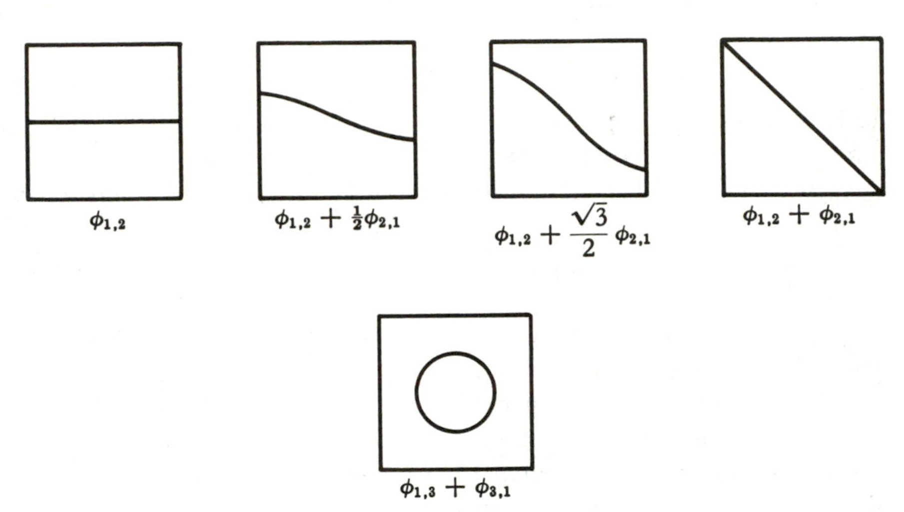

FIG. 10.2. Some nodal curves for a square membrane.

It will be useful in Section 5 to know that, in the sequence (10.3.11), each eigenvalue occurs only a finite number of times. To prove this, we let λ be a positive constant, and consider the number of ways of choosing the positive integers m and n such that $\lambda_{m,n} = \lambda$. Since $\lambda_{m,n} > m^2\pi^2/a^2$ it follows from $\lambda_{m,n} = \lambda$ that $m < a\sqrt{\lambda}/\pi$, and it follows similarly that $n < b\sqrt{\lambda}/\pi$. Hence there are at most a finite number of choices of the positive integers m and n consistent with the equation $\lambda_{m,n} = \lambda$.

EXERCISES 10a

1. Solve the initial-boundary value problem (evaluate all of the constants)

$$
\begin{array}{lllll}
\text{D.E.} & u_t = k(u_{xx} + u_{yy}), & 0 < x < a, & 0 < y < b, & t > 0, \\
\text{B.C.} & \left.\begin{array}{l} u(0,y,t) = 0 \\ u_x(a,y,t) = 0 \end{array}\right\} & & 0 < y < b, & t > 0, \\
 & \left.\begin{array}{l} u_y(x,0,t) = 0 \\ u_y(x,b,t) = 0 \end{array}\right\} & 0 < x < a, & & t > 0, \\
\text{I.C.} & u(x,y,0) = x + y & 0 < x < a, & 0 < y < b. &
\end{array}
$$

2. For the following problem (a) state the three-dimensional eigenvalue problem obtained by separating the time variable, (b) find the eigenvalues and eigenfunctions, (c) find the solution of the problem.

D.E. $u_t = k(u_{xx} + u_{yy} + u_{zz})$ $\quad 0 < x < a, \quad 0 < y < b, \quad 0 < z < c, \quad t > 0$

B.C. $u_x = 0$ $\quad$ if $x = 0$ or $x = a$

$u = 0$ $\quad$ if $y = 0$ or $y = b$

$u = 0$ $\quad$ if $z = 0$

$u_z = 0$ $\quad$ if $z = c$

I.C. $u(x,y,z,0) = f(x,y,z)$ $\quad 0 < x < a,\ 0 < y < b,\ 0 < z < c.$

3. If the natural frequencies of a membrane were integral multiples of a fixed frequency ν_0, then the difference between any two different natural frequencies would also be an integral multiple of ν_0, and so could not be numerically less than ν_0. Show that no rectangular membrane has this property, by evaluating

$$\lim_{m\to\infty} (\nu_{m,2} - \nu_{m,1})$$

4. For a *square* membrane of side $a = \pi$ the eigenvalues (10.3.9) are $\lambda_{m,n} = m^2 + n^2$. Since $\lambda_{m,n} = \lambda_{n,m}$ there are many eigenvalues of multiplicity at least two. Actually there are eigenvalues of arbitrarily high multiplicity.

(a) Use the relations $3^2 + 4^2 = 5^2$, $5^2 + 12^2 = 13^2$ to show that $65^2 = 5^2 \cdot 13^2$ is an eigenvalue of multiplicity at least four.

(b) Find an eigenvalue of multiplicity at least ten.
Hint: $(p^2 - q^2)^2 + (2pq)^2 = (p^2 + q^2)^2$.

(c) Find the lowest eigenvalue of multiplicity three.

5. Show that all the eigenvalues (10.3.9) are of multiplicity one if a^2/b^2 is irrational.

6. For the following problem (a) find the eigenvalues and eigenfunctions, (b) state the orthogonality relation satisfied by the eigenfunctions.

D.E. $x^2\varphi_{xx} + x\varphi_x + \varphi_{yy} + \lambda\varphi = 0$ $\quad 1 < x < 2, \quad 0 < y < \pi,$

B.C. $\varphi(1,y) = 0$ $\quad 0 < y < \pi,$

$\varphi(2,y) = 0$ $\quad 0 < y < \pi,$

$\varphi_y(x,0) = 0$ $\quad 1 < x < 2,$

$\varphi_y(x,\pi) = 0$ $\quad 1 < x < 2.$

7. Solve

D.E. $u_{tt} = c^2(u_{xx} + u_{yy}),$ $\quad 0 < x < a, \quad 0 < y < b, \quad t > 0$

B.C. $\left.\begin{array}{l} u(x,0,t) = 0 \\ u(x,b,t) = 0 \end{array}\right\}$ $\quad 0 < x < a, \quad t > 0$

$\left.\begin{array}{l} u_x(0,y,t) = 0 \\ u_x(a,y,t) = 0 \end{array}\right\}$ $\quad 0 < y < b, \quad t > 0$

I.C. $\left.\begin{array}{l} u(x,y,0) = f(x,y) \\ u_t(x,y,0) = g(x,y) \end{array}\right\}$ $\quad 0 < x < a, \quad 0 < y < b.$

10.4. *GREEN'S FORMULAS AND APPLICATIONS*

For the two-dimensional Laplace operator $\Delta = \partial^2/\partial x^2 + \partial^2/\partial y^2$, **Green's first and second formulas** are respectively

$$(10.4.1) \quad \iint_D (\Delta f) g \, dx \, dy = -\iint_D \left[\frac{\partial f}{\partial x}\frac{\partial g}{\partial x} + \frac{\partial f}{\partial y}\frac{\partial g}{\partial y}\right] dx \, dy + \int_C \frac{\partial f}{\partial \nu} g \, ds ,$$

$$(10.4.2) \quad \iint_D [(\Delta f) g - f(\Delta g)] \, dx \, dy = \int_C \left[\frac{\partial f}{\partial \nu} g - f \frac{\partial g}{\partial \nu}\right] ds .$$

In these formulas D is a finite domain with boundary C and $\partial/\partial\nu$ denotes the outward normal derivative on C. The formulas are valid provided that the domain D and the functions $f(x,y)$, $g(x,y)$ satisfy suitable restrictions. Roughly speaking, these restrictions require that the boundary C be sufficiently smooth and that the functions and their derivatives which appear in the formulas be appropriately continuous. We state below precise restrictions, which while far more stringent than required, will suffice for our purposes.

A domain will be called a **regular domain** if it is a finite domain and its boundary consists of a finite number of straight-line segments and circular arcs. Formula (10.4.1) is valid if D is a regular domain and if $f(x,y)$ and its first- and second-order partial derivatives, and $g(x,y)$ and its first-order partial derivatives, are continuous on the closed region consisting of D together with its boundary C. This closed region will hereafter be denoted by $D + C$. Formula (10.4.2) is valid if D is a regular domain and if both $f(x,y)$ and $g(x,y)$ have continuous first- and second-order partial derivatives on the closed region $D + C$.

We will prove Green's formulas only for the case when the domain D is a rectangle. For more general proofs the reader is referred to texts on advanced calculus. If D is a rectangle with sides a and b we can choose a cartesian coordinate system so that D consists of all points (x,y) such that $0 < x < a$, $0 < y < b$. The boundary C consists of four straight-line segments (see Fig. 10.3) OP, PQ, OR, RQ. Green's first formula will be proved by integrating the identity

$$(10.4.3) \quad (\Delta f) g = \left[\frac{\partial}{\partial x}\left(\frac{\partial f}{\partial x} g\right) + \frac{\partial}{\partial y}\left(\frac{\partial f}{\partial y} g\right)\right] - \left[\frac{\partial f}{\partial x}\frac{\partial g}{\partial x} + \frac{\partial f}{\partial y}\frac{\partial g}{\partial y}\right]$$

over the domain D. This identity is verified simply by application of the rule for differentiation of products to the first term on the right. We assume that f has continuous first- and second-order partial derivatives and g has continuous first-order partial derivatives in the closed region $D + C$. From (10.4.3) we obtain

$$\iint_D (\Delta f)g\,dx\,dy = -\iint_D \left[\frac{\partial f}{\partial x}\frac{\partial g}{\partial x} + \frac{\partial f}{\partial y}\frac{\partial g}{\partial y}\right] dx\,dy + J \tag{10.4.4}$$

where

$$J = \iint_D \left[\frac{\partial}{\partial x}\left(\frac{\partial f}{\partial x}g\right) + \frac{\partial}{\partial y}\left(\frac{\partial f}{\partial y}g\right)\right] dx\,dy\,. \tag{10.4.5}$$

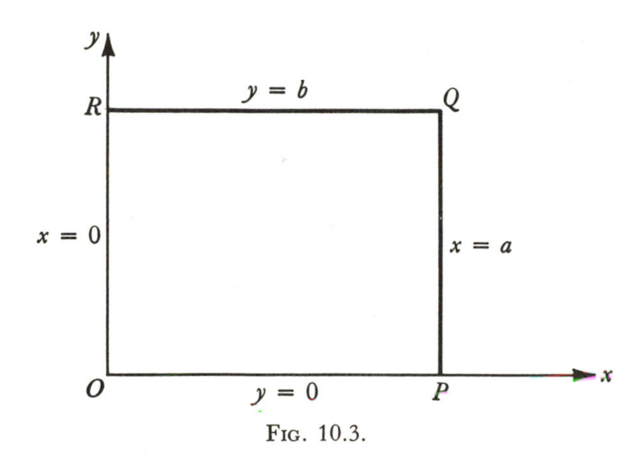

Fig. 10.3.

The integral J can be simplified by taking advantage of the fact that each term of the integrand is a derivative of a continuous function. If the integral of the first term is expressed as an iterated integral, we have

$$\begin{aligned}
\iint_D \frac{\partial}{\partial x}\left(\frac{\partial f}{\partial x}g\right) dx\,dy &= \int_0^b dy \int_0^a \frac{\partial}{\partial x}\left(\frac{\partial f}{\partial x}g\right) dx \\
&= \int_0^b dy \left[\frac{\partial f}{\partial x}g\right]_{x=0}^{x=a} \qquad (10.4.6)\\
&= \int_0^b [f_x(a,y)g(a,y) - f_x(0,y)g(0,y)]\,dy\,.
\end{aligned}$$

Now the point (a,y) is a typical point on the segment PQ of the boundary C. At this point the outward normal of D points in the positive x-direction so that

$$\frac{\partial f}{\partial x} = \frac{\partial f}{\partial \nu}$$

where $\partial/\partial\nu$ denotes the outward normal derivative. Further, the element of arc length ds is simply dy. Accordingly

$$\int_0^b f_x(a,y)g(a,y)\,dy = \int_{\mathrm{PQ}} \frac{\partial f}{\partial \nu} g\,ds\,. \tag{10.4.7}$$

Again, the point $(0,y)$ is a typical point on the segment RO, and at this point the outward normal points in the negative x-direction so that

$$\frac{\partial f}{\partial x} = -\frac{\partial f}{\partial \nu},$$

and again $ds = dy$. Accordingly,

$$-\int_0^b f_x(0,y)g(0,y)\,dy = \int_{\mathrm{RO}} \frac{\partial f}{\partial \nu} g\,ds\,. \tag{10.4.8}$$

From (10.4.6), (10.4.7), (10.4.8) we have

$$\iint_D \frac{\partial}{\partial x}\left(\frac{\partial f}{\partial x} g\right) dx\,dy = \int_{\mathrm{PQ}} \frac{\partial f}{\partial \nu} g\,ds + \int_{\mathrm{RO}} \frac{\partial f}{\partial \nu} g\,ds\,.$$

By a similar calculation,

$$\iint_D \frac{\partial}{\partial y}\left(\frac{\partial f}{\partial y} g\right) dx\,dy = \int_{\mathrm{QR}} \frac{\partial f}{\partial \nu} g\,ds + \int_{\mathrm{OP}} \frac{\partial f}{\partial \nu} g\,ds\,.$$

When these two equations are added, we obtain

$$J = \int_{\mathrm{PQ}} \frac{\partial f}{\partial \nu} g\,ds + \int_{\mathrm{RO}} \frac{\partial f}{\partial \nu} g\,ds + \int_{\mathrm{QR}} \frac{\partial f}{\partial \nu} g\,ds + \int_{\mathrm{OP}} \frac{\partial f}{\partial \nu} g\,ds\,. \tag{10.4.9}$$

But the four segments PQ, QR, RO, OP form the entire boundary C, and hence (10.4.9) is equivalent to

$$J = \int_C \frac{\partial f}{\partial \nu} g\,ds\,. \tag{10.4.10}$$

Now substituting (10.4.10) in (10.4.4), we obtain Green's first formula (10.4.1). Green's second formula for the rectangle can be obtained by first interchanging f and g in Green's first formula, to get

(10.4.11)

$$\iint_D f(\Delta g)\,dx\,dy = -\iint_D \left[\frac{\partial f}{\partial x}\frac{\partial g}{\partial x} + \frac{\partial f}{\partial y}\frac{\partial g}{\partial y}\right] dx\,dy + \int_C f\frac{\partial g}{\partial \nu}\,ds\,.$$

Then, subtraction of (10.4.11) from (10.4.1) leads to (10.4.2).

As an application, we will use Green's first formula to derive an energy equation satisfied by solutions of the membrane equation

$$\rho u_{tt} = T\,\Delta u + q_1(x,y,t,u,u_x,u_y,u_t)\,. \tag{10.4.12}$$

Let D be a domain in the (x,y)-plane with boundary C, and $u(x,y,t)$ a solution in D for $t > 0$. We assume that ρ and T are constants, that the external force function q_1 is a continuous function of its arguments, and that the first- and second-order partial derivatives of u with respect to x,y,t are continuous for (x,y) in $D + C$ and $t > 0$. We multiply (10.4.12) by u_t and integrate the resulting equation over D. Since $\rho u_{tt}u_t = \frac{1}{2}(\rho u_t^2)_t$ we obtain for $t > 0$,

$$\frac{d}{dt}\iint_D \frac{1}{2}\rho u_t^2\,dx\,dy = \iint_D T(\Delta u)u_t\,dx\,dy + \iint_D q_1 u_t\,dx\,dy\,. \tag{10.4.13}$$

The first integral on the right can be transformed by applying Green's first formula (10.4.1), with $f = u$, $g = u_t$. Thus

$$\iint_D T(\Delta u)u_t\,dx\,dy = -\iint_D T(u_x u_{xt} + u_y u_{yt})\,dx\,dy + \int_C T\frac{\partial u}{\partial \nu}u_t\,ds\,.$$

Since $T(u_x u_{xt} + u_y u_{yt}) = \frac{1}{2}T(u_x^2 + u_y^2)_t$, Equation (10.4.13) is therefore equivalent to

$$\frac{d}{dt}\left\{\iint_D \frac{1}{2}\rho u_t^2\,dx\,dy + \iint_D \frac{1}{2}T(u_x^2 + u_y^2)\,dx\,dy\right\} = \int_C T\frac{\partial u}{\partial \nu}u_t\,ds + \iint_D q_1 u_t\,dx\,dy\,, \qquad t > 0\,, \tag{10.4.14}$$

which is the energy equation for the membrane.

EXERCISES 10b

In Problems 1, 2, and 3 assume that every eigenfunction has continuous first- and second-order partial derivatives on the closed region $D + C$.

1. Let D be a regular domain with boundary C. Show that for the eigenvalue problem

D.E.	$\Delta\varphi + \lambda\varphi = 0$	in D,
B.C.	$\varphi = 0$	on C,

(i) all eigenvalues are real and nonnegative,
(ii) eigenfunctions belonging to different eigenvalues are orthogonal on D.

2. Let D be a regular domain with boundary C and let h be a smooth nonnegative function defined on C. Repeat Problem 1 for the eigenvalue problem

$$\text{D.E.} \qquad \Delta\varphi + \lambda\varphi = 0 \qquad \text{in } D\,,$$

$$\text{B.C.} \qquad \frac{\partial\varphi}{\partial\nu} + h\varphi = 0 \qquad \text{on } C\,.$$

3. Let D be a regular domain with boundary C and let $q(x,y)$ be real valued and continuous on $D + C$. Let M_1, M_2 be constants such that for all points (x,y) in $D + C$, we have $M_1 \leq q(x,y) \leq M_2$. Show that for the problem

$$\begin{array}{lll} \text{D.E.} & \Delta\varphi + q(x,y)\varphi + \lambda\varphi = 0 & \text{in } D\,, \\ \text{B.C.} & \varphi = 0 & \text{on } C\,, \end{array}$$

every eigenvalue λ satisfies $\lambda \geq -M_2$.

4. Show that if the function $f(x,y)$ in (10.3.16) has continuous first and second derivatives on the closed rectangle $D + C$, and satisfies the boundary condition $f = 0$ on C, then there is a constant M such that for all m,n

$$|A_{m,n}| \leq \frac{M}{m^2 + n^2}\,.$$

5. Use the energy equation (10.4.14) to show that within a suitably defined class of functions $u(x,y,t)$, the problem

$$\begin{array}{lll} \text{D.E.} & \rho u_{tt} = T\Delta u + q(x,y) & \text{in } D,\ t > 0 \\ \text{B.C.} & u = 0 & \text{on } C,\ t > 0 \\ \text{I.C.} & u(x,y,0) = f(x,y)\,, \quad u_t(x,y,0) = g(x,y) & \text{in } D \end{array}$$

has at most one solution. Here D is a regular domain, C is the boundary of D, and $q(x,y)$, $f(x,y)$, $g(x,y)$ are assumed continuous in $D + C$.

6. (i) Let $u(x,y,t)$ be a solution of the heat equation

$$\text{D.E.} \qquad u_t = k\Delta u + q(x,y)$$

in a regular domain D with boundary C, and assume that u and its first and second partial derivatives are continuous for (x,y) in $D + C$ and $t > 0$. Establish an integral identity for the heat equation analogous to the energy equation for the wave equation by multiplying the D.E. by u, integrating over D, and using Green's first formula.

(ii) Formulate a uniqueness theorem, analogous to that of Problem 5, for the initial-boundary value problem

$$\text{D.E.} \qquad u_t = k\Delta u + q(x,y) \qquad \text{in } D,\ t > 0$$

$$\text{B.C.} \qquad \frac{\partial u}{\partial \nu} = 0 \qquad \text{on } C,\ t > 0$$

$$\text{I.C.} \qquad u(x,y,0) = f(x,y) \qquad \text{in } D,$$

and prove the theorem using the result of part (i).

10.5. *DOUBLE FOURIER SERIES*

In the solution of the rectangular membrane problem we encountered the eigenvalue problem

$$\begin{aligned} \Delta\varphi + \lambda\varphi &= 0 \qquad \text{in } D\,, \\ \varphi &= 0 \qquad \text{on } C\,, \end{aligned} \tag{10.5.1}$$

where D is the rectangle $0 < x < a$, $0 < y < b$. The eigenfunctions form a double sequence

$$\varphi_{m,n}(x,y) = \sin\frac{m\pi x}{a}\sin\frac{n\pi y}{b}\,, \qquad m,n = 1, 2, \ldots$$

and the associated eigenfunction expansions

$$\begin{aligned} f(x,y) &\sim \sum_{m,n=1}^{\infty} A_{m,n}\varphi_{m,n}(x,y)\,, \\ A_{m,n} &= \frac{4}{ab}\iint_D f(x,y)\varphi_{m,n}(x,y)\,dx\,dy \end{aligned} \tag{10.5.2}$$

are called double Fourier series, more precisely, **sine-sine series.** In related problems we encounter **cosine-cosine series, cosine-sine series,** and other kinds of **double Fourier series.**

The **rectangular partial sums** of the double series (10.5.2) are the finite partial sums

$$s_{m,n}(x,y) = \sum_{i=1}^{m}\sum_{j=1}^{n} a_{i,j}\varphi_{i,j}(x,y)\,. \tag{10.5.3}$$

We say that a double sequence $s_{m,n}(x,y)$ converges to the limit $s(x,y)$, in symbols

$$\lim_{\substack{m\to\infty \\ n\to\infty}} s_{m,n}(x,y) = s(x,y) \tag{10.5.4}$$

if, to any positive number ϵ there correspond integers p,q such that

$$|s(x,y) - s_{m,n}(x,y)| < \epsilon \qquad \text{whenever } m \geq p \text{ and } n \geq q\,.$$

If the double sequence $s_{m,n}(x,y)$ of rectangular partial sums (10.5.3) converges to $s(x,y)$, then we say the **double series converges** to $s(x,y)$, and we write

$$s(x,y) = \sum_{m,n=1}^{\infty} A_{m,n}\varphi_{m,n}(x,y)\,. \tag{10.5.5}$$

The double series (10.5.2) is called **absolutely convergent** if

$$\sum_{m,n=1}^{\infty} |A_{m,n}\varphi_{m,n}(x,y)|$$

is convergent. When the double series is absolutley convergent, it is convergent and the two iterated series

$$\sum_{m=1}^{\infty}\sum_{n=1}^{\infty} A_{m,n}\varphi_{m,n}(x,y)\,, \tag{10.5.6}$$

$$\sum_{n=1}^{\infty}\sum_{m=1}^{\infty} A_{m,n}\varphi_{m,n}(x,y)\,, \tag{10.5.7}$$

also converge and have the same sum as that given by (10.5.5). The convergence of **iterated series** such as (10.5.6), (10.5.7) is defined in terms of the convergence of ordinary series. Thus, for example, the series (10.5.6) is convergent if the series

$$p_m(x,y) = \sum_{n=1}^{\infty} A_{m,n}\varphi_{m,n}(x,y)$$

is convergent for each m, $m = 1, 2, \ldots$, and the series

$$\sum_{m=1}^{\infty} p_m(x,y)$$

is also convergent. The sum (10.5.6) is then the sum of this latter series.

The representation of a function $f(x,y)$ in terms of the eigenfunctions $\varphi_{m,n}(x,y)$ by means of an iterated series is easy to establish.

If $f(x,y)$ is continuous and has continuous first-order partial derivatives on the closed region $D + C$, then the series (10.5.6) *and* (10.5.7) *converge to $f(x,y)$ at each point (x,y) in D.*

We shall prove the statement for (10.5.6), which is obviously sufficient. By the pointwise convergence theorem of Chapter 6, for each fixed y, $f(x,y)$ is a function of x whose Fourier sine series on $[0,a]$ converges to the function for $0 < x < a$; that is,

$$f(x,y) = \sum_{m=1}^{\infty} a_m(y) \sin\frac{m\pi x}{a} \qquad 0 < x < a\,, \tag{10.5.8}$$

where

$$a_m(y) = \frac{2}{a}\int_0^a f(x,y)\sin\frac{m\pi x}{a}\,dx \qquad m = 1, 2, \ldots. \tag{10.5.9}$$

The hypothesis on $f(x,y)$ implies that the functions $a_m(y)$ and their first derivatives are continuous on $[0,b]$, so that, again by the pointwise convergence theorem of Chapter 6, for $m = 1, 2, \ldots$

$$a_m(y) = \sum_{n=1}^{\infty} A_{m,n} \sin \frac{n\pi y}{b} \qquad 0 < y < b, \tag{10.5.10}$$

where

$$A_{m,n} = \frac{2}{b}\int_0^b a_m(y) \sin \frac{n\pi y}{b}\, dy. \tag{10.5.11}$$

Substitution of (10.5.10) in (10.5.8) and of (10.5.9) in (10.5.11) completes the proof.

The convergence of the iterated series implies nothing about the convergence of the corresponding double series, the latter being the sense in which the convergence of the eigenfunction expansion (10.5.2) should properly be understood. We will state two theorems which give conditions under which the eigenfunction expansion (10.5.2) converges in this sense to the function $f(x,y)$. Each of these theorems refers to the particular eigenvalue problem (10.5.1), but analogous theorems apply to other eigenvalue problems (see Exercises 10c). (The same statement applies to the preceding theorem on iterated series, as can be easily verified in many cases.) The proofs of the two theorems are not given, but a method of proof for similar theorems regarding double Fourier integrals is discussed in Section 9. See also Exercises 10f, Problem 3. The first theorem is a two-dimensional analogue of the uniform convergence theorem of Chapter 6.

Theorem. *If $f(x,y)$ is continuous and has continuous first- and second-order partial derivatives on the closed region $D + C$, and satisfies the boundary condition of* (10.5.1), *that is, $f = 0$ on C, then the series* (10.5.2) *is absolutely convergent and the rectangular partial sums of the series converge to $f(x,y)$ uniformly on $D + C$.*

The second theorem is concerned with two-dimensional analogues of mean square convergence and the Parseval equation.

Theorem. *If $f(x,y)$ is continuous on the closed rectangle $D + C$ then the rectangular partial sums of the series* (10.5.2) *converge in mean square on D to $f(x,y)$, that is,*

$$\lim_{\substack{m\to\infty \\ n\to\infty}} \iint_D [f(x,y) - s_{m,n}(x,y)]^2\, dx\, dy = 0, \tag{10.5.12}$$

and the Parseval equation

$$\iint_D f^2(x,y)\,dx\,dy = \frac{ab}{4}\sum_{m,n=1}^{\infty} A_{m,n}^2 \tag{10.5.13}$$

is valid.

Let $g(x,y)$ be another function continuous on $D + C$, with eigenfunction expansion

$$g(x,y) \sim \sum_{m,n=1}^{\infty} B_{m,n}\varphi_{m,n}(x,y)\,. \tag{10.5.14}$$

Then the bilinear version of Parseval's equation, namely

$$\iint_D f(x,y)g(x,y)\,dx\,dy = \frac{ab}{4}\sum_{m,n=1}^{\infty} A_{m,n}B_{m,n} \tag{10.5.15}$$

can be deduced from (10.5.13). This deduction is made, just as in the one-dimensional case in Chapter 6, by applying (10.5.13) to the two functions $f + g, f - g$ and using the identity

$$fg = \tfrac{1}{4}(f+g)^2 - \tfrac{1}{4}(f-g)^2\,.$$

We will use the Parseval equation to prove that *every eigenvalue of problem* (10.3.5) *is contained in the double sequence* (10.3.11). That is, all eigenvalues of problem (10.3.5) are obtained by the method of separation of the variables. We will also show that all the eigenfunctions of problem (10.3.5) were found by the method of separation of the variables; more precisely, *every eigenfunction of* (10.3.5) *is either a constant multiple of one of the eigenfunctions* $\varphi_{m,n}(x,y)$ *in the double sequence* (10.3.12), *or it is a linear combination of a finite number of the eigenfunctions* $\varphi_{m,n}(x,y)$ *belonging to the same eigenvalue.*

We assume that any eigenfunction is continuous and has continuous first- and second-order partial derivatives in the closed region $D + C$. Let λ be an eigenvalue. Then λ is real (Exercises 10b, Problem 1), and there will be a corresponding real-valued eigenfunction $\psi(x,y)$. We write the expansion of $\psi(x,y)$ in terms of the eigenfunctions (10.3.12),

$$\psi(x,y) = \sum_{m,n=1}^{\infty} A_{m,n}\varphi_{m,n}(x,y)\,. \tag{10.5.16}$$

The corresponding Parseval equation is

$$\iint_D \psi^2(x,y)\,dx\,dy = \frac{ab}{4}\sum_{m,n=1}^{\infty} A_{m,n}^2\,, \tag{10.5.17}$$

where

$$A_{m,n} = \frac{4}{ab} \iint_D \psi(x,y)\varphi_{m,n}(x,y)\, dx\, dy\,. \tag{10.5.18}$$

Since $\psi(x,y)$ is an eigenfunction, it is not identically zero and so the integral in (10.5.17) is not zero. Therefore, at least one term of the series in (10.5.17) is not zero; that is, $A_{m,n} \neq 0$ for some choice of m and n. With this choice of m and n we use (10.5.18) and Green's formula to obtain

$$\begin{aligned}(\lambda - \lambda_{m,n})A_{m,n} &= \frac{4}{ab} \iint_D [\psi\Delta\varphi_{m,n} - (\Delta\psi)\varphi_{m,n}]\, dx\, dy \\ &= \frac{4}{ab} \int_C \left[\psi \frac{\partial\varphi_{m,n}}{\partial\nu} - \frac{\partial\psi}{\partial\nu}\varphi_{m,n}\right] ds \\ &= 0\,,\end{aligned}$$

since both ψ and $\varphi_{m,n}$ are zero on C. Now we have

$$A_{m,n} \neq 0\,, \qquad (\lambda - \lambda_{m,n})A_{m,n} = 0$$

so $\lambda - \lambda_{m,n} = 0$; that is, $\lambda = \lambda_{m,n}$. This shows that the arbitrary eigenvalue λ is equal to one of those in the sequence (10.3.11), and our first proposition is proved.

Now consider the series (10.5.16). If only one of the coefficients is different from zero, say $A_{m,n} \neq 0$, then

$$\psi(x,y) = A_{m,n}\varphi_{m,n}(x,y)\,,$$

and ψ is a constant multiple of $\varphi_{m,n}$. Suppose that more than one coefficient is different from zero, say $A_{m,n} \neq 0$ and $A_{p,q} \neq 0$. Then, as we have shown above, $\lambda_{m,n} = \lambda$ and similarly $\lambda_{p,q} = \lambda$ so that $\lambda_{m,n} = \lambda_{p,q}$. Thus in the series (10.5.16) every nonzero term is associated with the eigenvalue λ. Now we saw, at the end of Section 3, that for a given value λ there are only a finite number of choices of the integers m and n such that $\lambda_{m,n} = \lambda$. Therefore the series (10.5.16) can have only a finite number of nonzero terms. Each of the nonzero terms is a multiple of an eigenfunction $\varphi_{m,n}(x,y)$ belonging to the fixed eigenvalue $\lambda_{m,n} = \lambda$. In other words the series (10.5.16) shows that any given eigenfunction $\psi(x,y)$ is a finite linear combination of eigenfunctions $\varphi_{m,n}(x,y)$ all belonging to the same eigenvalue.

EXERCISES 10c

Let D be the domain defined by the inequalities $0 < x < y < a$. The boundary C of D is an isosceles right-angled triangle. We shall consider the eigenvalue problems

(I) D.E. $\Delta\psi + \lambda\psi = 0$ in D,
B.C. $\psi = 0$ on C

and

(II) D.E. $\Delta\psi + \lambda\psi = 0$ in D,
B.C. $\dfrac{\partial\psi}{\partial\nu} = 0$ on C.

Let D' be the square $0 < x < a$, $0 < y < a$, and let C' be the boundary of D'. A function $f(x,y)$ defined in D' is called *symmetric* if

$$f(y,x) = f(x,y)$$

and is called *skew symmetric* if

$$f(y,x) = -f(x,y)\,.$$

1. Along the diagonal line where $y = x$ a skew symmetric function satisfies $f = 0$. Show that a differentiable symmetric function satisfies $\partial f/\partial\nu = 0$ on this line. Deduce that if φ is a solution of $\Delta\varphi + \lambda\varphi = 0$ in D', $\varphi = 0$ on C', then

$$\psi(x,y) = \varphi(x,y) - \varphi(y,x)$$

is a (possibly trivial) solution of I; and if φ is a solution of $\Delta\varphi + \lambda\varphi = 0$ in D', $\partial\varphi/\partial\nu = 0$ on C', then

$$\psi(x,y) = \varphi(x,y) + \varphi(y,x)$$

is a solution of II.

2. From the eigenfunctions of the square membrane

$$\varphi_{m,n}(x,y) = \sin\frac{m\pi x}{a}\sin\frac{n\pi y}{a}\,, \qquad m,n = 1, 2, \ldots$$

we obtain eigenfunctions of I,

$$\psi_{m,n}(x,y) = \varphi_{m,n}(x,y) - \varphi_{n,m}(x,y) = \begin{vmatrix} \sin\dfrac{m\pi x}{a} & \sin\dfrac{m\pi y}{a} \\ \sin\dfrac{n\pi x}{a} & \sin\dfrac{n\pi y}{a} \end{vmatrix},$$

with eigenvalues

$$\lambda_{m,n} = \left(\frac{\pi}{a}\right)^2 (m^2 + n^2)\,, \qquad 1 \le m < n < \infty\,.$$

The functions $\psi_{m,n}$ are zero if $m = n$ and those with $n < m$ are omitted because

$$\psi_{n,m} = -\psi_{m,n}\,.$$

Show that these eigenfunctions form an orthogonal system on D and evaluate

$$\iint\limits_D \psi_{m,n}^2\,dx\,dy\,.$$

3. Let $f(x,y)$ be continuous on $D' + C'$ and skew symmetric. Let the eigenfunction expansions of f with respect to the systems $\{\varphi_{m,n}\}$, $\{\psi_{m,n}\}$ be respectively the series

$$f(x,y) \sim \sum_{m=1}^{\infty} \sum_{n=1}^{\infty} A_{m,n} \varphi_{m,n}(x,y) \,,$$

$$f(x,y) \sim \sum_{n=2}^{\infty} \sum_{m=1}^{n-1} B_{m,n} \psi_{m,n}(x,y) \,.$$

Show that the Parseval equation for the second series can be deduced from the Parseval equation for the first series.

4. Find the eigenvalues and eigenfunctions of Problem II and establish the Parseval equation for the problem.

5. State a problem whose eigenvalues and eigenfunctions are

$$\lambda_{l,m,n} = l^2 + m^2 + n^2 \,, \qquad 1 \le l < m < n \,,$$

$$\psi_{l,m,n}(x,y,z) = \begin{vmatrix} \sin lx & \sin ly & \sin lz \\ \sin mx & \sin my & \sin mz \\ \sin nx & \sin ny & \sin nz \end{vmatrix} ,$$

and evaluate

$$\iiint_D \psi_{l,m,n}^2 \, dx \, dy \, dz$$

where D is the relevant domain.

10.6. *INHOMOGENEOUS PROBLEMS*

The methods of Chapter 5 for solving initial-boundary value problems, in which either the D.E. or the B.C. or both are inhomogeneous, can be adapted to the case when there are two or more space variables. In particular, the method of variation of parameters carries over with only minor changes. On the other hand it becomes more difficult to find equilibrium solutions and asymptotic solutions; the reason for this is evident if we note that an equilibrium solution of the one-dimensional heat equation is a function $u = u(x)$ which satisfies an ordinary differential equation, whereas an equilibrium solution of the two-dimensional heat equation is a function $u = u(x,y)$ which satisfies a partial differential equation.

The solution of equilibrium problems is discussed in general in the next chapter. Here and in the next section we will illustrate the method of variation of parameters by means of examples. As an example of a problem with an inhomogeneous D.E. we consider

$$
(10.6.1) \qquad
\begin{array}{lll}
\text{D.E.} & u_t = k\,\Delta u + q(x,y,t)\,, & 0 < x < a\,, \\
 & & 0 < y < b\,, \quad t > 0\,, \\
\text{B.C.} & u(x,0,t) = 0\,, & 0 < x < a\,, \quad t > 0\,, \\
 & u(x,b,t) = 0\,, & 0 < x < a\,, \quad t > 0\,, \\
 & u(0,y,t) = 0\,, & 0 < y < b\,, \quad t > 0\,, \\
 & u(a,y,t) = 0\,, & 0 < y < b\,, \quad t > 0\,, \\
\text{I.C.} & u(x,y,0) = f(x,y)\,, & 0 < x < a\,, \\
 & & 0 < y < b\,.
\end{array}
$$

Just as in the one-dimensional case, the first step in the method of variation of parameters is to find the eigenvalues and eigenfunctions of the related homogeneous problem. Since, in the case of problem (10.6.1), the related homogeneous problem leads to the eigenvalue problem (10.3.5), the eigenvalues $\lambda_{m,n}$ and eigenfunctions $\varphi_{m,n}(x,y)$ are those given by (10.3.9) and (10.3.10).

If $q(x,y,t)$ is continuous it is reasonable to assume that the solution $u(x,y,t)$, considered as a function of x and y for fixed $t > 0$, will be continuous and have continuous first- and second-order partial derivatives in the closed rectangle. Since $u(x,y,t)$ also satisfies the homogeneous boundary conditions for $t > 0$, it follows from the uniform convergence theorem of Section 5 that $u(x,y,t)$ can be represented by its eigenfunction expansion,

$$
(10.6.2) \qquad u(x,y,t) = \sum_{m,n=1}^{\infty} B_{m,n}(t)\varphi_{m,n}(x,y)\,,
$$

where

$$
(10.6.3) \qquad B_{m,n}(t) = \frac{4}{ab}\iint_D u(x,y,t)\varphi_{m,n}(x,y)\,dx\,dy\,,
$$

and D is the rectangular domain $0 < x < a$, $0 < y < b$. We differentiate (10.6.3) with respect to t and then use the D.E. in (10.6.1) to get

$$
(10.6.4) \qquad \frac{dB_{m,n}}{dt} = \frac{4}{ab}\iint_D \frac{\partial u}{\partial t}\varphi_{m,n}\,dx\,dy
$$

$$
= \frac{4}{ab}\iint_D k(\Delta u)\varphi_{m,n}\,dx\,dy + \frac{4}{ab}\iint_D q\varphi_{m,n}\,dx\,dy\,.
$$

The last integral on the right of (10.6.4) is

$$
(10.6.5) \qquad Q_{m,n}(t) = \frac{4}{ab}\iint_D q(x,y,t)\varphi_{m,n}(x,y)\,dx\,dy\,,
$$

which is a known function of t, since it is determined by (10.6.5) from the data of problem (10.6.1). The first integral on the right can be transformed by Green's formula, and because $\Delta\varphi_{m,n} = -\lambda_{m,n}\varphi_{m,n}$ we obtain

$$(10.6.6)\quad \frac{4}{ab}\iint_D k(\Delta u)\varphi_{m,n}\,dx\,dy = -\lambda_{m,n}k\,\frac{4}{ab}\iint_D u\varphi_{m,n}\,dx\,dy + \frac{4k}{ab}\int_C \left(\frac{\partial u}{\partial \nu}\,\varphi_{m,n} - u\,\frac{\partial \varphi_{m,n}}{\partial \nu}\right) ds\,,$$

where C is the boundary of D.

Since both u and $\varphi_{m,n}$ are zero on C, the integral over C is zero. The first integral on the right of (10.6.6) is equal to $-\lambda_{m,n}kB_{m,n}(t)$, and hence (10.6.4) is equivalent to

$$(10.6.7)\quad \frac{dB_{m,n}}{dt} = -\lambda_{m,n}kB_{m,n} + Q_{m,n}(t)\,.$$

If we let $t \to 0$ in (10.6.3) and use the I.C. in (10.6.1) we obtain

$$(10.6.8)\quad B_{m,n}(0) = A_{m,n}\,,$$

where

$$(10.6.9)\quad A_{m,n} = \frac{4}{ab}\iint_D f(x,y)\varphi_{m,n}(x,y)\,dx\,dy$$

is known. The solution of the differential equation (10.6.7) subject to the initial condition (10.6.8) is

$$(10.6.10)\quad B_{m,n}(t) = A_{m,n}e^{-\lambda_{m,n}kt} + \int_0^t e^{-\lambda_{m,n}k(t-\tau)}Q_{m,n}(\tau)\,d\tau\,.$$

Thus the coefficients $B_{m,n}(t)$ in (10.6.2) are determined and the formal solution of problem (10.6.1) is complete.

10.7. *SPECIAL METHOD FOR INHOMOGENEOUS BOUNDARY CONDITIONS*

In problem (10.6.1) the D.E. is inhomogeneous but the boundary conditions are homogeneous. Because the boundary conditions are homogeneous, it is clear from the beginning that the solution can be represented by a double series of the form (10.6.2). However, if we consider a problem with inhomogeneous boundary conditions, such as the problem

$$
(10.7.1)\qquad
\begin{array}{llll}
\text{D.E.} & u_t = k\,\Delta u & 0 < x < a\,, & \\
 & & 0 < y < b\,, & t > 0\,, \\
\text{B.C.} & u(x,0,t) = g(x)\,, & 0 < x < a\,, & t > 0\,, \\
 & u(x,b,t) = h(x)\,, & 0 < x < a\,, & t > 0\,, \\
 & u(0,y,t) = 0\,, & 0 < y < b\,, & t > 0\,, \\
 & u(a,y,t) = 0\,, & 0 < y < b\,, & t > 0\,, \\
\text{I.C.} & u(x,y,0) = 0\,, & 0 < x < a\,, & \\
 & & 0 < y < b\,, &
\end{array}
$$

then the results of Section 5 ensure only that an eigenfunction expansion of the solution is convergent to the solution as an iterated series. With this understanding, however, the method of variation of parameters is still applicable. Here we shall solve problem (10.7.1) using a modification of the method, which has the advantage that the equilibrium part of the solution is represented by an ordinary infinite series.

For fixed $t > 0$, and fixed y, $0 < y < b$, the solution $u(x,y,t)$ of (10.7.1) is a function of x defined for $0 \le x \le a$. We assume that the first and second derivatives of this function are continuous on $0 \le x \le a$. Since the function is zero at each end of the interval, it follows from the uniform convergence theorem of Chapter 6 that the function is equal to the sum of its Fourier sine series; that is,

$$
(10.7.2)\qquad u(x,y,t) = \sum_{m=1}^{\infty} u_m(y,t) \sin \frac{m\pi x}{a}\,,
$$

where

$$
(10.7.3)\qquad u_m(y,t) = \frac{2}{a}\int_0^a u(x,y,t) \sin \frac{m\pi x}{a}\, dx\,.
$$

Thus the solution of (10.7.1) can be represented by the series (10.7.2) for $t > 0$, $0 < y < b$ and $0 \le x \le a$. The unknown coefficients $u_m(y,t)$ are still to be determined. By differentiations of (10.7.3) we obtain

$$
(10.7.4)\qquad \frac{\partial u_m}{\partial t} = \frac{2}{a}\int_0^a k(u_{xx} + u_{yy})\varphi_m(x)\, dx
$$

where $\varphi_m(x) = \sin(m\pi x/a)$. Since

$$
k\,\frac{2}{a}\int_0^a u_{yy}\varphi_m(x)\, dx = k\,\frac{\partial^2 u_m}{\partial y^2}\,,
$$

and since

$$k\frac{2}{a}\int_0^a u_{xx}\varphi_m(x)\,dx = k\frac{2}{a}(u_x\varphi_m - u\varphi_m')\Big|_0^a + k\frac{2}{a}\int_0^a u\varphi_m''(x)\,dx$$

$$= -\left(\frac{m\pi}{a}\right)^2 k\frac{2}{a}\int_0^a u\varphi_m(x)\,dx$$

$$= -\left(\frac{m\pi}{a}\right)^2 ku_m\,,$$

we see that Equation (10.7.4) is equivalent to

$$\frac{\partial u_m}{\partial t} = k\left[\frac{\partial^2 u_m}{\partial y^2} - \left(\frac{m\pi}{a}\right)^2 u_m\right]. \tag{10.7.5}$$

If we let $y \to 0$ in (10.7.3) we obtain from the B.C. of (10.7.1)

$$u_m(0,t) = \frac{2}{a}\int_0^a g(x)\sin\frac{m\pi x}{a}\,dx = g_m\,, \tag{10.7.6}$$

and similarly, by letting $y \to b$ in (10.7.3) we get

$$u_m(b,t) = \frac{2}{a}\int_0^a h(x)\sin\frac{m\pi x}{a}\,dx = h_m\,, \tag{10.7.7}$$

where g_m and h_m are known constants. Finally, if we let $t \to 0$ in Equation (10.7.3) and use the I.C. of (10.7.1) we obtain

$$u_m(y,0) = 0\,. \tag{10.7.8}$$

By collecting the results (10.7.5) to (10.7.8) we get, for the determination of $u_m(y,t)$, the initial-boundary value problem

$$\begin{array}{lll} \text{D.E.} & \dfrac{\partial u_m}{\partial t} = k\left[\dfrac{\partial^2 u_m}{\partial y^2} - \left(\dfrac{m\pi}{a}\right)^2 u_m\right], & 0 < y < b\,,\ t > 0\,, \\ \text{B.C.} & u_m(0,t) = g_m,\ u_m(b,t) = h_m, & t > 0\,, \\ \text{I.C.} & u_m(y,0) = 0\,, & 0 < y < b\,. \end{array} \tag{10.7.9}$$

This problem can be solved by the methods of Chapter 5. We write the solution as the sum of an equilibrium solution $w_m(y)$ which satisfies the boundary conditions, and a transient term $v_m(y,t)$. That is,

$$u_m(y,t) = w_m(y) + v_m(y,t)\,, \tag{10.7.10}$$

where w_m and v_m satisfy

$$\begin{array}{lll} \text{D.E.} & \dfrac{d^2 w_m}{dy^2} - \left(\dfrac{m\pi}{a}\right)^2 w_m = 0\,, & 0 < y < b\,, \\ \text{B.C.} & w_m(0) = g_m\,,\ w_m(b) = h_m\,, & \end{array} \tag{10.7.11}$$

and

$$\text{(10.7.12)}\qquad \begin{array}{lll} \text{D.E.} & \dfrac{\partial v_m}{\partial t} = k\left[\dfrac{\partial^2 v_m}{\partial y^2} - \left(\dfrac{m\pi}{a}\right)^2 v_m\right], & 0 < y < b\,,\ t > 0\,, \\ \text{B.C.} & v_m(0,t) = 0\,,\ v_m(b,t) = 0\,, & t > 0\,, \\ \text{I.C.} & v_m(y,0) = -w_m(y)\,, & 0 < y < b\,. \end{array}$$

Problem (10.7.11), which involves only an ordinary differential equation, is easily solved to give

$$\text{(10.7.13)}\qquad w_m(y) = g_m \frac{\sinh \dfrac{m\pi(b-y)}{a}}{\sinh \dfrac{m\pi b}{a}} + h_m \frac{\sinh \dfrac{m\pi y}{a}}{\sinh \dfrac{m\pi b}{a}}\,.$$

Problem (10.7.12) is similar to problems solved in Chapter 4, and for its solution we find

$$\text{(10.7.14)}\qquad v_m(y,t) = \sum_{n=1}^{\infty} C_{m,n} e^{-\lambda_{m,n} kt} \sin \frac{n\pi y}{b}\,,$$

where $\lambda_{m,n}$ is given by (10.3.9) and where

$$\text{(10.7.15)}\qquad C_{m,n} = -\frac{2}{b}\int_0^b w_m(y) \sin \frac{n\pi y}{b}\, dy\,.$$

The integrals (10.7.15) can be evaluated explicitly, either by using formula (10.7.13), or more easily (see Exercises 10d, Problem 5) by using (10.7.11) and Green's formula. The result is

$$\text{(10.7.16)}\qquad C_{m,n} = \frac{2}{b\lambda_{m,n}}\left(\frac{n\pi}{b}\right)[g_m - (-1)^n h_m]\,.$$

Finally, when all these results are collected, we obtain the solution of problem (10.7.1),

(10.7.17)

$$u(x,y,t) = \sum_{m=1}^{\infty} w_m(y) \sin \frac{m\pi x}{a} + \sum_{m=1}^{\infty} v_m(y,t) \sin \frac{m\pi x}{a}$$

$$= \sum_{m=1}^{\infty} g_m \frac{\sinh \dfrac{m\pi(b-y)}{a}}{\sinh \dfrac{m\pi b}{a}} \sin \frac{m\pi x}{a} + \sum_{m=1}^{\infty} h_m \frac{\sinh \dfrac{m\pi y}{a}}{\sinh \dfrac{m\pi b}{a}} \sin \frac{m\pi x}{a}$$

$$+ \sum_{m=1}^{\infty}\sum_{n=1}^{\infty} \frac{2}{b\lambda_{m,n}}\left(\frac{n\pi}{b}\right) [g_m - (-1)^n h_m]\, e^{-\lambda_{m,n} kt} \sin \frac{m\pi x}{a} \sin \frac{n\pi y}{b}\,.$$

EXERCISES 10d

1. Solve

D.E.	$u_{tt} = c^2\Delta u + q(x,y,t)$,	$0 < x < a$, $0 < y < b$, $t > 0$,
B.C.	$u(x,0,t) = u_y(x,b,t) = 0$,	$0 < x < a$, $t > 0$,
	$u(0,y,t) = u_x(a,y,t) = 0$,	$0 < y < b$, $t > 0$,
I.C.	$u(x,y,0) = f(x,y)$,	$0 < x < a$, $0 < y < b$,
	$u_t(x,y,0) = g(x,y)$,	$0 < x < a$, $0 < y < b$.

In Problems 2, 3, and 4, D is the rectangle $0 < x < a$, $0 < y < b$, and C is the boundary of D.

2. (a) Find the solution of

D.E.	$u_t = k\Delta u + Q(x,y)$	in D, $t > 0$,
B.C.	$u = 0$	on C,
I.C.	$u(x,y,0) = f(x,y)$	in D.

(b) Find $\lim_{t\to\infty} u(x,y,t) = v(x,y)$ from your answer in part (a) and state the equilibrium problem which has $v(x,y)$ as its solution. How could you find the formula for $v(x,y)$ directly, without first solving part (a)? (Write a two-line explanation.)

3. Solve

D.E.	$\Delta u = -Q(x,y)$	in D,
B.C.	$\dfrac{\partial u}{\partial \nu} = 0$	on C,

assuming that

$$\iint_D Q(x,y)\,dx\,dy = 0 .$$

Explain, in terms of heat conduction, why this extra assumption is necessary for the problem to have a solution. At what stage in your computations was the assumption needed?

4. Solve (find all constants)

D.E.	$u_t = \Delta u + 2u_x$,	(x,y) in D,	$t > 0$,
B.C.	$u(x,0,t) = x$,	$0 < x < a$,	$t > 0$,
	$u(x,y,t) = 0$,	elsewhere on C,	$t > 0$,
I.C.	$u(x,y,0) = 0$,	in D.	

5. Let $\psi(y)$ be the solution of

D.E.	$\psi'' - \alpha^2\psi = 0$,	$0 < y < b$
B.C.	$\psi(0) = 1$,	$\psi(b) = 0$.

(a) Use Green's formula to evaluate the constants c_n in the series

$$\psi(y) \sim \sum_{n=1}^{\infty} c_n \sin \frac{n\pi y}{b} .$$

(b) For what values of y in $0 \leq y \leq b$ does the series in part (a) converge to $\psi(y)$?

(c) Deduce formula (10.7.16).

6. State the equilibrium problem satisfied by the function

$$w(x,y) = \sum_{m=1}^{\infty} w_m(y) \sin \frac{m\pi x}{a}$$

which appears in (10.7.17).

7. (a) Solve

$$\begin{array}{lll} \text{D.E.} & v_t = k\Delta v\,, & 0 < x < a\,,\ 0 < y < b\,,\ t > 0\,, \\ \text{B.C.} & v_y(x,0,t) = g(x)\,, & 0 < x < a\,,\ t > 0\,, \\ & v_y(x,b,t) = 0\,, & 0 < x < a\,,\ t > 0\,, \\ & v_x(0,y,t) = v_x(a,y,t) = 0\,, & 0 < y < b\,,\ t > 0\,, \\ \text{I.C.} & v(x,y,0) = 0\,, & 0 < x < a\,,\ 0 < y < b\,. \end{array}$$

(b) Find a constant A such that

$$\lim_{t\to\infty} [v(x,y,t) - At] = w(x,y)$$

exists, and find $w(x,y)$.

(c) State the equilibrium problem satisfied by $w(x,y)$.

8. Solve problem (10.7.1) with $g(x)$ and $h(x)$ replaced by $g(x,t)$ and $h(x,t)$.

9. Solve

$$\begin{array}{lll} \text{D.E.} & u_t = k\Delta u\,, & 0 < x < a\,,\ 0 < y < b\,,\ t > 0\,, \\ \text{B.C.} & u(x,0,t) = g(x)\,, & 0 < x < a\,,\ t > 0\,, \\ & u(x,b,t) = 0\,, & 0 < x < a\,,\ t > 0\,, \\ & u(0,y,t) = h(y)\,, & 0 < y < b\,,\ t > 0\,, \\ & u(a,y,t) = 0\,, & 0 < y < b\,,\ t > 0\,, \\ \text{I.C.} & u(x,y,0) = 0\,, & 0 < x < a\,,\ 0 < y < b\,. \end{array}$$

10.8. *PROBLEMS WITH INFINITE DOMAINS*

We have seen that it is easier to study the propagation of disturbances on an infinite string than on a finite string; similarly, heat conduction in an infinite rod has a simpler description than that in a finite rod. A problem involving an infinite string or infinite rod can usually be interpreted as a limiting case of a problem for a very long but finite string or rod. The gain in simplicity in the limiting case may be attributed to the elimination of end effects.

In problems with more than one space variable, we can expect a similar gain in simplicity for infinite domains as compared to finite domains, due to the elimination of boundary effects. In this section we will discuss some examples of problems in infinite domains, namely those in which the domain is the limiting case of a rectangular domain, when one or both of the dimensions of the rectangle become infinite. We expect that, in analogy with the results of Chapter 9, in the passage to the limit the representations of functions by eigenfunction expansions appropriate for various problems in finite rectangles will pass into integral representation formulas. This is indeed the case, but instead of imitating the discussion in Chapter 9, we choose the alternative of discussing at the outset the initial-boundary value problems in infinite domains directly. In the course of this discussion we will see that appropriate integral representation formulas for functions of several variables can be obtained formally from those for functions of a single variable. In the next section we will discuss independently these integral representation formulas and their validity.

As our first example, we consider the limiting form of problem (10.7.1) as $b \to \infty$ with a fixed. Although this is not a particularly typical example, it has the merit that the discussion of the limiting form very closely parallels that of the finite form of the problem. As $b \to \infty$ the rectangle of problem (10.7.1) passes into an infinite domain D, namely the semi-infinite strip $0 < x < a$, $0 < y < \infty$, whose boundary C consists of the segment $y = 0$, $0 < x < a$, and the two half-lines $x = 0$, $0 < y < \infty$ and $x = a$, $0 < y < \infty$. The limiting form of the problem is

$$
(10.8.1)\qquad
\begin{array}{llll}
\text{D.E.} & u_t = k\,\Delta u\,, & 0 < x < a\,, & 0 < y < \infty\,, \\
 & & & t > 0\,, \\
\text{B.C.} & u(x,0,t) = g(x)\,, & 0 < x < a\,, & t > 0\,, \\
 & u(0,y,t) = 0\,, & 0 < y < \infty\,, & t > 0\,, \\
 & u(a,y,t) = 0\,, & 0 < y < \infty\,, & t > 0\,, \\
\text{I.C.} & u(x,y,0) = 0\,, & 0 < x < a\,, & 0 < y < \infty\,.
\end{array}
$$

The solution of problem (10.7.1) can be interpreted as the temperature distribution in a rectangular plate, initially at temperature zero, when the edges $x = 0$, $x = a$ are at temperature zero and the edges $y = 0$, $y = b$ are at temperatures $g(x)$, $h(x)$, respectively. Suppose for the moment that $g(x)$ is a positive function. Then heat will enter the plate through the edge $y = 0$, flow through the plate, and leave at the cooler edges $x = 0$, $x = a$, and possibly at $y = b$. It seems plausible from physical considerations that if b is very large compared to a, then the temperature in the part of the plate near $y = 0$ will be almost independent of the temperature distribution $h(x)$ along the distant boundary $y = b$, and that most of the heat lost by the plate will result from flow through the edges $x = 0$ and $x = a$.

In our solution of (10.8.1) we shall find it necessary to impose conditions on $u(x,y,t)$ as $y \to \infty$. The considerations above suggest that the conditions $u(x,y,t) \to 0$ and $u_y(x,y,t) \to 0$ as $y \to \infty$ are reasonable if we seek the solution of (10.8.1) which approximates the solution of (10.7.1) for b very large compared to a.

To solve (10.8.1) we proceed initially just as we did to solve (10.7.1). For every fixed $t > 0$ and fixed $y > 0$, the solution $u(x,y,t)$ of (10.8.1) is a function of x, with continuous first and second derivatives on $0 \leq x \leq a$, which vanishes at $x = 0$ and $x = a$. It follows that

$$u(x,y,t) = \sum_{m=1}^{\infty} u_m(y,t) \sin \frac{m\pi x}{a}, \tag{10.8.2}$$

where

$$u_m(y,t) = \frac{2}{a} \int_0^a u(x,y,t) \sin \frac{m\pi x}{a}\, dx. \tag{10.8.3}$$

We find that $u_m(y,t)$ is the solution of the initial-boundary value problem

$$\begin{array}{llr} \text{D.E.} & \dfrac{\partial u_m}{\partial t} = k\left[\dfrac{\partial^2 u_m}{\partial y^2} - \left(\dfrac{m\pi}{a}\right)^2 u_m\right], & 0 < y < \infty,\ t > 0, \\ \text{B.C.} & u_m(0,t) = g_m, & t > 0, \\ \text{I.C.} & u_m(y,0) = 0, & 0 < y < \infty, \end{array} \tag{10.8.4}$$

where

$$g_m = \frac{2}{a} \int_0^a g(x) \sin \frac{m\pi x}{a}\, dx. \tag{10.8.5}$$

To solve (10.8.4) we write the solution as the sum of an equilibrium solution $w_m(y)$ which satisfies the boundary condition and a transient term $v_m(y,t)$; that is,

$$u_m(y,t) = w_m(y) + v_m(y,t), \tag{10.8.6}$$

where w_m and v_m satisfy

$$\begin{array}{llr} \text{D.E.} & \dfrac{d^2 w_m}{dy^2} - \left(\dfrac{m\pi}{a}\right)^2 w_m = 0, & 0 < y < \infty, \\ \text{B.C.} & w_m(0) = g_m, & \end{array} \tag{10.8.7}$$

and

$$\begin{array}{llr} \text{D.E.} & \dfrac{\partial v_m}{\partial t} = k\left[\dfrac{\partial^2 v_m}{\partial y^2} - \left(\dfrac{m\pi}{a}\right)^2\right], & 0 < y < \infty,\ t > 0, \\ \text{B.C.} & v_m(0,t) = 0, & t > 0, \\ \text{I.C.} & v_m(y,0) = -w_m(y), & 0 < y < \infty. \end{array} \tag{10.8.8}$$

Up to this point our discussion has been essentially identical with that of Section 7. Now, however, the fact that problems (10.8.7) and (10.8.8), which are the correspondents of (10.7.11) and (10.7.12) respectively, are problems on semi-infinite intervals is responsible for two significant differences. First, the solution of (10.8.7) is not unique, as was that of (10.7.11); second, in solving (10.8.8) we must employ the Fourier sine integral formula instead of the Fourier sine series formula which was used for (10.7.12).

Although the solution of (10.8.7) is not unique, there is only one solution which has the property $w_m(y) \to 0$ as $y \to \infty$, namely the solution

$$w_m(y) = g_m e^{-m\pi y/a} . \tag{10.8.9}$$

Our assumption that $u(x,y,t) \to 0$ as $y \to \infty$ leads us to choose this solution. With this choice of $w_m(y)$, which appears as part of the data in problem (10.8.7), it is plausible that $v_m(y,t) \to 0$ as $y \to \infty$ and even that $v_m(y,t)$ can be represented by its Fourier sine integral,

$$v_m(y,t) = \frac{2}{\pi} \int_0^\infty V_m(\lambda,t) \sin \lambda y \, d\lambda , \tag{10.8.10}$$

where

$$V_m(\lambda,t) = \int_0^\infty v_m(y,t) \sin \lambda y \, dy . \tag{10.8.11}$$

If we assume that $v_m \to 0$ and $\partial v_m/\partial y \to 0$ as $y \to \infty$, corresponding to the assumptions $u(x,y,t) \to 0$, $u_y(x,y,t) \to 0$ as $y \to \infty$, made above, then from (10.8.11) we obtain

$$\text{D.E.} \qquad \frac{\partial V_m}{\partial t} = -\left[\left(\frac{m\pi}{a}\right)^2 + \lambda^2\right] k V_m ,$$

$$\text{I.C.} \qquad V_m(\lambda,0) = \int_0^\infty - w_m(y) \sin \lambda y \, dy = -g_m \frac{\lambda}{\left(\frac{m\pi}{a}\right)^2 + \lambda^2} . \tag{10.8.12}$$

The solution of (10.8.12) is

$$V_m(\lambda,t) = -g_m \frac{\lambda}{\left(\frac{m\pi}{a}\right)^2 + \lambda^2} e^{-[(m\pi/a)^2 + \lambda^2]kt} \tag{10.8.13}$$

From (10.8.2), (10.8.5), (10.8.6), (10.8.9), (10.8.10) and (10.8.13) we finally have, for the solution of problem (10.8.1),

(10.8.14)

$$u(x,y,t) = \sum_{m=1}^{\infty} g_m e^{-(m\pi y/a)} \sin \frac{m\pi x}{a}$$

$$- \frac{2}{\pi} \sum_{m=1}^{\infty} \int_0^{\infty} g_m \frac{\lambda}{\left(\frac{m\pi}{a}\right)^2 + \lambda^2} e^{-[(m\pi/a)^2+\lambda^2]kt} \sin \frac{m\pi x}{a} \sin \lambda y \, d\lambda \, .$$

It is interesting to compare (10.8.14), the solution of (10.8.1), with (10.7.17), the solution of (10.7.1). As we expected, the solution of the problem in the infinite domain is analytically simpler than the solution of the problem in the finite domain.

We next consider the more typical problem

$$\text{(10.8.15)} \quad \begin{array}{lll} \text{D.E.} & u_{tt} = c^2 \Delta u & \text{in } D\,, \quad t > 0\,, \\ \text{B.C.} & u = 0 & \text{on } C\,, \quad t > 0\,, \\ \text{I.C.} & u(x,y,0) = f(x,y)\,, & \\ & u_t(x,y,0) = g(x,y) & \text{in } D\,, \end{array}$$

where the domain D is the quarter-plane $0 < x < \infty$, $0 < y < \infty$, and the boundary C of D consists of the two half-lines $0 \leq x < \infty$, $y = 0$, and $x = 0$, $0 \leq y < \infty$. In a formal way, problem (10.8.15) is the limiting case of problem (10.3.1) when both $a \to \infty$ and $b \to \infty$.

The solution $u(x,y,t)$ of (10.8.15) can be interpreted as the displacement function of an infinite membrane, subject to an initial disturbance. If the initial disturbance is confined to a finite region near the origin, that is if $f(x,y)$ and $g(x,y)$ are zero except near the origin, then, since it can be shown as in Section 8.7 that disturbances on the membrane travel with finite velocity c, the solution u will, at each time $t > 0$, also be zero outside of a finite region. To determine a solution of (10.8.15) we will assume that u and its first- and second-order partial derivatives are continuous in D and that all of these functions are zero outside of some finite region, at each instant $t \geq 0$.

The requirements of the problem (10.8.15) that for fixed $t > 0$ and $y > 0$, the solution $u(x,y,t)$ be a function of x on the semi-infinite interval $x \geq 0$, which satisfies a boundary condition of the first kind at $x = 0$, suggests the use of the Fourier sine transform of u with respect to x. The conditions satisfied by u, stated in the preceding paragraph, ensure that it can be represented by its Fourier sine integral; that is,

$$\text{(10.8.16)} \qquad u(x,y,t) = \frac{2}{\pi} \int_0^{\infty} \Phi(\lambda,y,t) \sin \lambda x \, d\lambda \, ,$$

where

$$\Phi(\lambda,y,t) = \int_0^\infty u(x,y,t) \sin \lambda x \, dx \, . \tag{10.8.17}$$

Because of our assumptions, Φ is a continuous function of y, has continuous first and second derivatives, and is zero at $y = 0$. Moreover Φ is zero for all sufficiently large y since u is zero outside of a finite region. Therefore Φ can be represented by its Fourier sine integral; that is,

$$\Phi(\lambda,y,t) = \frac{2}{\pi} \int_0^\infty U(\lambda,\mu,t) \sin \mu y \, d\mu \, , \tag{10.8.18}$$

where

$$U(\lambda,\mu,t) = \int_0^\infty \Phi(\lambda,y,t) \sin \mu y \, dy \, . \tag{10.8.19}$$

If we substitute (10.8.18) into (10.8.16) we obtain

$$u(x,y,t) = \left(\frac{2}{\pi}\right)^2 \int_0^\infty \int_0^\infty U(\lambda,\mu,t) \sin \lambda x \sin \mu y \, d\mu \, d\lambda. \tag{10.8.20}$$

Similarly if we substitute (10.8.17) into (10.8.19) we obtain

$$U(\lambda,\mu,t) = \int_0^\infty \int_0^\infty u(x,y,t) \sin \lambda x \sin \mu y \, dx \, dy \, . \tag{10.8.21}$$

The function $U(\lambda,\mu,t)$ is called the **double Fourier sine transform** of $u(x,y,t)$. Since u is represented in terms of U by formula (10.8.20), we can solve our problem by finding U.

Let $\varphi(x,y;\lambda,\mu) = \sin \lambda x \sin \mu y$. Then (10.8.21) can be written as

$$U(\lambda,\mu,t) = \iint_D u\varphi \, dx \, dy \, . \tag{10.8.22}$$

By differentiation of (10.8.22) we obtain

$$\frac{\partial^2 U}{\partial t^2} = \iint_D \frac{\partial^2 u}{\partial t^2} \varphi \, dx \, dy = c^2 \iint_D (\Delta u)\varphi \, dx \, dy \, . \tag{10.8.23}$$

Now we would like to apply Green's formula, but this holds only in a finite domain. Hence a limit process is needed. Let D_a denote the part of D which lies inside the circle with radius a and center at the origin. The boundary C_a of D_a consists of the two segments $y = 0$, $0 \leq x \leq a$, and $x = 0$, $0 \leq y \leq a$, and a circular arc. Then, applying Green's formula to D_a,

(10.8.24)

$$\iint_{D_a} (\Delta u)\varphi \, dx \, dy = \iint_{D_a} u(\Delta\varphi) \, dx \, dy + \int_{C_a} \left(\frac{\partial u}{\partial \nu} \varphi - u \frac{\partial \varphi}{\partial \nu}\right) ds \, .$$

Both u and φ are zero on the two line segments of C_a. Moreover, since u and its partial derivatives are zero outside of a finite region, it follows that if a is large enough both u and $\partial u/\partial \nu$ are zero on the circular arc of C_a. Therefore the line integral in (10.8.24) is zero for all large a, and we have

$$\iint\limits_{D_a} (\Delta u)\varphi \, dx \, dy = \iint\limits_{D_a} u(\Delta\varphi) \, dx \, dy \, . \tag{10.8.25}$$

Now we let $a \to \infty$ in (10.8.25) to get

$$\iint\limits_{D} (\Delta u)\varphi \, dx \, dy = \iint\limits_{D} u(\Delta\varphi) \, dx \, dy \, . \tag{10.8.26}$$

We substitute (10.8.26) in (10.8.23) and note that $\Delta\varphi = -(\lambda^2 + \mu^2)\varphi$. The result is

$$\frac{\partial^2 U}{\partial t^2} = -c^2(\lambda^2 + \mu^2)U \, . \tag{10.8.27}$$

From (10.8.21) and the I.C. of (10.8.15), we get

$$U(\lambda,\mu,0) = \iint\limits_{D} f(x,y) \sin \lambda x \sin \mu y \, dx \, dy = F(\lambda,\mu) \, , \tag{10.8.28}$$

and

$$U_t(\lambda,\mu,0) = \iint\limits_{D} g(x,y) \sin \lambda x \sin \mu y \, dx \, dy = G(\lambda,\mu) \, , \tag{10.8.29}$$

where $F(\lambda,\mu)$, $G(\lambda,\mu)$ are the double sine transforms of $f(x,y)$, $g(x,y)$, respectively. The solution of the differential equation (10.8.27) subject to the initial conditions (10.8.28), (10.8.29) is

$$U(\lambda,\mu,t) = F(\lambda,\mu) \cos ct\sqrt{\lambda^2 + \mu^2} + G(\lambda,\mu) \frac{\sin ct\sqrt{\lambda^2 + \mu^2}}{c\sqrt{\lambda^2 + \mu^2}} \, . \tag{10.8.30}$$

Therefore the solution $u(x,y,t)$ of (10.8.15) is

$$\begin{aligned} &\left(\frac{2}{\pi}\right)^2 \int_0^\infty \int_0^\infty F(\lambda,\mu) \cos ct\sqrt{\lambda^2 + \mu^2} \sin \lambda x \sin \mu y \, d\lambda \, d\mu \\ &+ \left(\frac{2}{\pi}\right)^2 \int_0^\infty \int_0^\infty G(\lambda,\mu) \frac{\sin ct\sqrt{\lambda^2 + \mu^2}}{c\sqrt{\lambda^2 + \mu^2}} \sin \lambda x \sin \mu y \, d\lambda \, d\mu \, . \end{aligned} \tag{10.8.31}$$

The formula (10.8.31), with (10.8.28) and (10.8.29), is analogous to (10.3.15), with (10.3.18) and (10.3.19), and is formally similar.

As our last and again typical example in this section we will solve the initial value problem for the heat equation in the plane. Thus D is the domain $-\infty < x < \infty$, $-\infty < y < \infty$, and our problem is

$$\text{(10.8.32)} \quad \begin{array}{lll} \text{D.E.} & u_t = k\,\Delta u & \text{in } D\,, \quad t > 0\,, \\ \text{I.C.} & u(x,y,0) = f(x,y) & \text{in } D\,. \end{array}$$

Problem (10.8.32) may be regarded as the limiting case of the initial-boundary value problem for the heat equation in the square $-L \leq x \leq L$, $-L \leq y \leq L$, as $L \to \infty$. This view suggests the use of the Fourier transform in solving the problem.

Since heat travels with infinite velocity we cannot now assume that $u(x,y,t)$ is zero outside of some finite region, even if $f(x,y)$ has that property. However it is plausible that $u(x,y,t)$ is very small at large distances from the origin. (Compare Section 9.7.)

Explicitly, we suppose that for any fixed y and t, $u(x,y,t)$ is a function of x which can be represented by its Fourier integral; that is,

$$\text{(10.8.33)} \qquad u(x,y,t) = \frac{1}{2\pi} \int_{-\infty}^{\infty} \Phi(\lambda,y,t) e^{-i\lambda x}\, d\lambda\,,$$

where

$$\text{(10.8.34)} \qquad \Phi(\lambda,y,t) = \int_{-\infty}^{\infty} u(x,y,t) e^{i\lambda x}\, dx\,.$$

Next we suppose that as a function of y, $\Phi(\lambda,y,t)$ can be represented by its Fourier integral,

$$\text{(10.8.35)} \qquad \Phi(\lambda,y,t) = \frac{1}{2\pi} \int_{-\infty}^{\infty} U(\lambda,\mu,t) e^{-i\mu y}\, d\mu\,,$$

where

$$\text{(10.8.36)} \qquad U(\lambda,\mu,t) = \int_{-\infty}^{\infty} \Phi(\lambda,\mu,t) e^{i\mu y}\, dy\,.$$

By substituting (10.8.34) in (10.8.36), and (10.8.35) in (10.8.33) we obtain the formulas

$$\text{(10.8.37)} \quad U(\lambda,\mu,t) = \int_{-\infty}^{\infty} \int_{-\infty}^{\infty} u(x,y,t) e^{i(\lambda x + \mu y)}\, dx\, dy\,,$$

$$\text{(10.8.38)} \quad u(x,y,t) = \frac{1}{(2\pi)^2} \int_{-\infty}^{\infty} \int_{-\infty}^{\infty} U(\lambda,\mu,t) e^{-i(\lambda x + \mu y)}\, d\mu\, d\lambda\,.$$

The function $U(\lambda,\mu,t)$ defined by (10.8.37) is called the **double Fourier transform** of $u(x,y,t)$. We will discuss double Fourier transforms in general in the next section. For the present, let us assume that the solution of problem (10.8.32) can be represented in terms of its Fourier transform by formula (10.8.38). Then to solve the problem we need to find $U(\lambda,\mu,t)$. To do this we will proceed formally by the method of variation of parameters.

Let $\varphi(x,y;\lambda,\mu) = e^{i(\lambda x+\mu y)}$, and write (10.8.37) in the form

$$U = \iint_D u\varphi \, dx \, dy \, . \tag{10.8.39}$$

By differentiation of (10.8.39) we get

$$\frac{\partial U}{\partial t} = k \iint_D (\Delta u)\varphi \, dx \, dy \, . \tag{10.8.40}$$

To apply Green's formula we consider the finite domain D_a consisting of the interior of the circle $x^2 + y^2 = a^2$. Let C_a be the boundary of D_a. Then

$$\iint_{D_a} (\Delta u)\varphi \, dx \, dy = \iint_{D_a} u(\Delta\varphi) \, dx \, dy + \int_{C_a} \left(\frac{\partial u}{\partial \nu} \varphi - u \frac{\partial \varphi}{\partial \nu} \right) ds \, . \tag{10.8.41}$$

If the initial temperature $f(x,y)$ is zero outside of some finite region, or even if $f(x,y) \to 0$ as $x^2 + y^2 \to \infty$, we can expect that u and its derivatives will also be small at large distances from the origin. We will assume that

$$\lim_{a\to\infty} \int_{C_a} |u| \, ds = 0 \, , \quad \lim_{a\to\infty} \int_{C_a} \left| \frac{\partial u}{\partial \nu} \right| ds = 0 \, . \tag{10.8.42}$$

The boundary integral in (10.8.41) will then have the limiting value zero when $a \to \infty$, and hence by letting $a \to \infty$ in (10.8.41) we obtain

$$\iint_D (\Delta u)\varphi \, dx \, dy = \iint_D u(\Delta\varphi) \, dx \, dy \, . \tag{10.8.43}$$

If we substitute (10.8.43) in (10.8.40) and note that $\Delta\varphi = -(\lambda^2 + \mu^2)\varphi$, the result is

$$\frac{\partial U}{\partial t} = -(\lambda^2 + \mu^2)kU \, . \tag{10.8.44}$$

When we let $t \to 0$ in (10.8.37) and use the I.C. of (10.8.32), we find

$$U(\lambda,\mu,0) = F(\lambda,\mu) \, , \tag{10.8.45}$$

where

$$F(\lambda,\mu) = \int_{-\infty}^{\infty} \int_{-\infty}^{\infty} f(x,y) e^{i(\lambda x+\mu y)} \, dx \, dy \, . \tag{10.8.46}$$

From (10.8.44) and (10.8.45) it follows that

$$U(\lambda,\mu,t) = F(\lambda,\mu) e^{-(\lambda^2+\mu^2)kt} \tag{10.8.47}$$

and hence, from (10.8.38), that the solution of (10.8.32) is

$$u(x,y,t) = \frac{1}{(2\pi)^2}\int_{-\infty}^{\infty}\int_{-\infty}^{\infty} F(\lambda,\mu)e^{-(\lambda^2+\mu^2)kt}e^{-i(\lambda x+\mu y)}\,d\lambda\,d\mu \tag{10.8.48}$$

with $F(\lambda,\mu)$ given by (10.8.46).

EXERCISES 10e

1. Solve

D.E.	$u_t = k\Delta u$,	$0 < x < \infty$,	$0 < y < \infty$,	$t > 0$,
B.C.	$u(x,0,t) = g(x)$,	$0 < x < \infty$,		$t > 0$,
	$u(0,y,t) = 0$,		$0 < y < \infty$,	$t > 0$,
I.C.	$u(x,y,0) = 0$,	$0 < x < \infty$,	$0 < y < \infty$.	

2. Solve for $u(x,y,z,t)$:

D.E.	$u_t = k\Delta u$,	$-\infty < x < \infty$,	$-\infty < y < \infty$,
		$0 < z < \infty$,	$t > 0$,
B.C.	$u(x,y,0,t) = g(x,y)$,	$-\infty < x < \infty$,	$-\infty < y < \infty$,
			$t > 0$,
I.C.	$u(x,y,z,0) = f(x,y,z)$,	$-\infty < x < \infty$,	$-\infty < y < \infty$,
		$0 < z < \infty$.	

3. Solve

D.E.	$u_{tt} = c^2\Delta u$,	$0 < x < \infty$,	$0 < y < \infty$,	$t > 0$,
B.C.	$u_y(x,0,t) = g(x)$,	$0 < x < \infty$,		$t > 0$,
	$u(0,y,t) = h(y)$,		$0 < y < \infty$,	$t > 0$,
I.C.	$u(x,y,0) = 0$,	$0 < x < \infty$,	$0 < y < \infty$,	
	$u_t(x,y,0) = f(x,y)$,	$0 < x < \infty$,	$0 < y < \infty$.	

4. Show that if $u(x,t)$ and $v(x,t)$ are solutions of $u_t = ku_{xx}$ and $v_t = kv_{xx}$ then $w(x,y,t) = u(x,t)v(y,t)$ is a solution of $w_t = k(w_{xx} + w_{yy})$.

10.9. *DOUBLE FOURIER TRANSFORMS*

In Section 8 we were led, by formal repeated use of Fourier integral formulas for functions of one variable, to certain integral representations of functions of two variables which we called double Fourier integral formulas. Here we consider these representations generally.

We define the **double Fourier transform** of a function $f(x,y)$, $-\infty < x < \infty$, $-\infty < y < \infty$, to be the function $F(\lambda,\mu)$ given by

$$F(\lambda,\mu) = \int_{-\infty}^{\infty}\int_{-\infty}^{\infty} f(x,y)e^{i(\lambda x+\mu y)}\,dx\,dy\,. \tag{10.9.1}$$

The **inversion formula,** which expresses $f(x,y)$ in terms of $F(\lambda,\mu)$ is

$$(10.9.2)\qquad f(x,y) = \frac{1}{(2\pi)^2}\int_{-\infty}^{\infty}\int_{-\infty}^{\infty} F(\lambda,\mu)e^{-i(\lambda x+\mu y)}\,d\lambda\,d\mu\,.$$

To assure that the integral (10.9.1) is convergent, and thus that the transform $F(\lambda,\mu)$ is actually defined, we assume that $f(x,y)$ is continuous and absolutely integrable; that is,

$$(10.9.3)\qquad \int_{-\infty}^{\infty}\int_{-\infty}^{\infty} |f(x,y)|\,dx\,dy < \infty\,.$$

With this assumption, the integral (10.9.1) is even uniformly convergent with respect to λ and μ, and $F(\lambda,\mu)$ is continuous and bounded.

Still further restrictions must be imposed on $f(x,y)$ to ensure that (10.9.2) is valid. A sufficient condition for this is that $F(\lambda,\mu)$ also be absolutely integrable. That is, *if* $f(x,y)$ *is continuous and satisfies* (10.9.3) *and if*

$$(10.9.4)\qquad I = \int_{-\infty}^{\infty}\int_{-\infty}^{\infty} |F(\lambda,\mu)|\,d\lambda\,d\mu < \infty$$

then Equation (10.9.2) *holds.* A method of proving this, based on theorems about Fourier transforms of functions of one variable, is indicated in Exercises 10f, Problem 1.

The criterion (10.9.4) for the validity of (10.9.2) is sometimes difficult to verify. Moreover, it is often preferable to have a criterion which is expressed more directly in terms of properties of $f(x,y)$. Just as in the one-dimensional case, it can be shown that (10.9.2) holds if $f(x,y)$ is sufficiently smooth. To find the appropriate condition in the present two-dimensional case, we will investigate the transforms of partial derivatives of $f(x,y)$.

Formal differentiation of (10.9.2) with respect to x leads to

$$(10.9.5)\qquad f_x(x,y) = \frac{1}{(2\pi)^2}\int_{-\infty}^{\infty}\int_{-\infty}^{\infty} -\,i\lambda F(\lambda,\mu)e^{-i(\lambda x+\mu y)}\,d\lambda\,d\mu$$

which suggests the conclusion that the transform of $f_x(x,y)$ is $-i\lambda F(\lambda,\mu)$. The conclusion is correct provided that $f_x(x,y)$ is continuous and absolutely integrable. We will omit the proof, which requires writing down the formula for the transform of $f_x(x,y)$ and using integration by parts. Similar results hold for $f_y(x,y)$ and for higher derivatives. The general rule is the following. If $f(x,y)$ is continuous and has continuous partial derivatives up to the order $m + n$, and if $f(x,y)$ and all of these derivatives are absolutely integrable, then the transform of

$$\frac{\partial^{m+n} f}{\partial x^m\,\partial y^n}$$

is

$$(-i\lambda)^m(-i\mu)^n F(\lambda,\mu) .$$

As a particular consequence, the transform of Δf is $-(\lambda^2 + \mu^2)F(\lambda,\mu)$.

We can now show that condition (10.9.4) holds, and therefore that (10.9.2) is valid, provided $f(x,y)$ has sufficiently many continuous and absolutely integrable derivatives. The integral (10.9.4) can be written as the sum of the integral over the unit disc $\lambda^2 + \mu^2 \leq 1$ and the integral over the remainder of the plane,

$$I = I_1 + I_2 ,$$

where

$$I_1 = \iint\limits_{\lambda^2+\mu^2 \leq 1} |F(\lambda,\mu)| \, d\lambda \, d\mu , \quad I_2 = \iint\limits_{\lambda^2+\mu^2 > 1} |F(\lambda,\mu)| \, d\lambda \, d\mu .$$

Since $F(\lambda,\mu)$ is continuous, I_1 is finite, so that we must show that I_2 is finite.

We first attempt to do this under the assumption that the first- and second-order derivatives of $f(x,y)$ are continuous and absolutely integrable. The function $g(x,y) = f_{xx}(x,y) + f_{yy}(x,y)$ is then a continuous, absolutely integrable function, and its transform is

$$G(\lambda,\mu) = -(\lambda^2 + \mu^2)F(\lambda,\mu) .$$

Hence

$$I_2 = \iint\limits_{\lambda^2+\mu^2 > 1} \frac{|G(\lambda,\mu)|}{\lambda^2 + \mu^2} \, d\lambda \, d\mu . \tag{10.9.6}$$

Since $G(\lambda,\mu)$ is continuous and bounded, there is a constant M such that $|G(\lambda,\mu)| \leq M$. Therefore,

$$I_2 \leq \iint\limits_{\lambda^2+\mu^2 > 1} \frac{M}{\lambda^2 + \mu^2} \, d\lambda \, d\mu , \tag{10.9.7}$$

and we see that I_2 will be finite if the integral

$$J_2 = \iint\limits_{\lambda^2+\mu^2 > 1} \frac{d\lambda \, d\mu}{\lambda^2 + \mu^2} \tag{10.9.8}$$

is convergent. We introduce polar coordinates ρ, φ defined by

$$\begin{array}{ll} \lambda = \rho \cos \varphi & \rho = (\lambda^2 + \mu^2)^{1/2} \\ \mu = \rho \sin \varphi & \varphi = \arctan (\mu/\lambda) \\ 0 \leq \rho < \infty , & -\pi \leq \varphi < \pi . \end{array} \tag{10.9.9}$$

Since the element of area in polar coordinates is $\rho \, d\rho \, d\varphi$ we have

(10.9.10) $$J_2 = \int_{-\pi}^{\pi} d\varphi \int_1^\infty \frac{\rho\, d\rho}{\rho^2} = 2\pi \int_1^\infty \frac{d\rho}{\rho}$$

which is divergent. Thus our first attempt to show that I_2 is finite has failed.

It is easy to see however that *if* $f(x,y)$ *has continuous partial derivatives up to the fourth order, and all of these partial derivatives are absolutely integrable then* I_2 *is finite and hence* (10.9.2) *is valid*. For in this case the function

$$h(x,y) = \Delta\Delta f = f_{xxxx} + 2f_{xxyy} + f_{yyyy}$$

has the Fourier transform

$$H(\lambda,\mu) = (\lambda^2 + \mu^2)^2 F(\lambda,\mu)\ .$$

Therefore

(10.9.11) $$I_2 = \iint\limits_{\lambda^2+\mu^2>1} \frac{|H(\lambda,\mu)|}{(\lambda^2+\mu^2)^2}\, d\lambda\, d\mu$$

and this is convergent by comparison with

(10.9.12) $$J_4 = \iint\limits_{\lambda^2+\mu^2>1} \frac{d\lambda\, d\mu}{(\lambda^2+\mu^2)^2} = 2\pi \int_1^\infty \frac{d\rho}{\rho^3} < \infty\ ,$$

which proves our statement.

If $f(x,y)$ is an odd function of both x and y, that is,

$$f(-x,y) = -f(x,y)$$
$$f(x,-y) = -f(x,y)\ ,$$

then from (10.9.1) we can deduce that $F(\lambda,\mu) = (2i)^2 B(\lambda,\mu)$ where

(10.9.13) $$B(\lambda,\mu) = \int_0^\infty \int_0^\infty f(x,y) \sin \lambda x \sin \mu y\, dx\, dy\ .$$

This function is clearly an odd function of both λ and μ, and hence from (10.9.2) we obtain

(10.9.14) $$f(x,y) = \left(\frac{2}{\pi}\right)^2 \int_0^\infty \int_0^\infty B(\lambda,\mu) \sin \lambda x \sin \mu y\, d\lambda\, d\mu\ .$$

The function $B(\lambda,\mu)$ defined by (10.9.13) is called the **double Fourier sine transform** of $f(x,y)$ and (10.9.14) is the corresponding **inversion formula.**

Two important theorems for Fourier transforms of one variable, namely Parseval's equation and the theorem on convolutions, have two-dimensional analogues. We state and prove these below.

The **Parseval equation** for double Fourier transforms is

$$(10.9.15)\quad \int_{-\infty}^{\infty}\int_{-\infty}^{\infty} f(x,y)\overline{g(x,y)}\,dx\,dy = \frac{1}{(2\pi)^2}\int_{-\infty}^{\infty}\int_{-\infty}^{\infty} F(\lambda,\mu)\overline{G(\lambda,\mu)}\,d\lambda\,d\mu$$

and the special case $g = f$ gives

$$(10.9.16)\quad \int_{-\infty}^{\infty}\int_{-\infty}^{\infty} |f(x,y)|^2\,dx\,dy = \frac{1}{(2\pi)^2}\int_{-\infty}^{\infty}\int_{-\infty}^{\infty} |F(\lambda,\mu)|^2\,d\lambda\,d\mu\,.$$

We will establish (10.9.15) under the assumption that f, g and F are continuous and absolutely integrable. Then $f(x,y)$ is given by (10.9.2). Substitution of (10.9.2) in (10.9.15) yields

$$\begin{aligned}\iint f\bar{g}\,dx\,dy &= \iint \bar{g}\left[\frac{1}{(2\pi)^2}\iint Fe^{-i(\lambda x+\mu y)}\,d\lambda\,d\mu\right]dx\,dy\\ &= \frac{1}{(2\pi)^2}\iint F\left[\iint \overline{ge^{i(\lambda x+\mu y)}}\,dx\,dy\right]d\lambda\,d\mu\\ &= \frac{1}{(2\pi)^2}\iint F\overline{G}\,d\lambda\,d\mu\,,\end{aligned}$$

where the interchange of the order of integration is justified by the fact that both g and F are absolutely integrable.

The **convolution** of two functions $f(x,y)$, $g(x,y)$ is the function

$$(10.9.17)\quad \begin{aligned}h(x,y) &= \int_{-\infty}^{\infty}\int_{-\infty}^{\infty} f(x-\xi,y-\eta)g(\xi,\eta)\,d\xi\,d\eta\\ &= \int_{-\infty}^{\infty}\int_{-\infty}^{\infty} g(x-\xi,y-\eta)f(\xi,\eta)\,d\xi\,d\eta\,.\end{aligned}$$

By methods resembling those used in the one-dimensional case, it can be shown that if f, g, and h are continuous and absolutely integrable, if the Fourier transform of g is absolutely integrable, and if the transform of h is the product of the transforms of f and g, that is,

$$(10.9.18)\qquad H(\lambda,\mu) = F(\lambda,\mu)G(\lambda,\mu)\,,$$

then h is the convolution (10.9.17).

To conclude this section we use the preceding theorem on convolutions to express the solution of problem (10.8.32) in an elegant form. The transform of the solution is given by (10.8.47) as a product

$$(10.9.19)\qquad U(\lambda,\mu,t) = F(\lambda,\mu)e^{-(\lambda^2+\mu^2)kt}\,.$$

Now the function

$$w(x,t) = \frac{1}{\sqrt{4\pi kt}}e^{-x^2/4kt}$$

was shown, in Chapter 9, to have the transform $e^{-\lambda^2kt}$. Therefore, as is easily verified, the function

$$w_2(x,y,t) = w(x,t)w(y,t) = \frac{1}{4\pi kt}e^{-(x^2+y^2)/4kt} \tag{10.9.20}$$

has the transform $e^{-(\lambda^2+\mu^2)kt}$. It follows that (10.9.19) is the transform of the convolution of $f(x,y)$ and $w_2(x,y,t)$; that is, the solution of problem (10.8.32) is

$$u(x,y,t) = \int_{-\infty}^{\infty}\int_{-\infty}^{\infty}\frac{1}{4\pi kt}e^{-[(x-\xi)^2+(y-\eta)^2]/4kt}f(\xi,\eta)\,d\xi\,d\eta\,. \tag{10.9.21}$$

EXERCISES 10f

1. Let $f(x,y)$ be a continuous, absolutely integrable function and suppose its Fourier transform $F(\lambda,\mu)$ is absolutely integrable. We wish to show that (10.9.2) is valid. The integral in (10.9.2) is uniformly convergent because of (10.9.4), so the value of the integral is a continuous function $f_1(x,y)$. We have to show that $f_1(x,y) = f(x,y)$.

   Let

$$g(x;\epsilon) = \begin{cases} 1 - \dfrac{|x|}{\epsilon}, & |x| \le \epsilon \\ 0, & |x| > \epsilon \end{cases}$$

   where ϵ is a positive constant.

   (a) Draw the graph of $g(x;\epsilon)$.

   (b) Show that if $\varphi(x)$ is a real-valued continuous function such that

$$\int_{-\infty}^{\infty}\varphi(x)g(x-a;\epsilon)\,dx = 0$$

   for all $\epsilon > 0$ and all real a, then $\varphi(x) = 0$ for all x.

   (c) Show by a similar argument that if $\varphi(x,y)$ is a real-valued continuous function and

$$\int_{-\infty}^{\infty}\int_{-\infty}^{\infty}\varphi(x,y)g(x-a;\epsilon)g(y-b;\epsilon)\,dx\,dy = 0$$

   for all $\epsilon > 0$ and all real a,b then $\varphi(x,y) = 0$ for all x,y.

   (d) Calculate the Fourier transform $H(\lambda,\mu)$ of $h(x,y) = g(x-a;\epsilon)g(y-b;\epsilon)$. Observe that $H(\lambda,\mu)$ is an absolutely integrable, bounded, and even function of (λ,μ) and use the theory of Chapter 9 to show that

$$h(x,y) = \frac{1}{(2\pi)^2}\int_{-\infty}^{\infty}\int_{-\infty}^{\infty}H(\lambda,\mu)e^{-i(\lambda x+\mu y)}\,d\lambda\,d\mu\,.$$

   (e) Use (d) to show that

$$\int_{-\infty}^{\infty}\int_{-\infty}^{\infty}f_1(x,y)h(x,y)\,dx\,dy = \int_{-\infty}^{\infty}\int_{-\infty}^{\infty}f(x,y)h(x,y)\,dx\,dy\,.$$

   (f) From (e) and (c) conclude that $f_1(x,y) = f(x,y)$.

2. (a) Use the inequality $2ab \le a^2 + b^2$ to show that if both of the series

$$\sum_{m=1}^{\infty} |a_m|^2, \qquad \sum_{m=1}^{\infty} |b_m|^2$$

are convergent, then the series

$$\sum_{m=1}^{\infty} a_m b_m$$

is absolutely convergent.

(b) Let $f(x)$, $g(x)$ be continuous and assume that both of the integrals

$$\int_{-\infty}^{\infty} |f(x)|^2\, dx, \qquad \int_{-\infty}^{\infty} |g(x)|^2\, dx$$

are convergent. Show that

$$\int_{-\infty}^{\infty} f(x)g(x)\, dx$$

is absolutely convergent.

(c) Show that if $Q(\lambda,\mu)$ is continuous and

$$\int_{-\infty}^{\infty} \int_{-\infty}^{\infty} |Q(\lambda,\mu)|^2\, d\lambda\, d\mu < \infty$$

then

$$\int_{-\infty}^{\infty} \int_{-\infty}^{\infty} \frac{Q(\lambda,\mu)}{\lambda^2 + \mu^2 + 1}\, d\lambda\, d\mu$$

is absolutely convergent.

(d) Show by comparison with (10.9.8) and (10.9.12) that

$$\sum_{m=1}^{\infty} \sum_{n=1}^{\infty} \frac{1}{m^2 + n^2}$$

diverges, but if $\sum_{m,n=1}^{\infty} |C_{m,n}|^2 < \infty$ then

$$\sum_{m,n=1}^{\infty} \left| \frac{C_{m,n}}{m^2 + n^2} \right| < \infty .$$

3. Let $f(x,y)$ be real-valued and continuous on the closed square $0 \le x \le \pi$, $0 \le y \le \pi$. Let

$$f(x,y) \sim \sum_{m,n=1}^{\infty} A_{m,n} \sin mx \sin ny$$

be the double Fourier sine series of $f(x,y)$.

(a) Show that the coefficients satisfy *Bessel's inequality* (see Exercises 6g, Problem 4)

$$\left(\frac{\pi}{2}\right)^2 \sum_{m,n=1}^{\infty} A_{m,n}^2 \le \int_0^{\pi} \int_0^{\pi} f^2(x,y)\, dx\, dy .$$

(b) Suppose that $f(x,y)$ and its first- and second-order partial derivatives are continuous on the closed square and $f = 0$ on the boundary of the square. Use part (a) applied to the coefficients of Δf, and Problem 2(d) to show that

$$\sum_{m,n=1}^{\infty} |A_{m,n}| < \infty ,$$

so that the double Fourier sine series of f is uniformly and absolutely convergent. Show that the sum of the series is $f(x,y)$ by considering the associated iterated series.

4. Let $f(x,y)$ be periodic with period 2π in each variable; that is, $f(x + 2\pi, y) = f(x,y + 2\pi) = f(x,y)$. The double Fourier series of $f(x,y)$ in complex form is

$$f(x,y) \sim \sum_{m=-\infty}^{\infty} \sum_{n=-\infty}^{\infty} C_{m,n} e^{-i(mx+ny)} ,$$

where

$$C_{m,n} = \frac{1}{(2\pi)^2} \int_{-\pi}^{\pi} \int_{-\pi}^{\pi} f(x,y) e^{i(mx+ny)}\, dx\, dy .$$

(a) Show that if $f(x,y)$ is continuous, then the coefficients satisfy Bessel's inequality,

$$\sum_{m,n=1}^{\infty} |C_{m,n}|^2 \leq \frac{1}{(2\pi)^2} \int_{-\pi}^{\pi} \int_{-\pi}^{\pi} |f(x,y)|^2\, dx\, dy .$$

(b) Use part (a) applied to Δf, and Problem 2(d) to show that if $f(x,y)$ and its first- and second-order derivatives are continuous and periodic then the Fourier series of f is absolutely and uniformly convergent.

11

Laplace's equation and related equations

11.1. *BOUNDARY VALUE PROBLEMS*

In this chapter we shall discuss the solution of problems for **Laplace's equation,**

$$\Delta u = 0 \,, \tag{11.1.1}$$

and for the related inhomogeneous equation

$$\Delta u = -q \,, \tag{11.1.2}$$

which is called **Poisson's equation.** We shall also consider the general *elliptic* linear equation with constant coefficients in standard form,

$$\Delta u + \gamma u = -q \,. \tag{11.1.3}$$

Here q is a function of (x,y), or of (x,y,z), and γ is a real constant. For the most part our discussion will be confined to the two-dimensional case, although much of it can be extended immediately to the case of three dimensions.

The problems we shall consider can be viewed as problems concerning equilibrium temperature distributions, and this viewpoint provides us with valuable heuristic insight. The importance of these problems, however, goes far beyond the theory of heat conduction because exactly the same problems occur in a vast array of physical and mathematical applications.

The physical interpretation of the equations we consider indicates that the appropriate problems for them are **boundary value problems,** rather than initial-boundary value or initial value problems. The first such problem for Laplace's equation which we encounter is the **Dirichlet problem** for a domain D, that is, the problem

$$\text{(11.1.4)} \qquad \begin{array}{lll} \text{D.E.} & \Delta u = 0 & \text{in } D\,, \\ \text{B.C.} & u = f & \text{on } C\,, \end{array}$$

where f is a known function defined on the boundary C of D. We may interpret the solution u as the equilibrium temperature distribution in a uniform heat-conducting body occupying the domain D, when the temperature distribution on the boundary of the body is kept fixed. On the basis of this physical model, it is plausible that the problem has a solution if D is a finite regular domain and if the data function f is sufficiently smooth. That the solution is unique is even more plausible, for we can argue that if u_1 and u_2 were any two solutions of (11.1.4), then their difference $v = u_2 - u_1$ would be a solution of

$$\text{(11.1.5)} \qquad \begin{array}{lll} \text{D.E.} & \Delta v = 0 & \text{in } D\,, \\ \text{B.C.} & v = 0 & \text{on } C\,. \end{array}$$

Hence v could be interpreted as the equilibrium temperature in D when the boundary is kept at temperature zero. From the physical model it is apparent that v must be identically zero, hence $u_2 = u_1$ and problem (11.1.4) has only one solution.

The **Neumann problem** for the domain D is

$$\text{(11.1.6)} \qquad \begin{array}{lll} \text{D.E.} & \Delta u = 0 & \text{in } D\,, \\ \text{B.C.} & \dfrac{\partial u}{\partial \nu} = f & \text{on } C\,, \end{array}$$

where $\partial u/\partial \nu$ denotes the outward normal derivative on the boundary. Here the solution u may be interpreted as the equilibrium temperature distribution in D when the flux on the boundary is prescribed. For such a temperature distribution, the heat content of D, which is a constant multiple of the integral of u over D, must be independent of the time, and this requires that the net flux across the boundary of D be zero. In other words, problem (11.1.6) cannot have a solution unless the integral of f over the boundary is zero. When D is a two-dimensional domain with boundary curve C, this **compatibility condition** that f must satisfy is

$$\text{(11.1.7)} \qquad \int_C f\, ds = 0\,.$$

It is easily verified that if u_1 is a solution of (11.1.6) and A is any constant, then $u_2 = u_1 + A$ is also a solution. Thus, when (11.1.6) has a solution, the solution is not unique. It is, however, plausible that the solution is unique except for such an additive constant.

Poisson's equation is satisfied by the equilibrium temperature distribution in D when there is a known distribution of sources in D. If the boundary is maintained at temperature zero, the equilibrium temperature satisfies

$$\text{(11.1.8)} \qquad \begin{array}{lll} \text{D.E.} & \Delta u = -q & \text{in } D\,, \\ \text{B.C.} & u = 0 & \text{on } C\,. \end{array}$$

The corresponding problem for the case of an insulated boundary is

$$\text{(11.1.9)} \qquad \begin{array}{lll} \text{D.E.} & \Delta u = -q & \text{in } D\,, \\ \text{B.C.} & \dfrac{\partial u}{\partial \nu} = 0 & \text{on } C\,. \end{array}$$

For this problem, as for problem (11.1.6), there is a compatibility condition indicated by physical considerations. Since the net rate of production of heat in D is proportional to

$$\text{(11.1.10)} \qquad \iint_D q\,dx\,dy\,,$$

there can be no solution of (11.1.9) unless the integral (11.1.10) is zero. When the problem has a solution, the solution is unique only to within an additive constant.

Problem (11.1.8) is a special case of the problem

$$\text{(11.1.11)} \qquad \begin{array}{lll} \text{D.E.} & \Delta u = -q & \text{in } D \\ \text{B.C.} & u = f & \text{on } C\,. \end{array}$$

If u_1 is a solution of (11.1.4) and u_2 a solution of (11.1.8), then $u = u_1 + u_2$ is a solution of (11.1.11). Thus, (11.1.11) can be solved by solving the apparently simpler problems (11.1.4) and (11.1.8).

It is often practicable to reduce problem (11.1.8) to problem (11.1.4). The following procedure is one method of effecting this reduction. The equation

$$\text{(11.1.12)} \qquad \Delta v = -q \qquad \text{in } D\,,$$

has many solutions, and it may be easy to find one which is continuous in the closed region $D + C$. If v is known, we can use the values of v on C as boundary data in problem (11.1.4), that is, we seek w such that

$$\text{(11.1.13)} \qquad \begin{array}{lll} \text{D.E.} & \Delta w = 0 & \text{in } D\,, \\ \text{B.C.} & w = v & \text{on } C\,. \end{array}$$

The function $u = v - w$ is then a solution of (11.1.8).

11.2. UNIQUENESS STATEMENTS AND COMPATIBILITY CONDITIONS

The uniqueness statements and compatibility conditions, which were formulated in Section 1 on the basis of physical considerations, can be established mathematically, provided the domain D and all functions considered are such that Green's formulas can be applied. We make this assumption throughout the following discussion.

Consider first the uniqueness question for the general *Neumann problem*

$$(11.2.1) \qquad \begin{aligned} &\text{D.E.} \qquad \Delta u = -q \qquad &&\text{in } D\,, \\ &\text{B.C.} \qquad \frac{\partial u}{\partial \nu} = f \qquad &&\text{on } C\,, \end{aligned}$$

where q or f may be zero. If u_1 and u_2 were two solutions of this inhomogeneous problem, then $v = u_2 - u_1$ would be a solution of the related homogeneous problem

$$(11.2.2) \qquad \begin{aligned} &\text{D.E.} \qquad \Delta v = 0 \qquad &&\text{in } D\,, \\ &\text{B.C.} \qquad \frac{\partial v}{\partial \nu} = 0 \qquad &&\text{on } C\,. \end{aligned}$$

In Green's first formula,

$$\iint_D w\Delta v \, dx\, dy + \iint_D (w_x v_x + w_y v_y)\, dx\, dy = \int_C w \frac{\partial v}{\partial \nu}\, ds\,,$$

we set $w = v$ to obtain

$$(11.2.3) \qquad \iint_D v\Delta v \, dx\, dy + \iint_D (v_x^2 + v_y^2)\, dx\, dy = \int_C v \frac{\partial v}{\partial \nu}\, ds\,.$$

Now let v be a solution of (11.2.2). Then the first integral on the left and the integral on the right vanish because of the D.E. and B.C. of problem (11.2.2), respectively, and (11.2.3) becomes

$$(11.2.4) \qquad \iint_D (v_x^2 + v_y^2)\, dx\, dy = 0\,.$$

Since $v_x^2 + v_y^2$ is nonnegative and the integral of a continuous nonnegative function can be zero only if the function is the constant function zero, we conclude that $v_x^2 + v_y^2 = 0$ in D, and hence $v_x = 0$ and $v_y = 0$ in D. Thus v is a constant function, say A, and $u_2 = u_1 + A$.

In exactly the same way we show that if u_1 and u_2 were two solutions of the general *Dirichlet problem*

$$\text{(11.2.5)} \qquad \begin{array}{lll} \text{D.E.} & \Delta u = -q & \text{in } D\,, \\ \text{B.C.} & u = f & \text{on } C\,, \end{array}$$

then the difference $v = u_2 - u_1$, which would satisfy the related homogeneous problem (11.1.5), is a constant function in D. Now, however, from the B.C. of (11.1.5) and the continuity of v in $D + C$, we conclude that the value of this constant must be zero, so that $v = 0$, $u_2 = u_1$ in D, and the solution of the Dirichlet problem is unique. In Section 8 we shall establish this uniqueness statement under assumptions which are much less restrictive, and more appropriate to the Dirichlet problem, than those made at the beginning of this section.

To establish the compatibility condition for the Neumann problem (11.2.1) we use Green's second formula,

$$\text{(11.2.6)} \qquad \iint\limits_D (w\Delta u - u\Delta w)\, dx\, dy = \int_C \left(w\,\frac{\partial u}{\partial \nu} - u\,\frac{\partial w}{\partial \nu} \right) ds\,.$$

Choosing w to be the constant function 1, so that $\Delta w = 0$ in D and $\partial w/\partial \nu = 0$ on C, we get from (11.2.6),

$$\text{(11.2.7)} \qquad \iint\limits_D \Delta u\, dx\, dy = \int_C \frac{\partial u}{\partial \nu}\, ds\,.$$

Finally, supposing u to be a solution of (11.2.1) and substituting the D.E. and B.C. of (11.2.1) in (11.2.7), we have the compatibility condition,

$$\text{(11.2.8)} \qquad -\iint\limits_D q\, dx\, dy = \int_C f\, ds\,.$$

The same techniques can be employed to establish uniqueness statements and compatibility conditions for boundary value problems for the differential equation (11.1.3) with $\gamma \neq 0$. The discussion of these problems is left for Exercises 11a. The statement and interpretation of these results are essentially related to the eigenvalue problems considered in Chapter 10, namely,

$$\text{(11.2.9)} \qquad \begin{array}{lll} \text{D.E.} & \Delta\varphi + \lambda\varphi = 0 & \text{in } D\,, \\ \text{B.C.} & \varphi = 0 & \text{on } C\,, \end{array}$$

and

$$\text{(11.2.10)} \qquad \begin{array}{lll} \text{D.E.} & \Delta\varphi + \lambda\varphi = 0 & \text{in } D\,, \\ \text{B.C.} & \dfrac{\partial \varphi}{\partial \nu} = 0 & \text{on } C\,, \end{array}$$

and the physical problems with which they are associated.

EXERCISES 11a

1. State a physical interpretation of the condition (11.2.8).

2. Use (11.2.3) to prove a uniqueness statement for *Robin's problem*

$$\text{D.E.} \qquad \Delta u = -q \qquad \text{in } D\,,$$

$$\text{B.C.} \qquad \frac{\partial u}{\partial \nu} + hu = f \qquad \text{on } C\,,$$

if h is a positive constant.

3. Use (11.2.3) to prove that both the Dirichlet problem,

$$\begin{array}{lll} \text{D.E.} & \Delta u + \gamma u = -q & \text{in } D\,, \\ \text{B.C.} & u = f & \text{on } C\,, \end{array}$$

and the Neumann problem,

$$\text{D.E.} \qquad \Delta u + \gamma u = -q \qquad \text{in } D\,,$$

$$\text{B.C.} \qquad \frac{\partial u}{\partial \nu} = f \qquad \text{on } C\,,$$

have unique solutions provided $\gamma < 0$.

4. (i) Using the definition of an eigenvalue problem, show that the Dirichlet problem of Problem 3 has a unique solution if and only if γ is not an eigenvalue of the eigenvalue problem (11.2.9).

(ii) Use the result of part (i) and a property of the eigenvalues of (11.2.9) to prove the first uniqueness statement of Problem 3, and to account for the failure of the procedure employed there if $\gamma > 0$.

5. (i) Use (11.2.6) to show that if γ is an eigenvalue of the eigenvalue problem (11.2.10), then the Neumann problem of Problem 3 has a solution only if q and f satisfy the compatibility conditions

$$-\iint_D \varphi q \, dx \, dy = \int_C \varphi f \, ds$$

for every eigenfunction φ of (11.2.10) belonging to the eigenvalue γ.

(ii) Show that the compatibility condition (11.2.8) is a special case of the result of part (i).

11.3. *DIRICHLET'S PROBLEM FOR A RECTANGLE*

The problem (11.1.4) for a rectangle requires us to find a function which satisfies Laplace's equation at every point inside the rectangle and has prescribed values along the boundary of the rectangle. If we choose a rectangular cartesian coordinate system with origin at a vertex of the rectangle and with the axes parallel to the sides of the rectangle, the problem is of the form

$$
(11.3.1)\qquad
\begin{aligned}
&\text{D.E.} && u_{xx} + u_{yy} = 0\,, && 0 < x < a\,, \quad 0 < y < b\,,\\
&\text{B.C.} && \begin{cases} u(x,0) = g(x)\,, & 0 < x < a\,,\\ u(x,b) = h(x)\,, & 0 < x < a\,,\\ u(0,y) = f(y)\,, & 0 < y < b\,,\\ u(a,y) = k(y)\,, & 0 < y < b\,,\end{cases}
\end{aligned}
$$

where g, h, f, k are given functions.

We will first solve a special case of problem (11.3.1) in which the prescribed values on a pair of parallel sides of the rectangle are zero. More precisely, we will first solve the problem

$$
(11.3.2)\qquad
\begin{aligned}
&\text{D.E.} && w_{xx} + w_{yy} = 0\,, && 0 < x < a\,, \quad 0 < y < b\,,\\
&\text{B.C.} && \begin{cases} w(x,0) = g(x)\,, & 0 < x < a\,,\\ w(x,b) = h(x)\,, & 0 < x < a\,,\\ w(0,y) = 0\,, & 0 < y < b\,,\\ w(a,y) = 0\,, & 0 < y < b\,.\end{cases}
\end{aligned}
$$

We attempt to solve problem (11.3.2) by following the procedure used for the solution of homogeneous initial-boundary value problems. The first step in the procedure is to seek functions of the form

$$w(x,y) = \varphi(x)\psi(y)\,,$$

which satisfy the D.E. and the homogeneous equations of the B.C. This leads to

$$
(11.3.3)\qquad
\begin{aligned}
&\text{D.E.} && \varphi'' + \lambda\varphi = 0\,, && 0 < x < a\,,\\
&\text{B.C.} && \varphi(0) = \varphi(a) = 0\,,
\end{aligned}
$$

and

$$(11.3.4)\qquad \psi'' - \lambda\psi = 0\,, \qquad 0 < y < b\,,$$

where λ is a constant. Now (11.3.3) is a familiar eigenvalue problem. The eigenvalues and eigenfunctions are

$$(11.3.5)\qquad \lambda_m = \left(\frac{m\pi}{a}\right)^2, \quad \varphi_m(x) = \sin\left(\frac{m\pi x}{a}\right), \qquad m = 1, 2, \ldots .$$

Hence, if $\psi_m(y)$ denotes any solution of (11.3.4) with $\lambda = \lambda_m$, that is, any solution of

$$(11.3.6)\qquad \psi_m'' - \lambda_m\psi_m = 0\,,$$

then

$$(11.3.7)\qquad w_m(x,y) = \psi_m(y)\sin\frac{m\pi x}{a}$$

satisfies the D.E. of (11.3.2) and also satisfies the B.C.

$$w_m(0,y) = w_m(a,y) = 0\,, \qquad 0 < y < b\,.$$

We now ask if the solution of (11.3.2) can be obtained by superposition of the functions (11.3.7), that is, if the functions $\psi_m(y)$, which must satisfy (11.3.6), can be chosen so that

$$w(x,y) = \sum_{m=1}^{\infty} \psi_m(y) \sin \frac{m\pi x}{a} \tag{11.3.8}$$

is a solution of (11.3.2). In a formal sense, the series (11.3.8) satisfies the D.E. of (11.3.2) and also the homogeneous equations of the B.C. of (11.3.2). The additional requirements imposed on (11.3.8) by (11.3.2) are

$$g(x) = \sum_{m=1}^{\infty} \psi_m(0) \sin \frac{m\pi x}{a}, \qquad 0 < x < a, \tag{11.3.9}$$

$$h(x) = \sum_{m=1}^{\infty} \psi_m(b) \sin \frac{m\pi x}{a}, \qquad 0 < x < a. \tag{11.3.10}$$

Since the series in (11.3.9) and (11.3.10) are Fourier sine series, these equations will be satisfied if

$$\psi_m(0) = g_m, \quad \psi_m(b) = h_m, \tag{11.3.11}$$

where

$$g_m = \frac{2}{a}\int_0^a g(x) \sin \frac{m\pi x}{a}\, dx, \quad h_m = \frac{2}{a}\int_0^a h(x) \sin \frac{m\pi x}{a}\, dx. \tag{11.3.12}$$

Therefore, the function $\psi_m(y)$ must be the solution of

$$\begin{array}{lll} \text{D.E.} & \psi_m'' - \lambda_m \psi_m = 0, & 0 < y < b, \\ \text{B.C.} & \psi_m(0) = g_m, \quad \psi_m(b) = h_m. & \end{array} \tag{11.3.13}$$

This is a boundary value problem for an ordinary differential equation. It can be solved by elementary methods, and we find the unique solution

$$\psi_m(y) = g_m \frac{\sinh \dfrac{m\pi(b-y)}{a}}{\sinh \dfrac{m\pi b}{a}} + h_m \frac{\sinh \dfrac{m\pi y}{a}}{\sinh \dfrac{m\pi b}{a}}. \tag{11.3.14}$$

Hence, for the formal solution of problem (11.3.2) we obtain

$$w(x,y) = \sum_{m=1}^{\infty} g_m \frac{\sinh \dfrac{m\pi(b-y)}{a}}{\sinh \dfrac{m\pi b}{a}} \sin \frac{m\pi x}{a} + \sum_{m=1}^{\infty} h_m \frac{\sinh \dfrac{m\pi y}{a}}{\sinh \dfrac{m\pi b}{a}} \sin \frac{m\pi x}{a}, \tag{11.3.15}$$

where g_m and h_m are given by (11.3.12).

By a similar calculation, the solution of the problem

$$
\text{(11.3.16)}\qquad
\begin{array}{lll}
\text{D.E.} & v_{xx} + v_{yy} = 0\,, & 0 < x < a\,,\quad 0 < y < b\,,\\
\text{B.C.} & \begin{cases} v(x,0) = 0\,, & 0 < x < a\,,\\ v(x,b) = 0\,, & 0 < x < a\,,\\ v(0,y) = f(y)\,, & 0 < y < b\,,\\ v(a,y) = k(y)\,, & 0 < y < b\,, \end{cases} &
\end{array}
$$

is found to be

$$
\text{(11.3.17)}\qquad v(x,y) = \sum_{n=1}^{\infty} f_n \frac{\sinh \dfrac{n\pi(a-x)}{b}}{\sinh \dfrac{n\pi a}{b}} \sin \frac{n\pi y}{b}
+ \sum_{n=1}^{\infty} k_n \frac{\sinh \dfrac{n\pi x}{b}}{\sinh \dfrac{n\pi a}{b}} \sin \frac{n\pi y}{b}\,,
$$

where

$$
\text{(11.3.18)}\qquad f_n = \frac{2}{b}\int_0^b f(y) \sin \frac{n\pi y}{b}\,dy\,,\quad k_n = \frac{2}{b}\int_0^b k(y) \sin \frac{n\pi y}{b}\,dy\,.
$$

The solution of problem (11.3.1) is clearly the sum of the solutions of problems (11.3.2) and (11.3.16), that is, $u(x,y) = w(x,y) + v(x,y)$.

EXERCISES 11b

1. Solve

$$
\begin{array}{lll}
\text{D.E.} & u_{xx} + u_{yy} + u_{zz} = 0\,, & 0 < x < a\,,\quad 0 < y < b\,,\quad 0 < z < c\,,\\
\text{B.C.} & u(x,y,0) = f(x,y)\,, & 0 < x < a\,,\quad 0 < y < b\,,\\
 & u(x,y,c) = 0\,, & 0 < x < a\,,\quad 0 < y < b\,,\\
 & u(x,0,z) = u(x,b,z) = 0\,, & 0 < x < a\,,\quad 0 < z < c\,,\\
 & u(0,y,z) = u(a,y,z) = 0\,, & 0 < y < b\,,\quad 0 < z < c\,.
\end{array}
$$

2. Obtain the solution of (11.3.2) from the solution of (10.7.1).

3. (i) Find the solution of

$$
\begin{array}{lll}
\text{D.E.} & \dfrac{d^2 w(x)}{dx^2} = 0\,, & 0 < x < a\,,\\
\text{B.C.} & w(0) = A\,, & w(a) = B
\end{array}
$$

by assuming a series solution

$$
w(x) = \sum c_n \varphi_n(x)\,,
$$

where $\varphi_n(x)$ are the eigenfunctions of the problem

$$\text{D.E.} \qquad \varphi'' + \lambda\varphi = 0\,, \qquad 0 < x < a\,,$$
$$\text{B.C.} \qquad \varphi(0) = 0\,, \qquad \varphi(a) = 0\,.$$

Proceed by observing that in the equation

$$0 = \int_0^a w(\varphi_n'' + \lambda_n\varphi_n)\,dx = \int_0^a w\varphi_n''\,dx + \lambda_n \int_0^a w\varphi_n\,dx\,,$$

the second integral on the right is c_n, and the first integral can be transformed with the aid of Green's second formula.

(ii) Solve problem (11.3.2) by assuming a series solution

$$w(x,y) = \sum c_{m,n}\varphi_{m,n}(x,y)\,,$$

where $\varphi_{m,n}(x,y)$ are the eigenfunctions of problem (10.3.4).

4. (i) Solve the Neumann problem

$$\begin{array}{lllll} \text{D.E.} & v_{xx} + v_{yy} = 0\,, & 0 < x < a\,, & 0 < y < b\,, \\ \text{B.C.} & v_y(x,0) = g(x)\,, & 0 < x < a\,, & \\ & v_y(x,b) = h(x)\,, & 0 < x < a\,, & \\ & v_x(0,y) = 0\,, & & 0 < y < b\,, \\ & v_x(a,y) = 0\,, & & 0 < y < b\,, \end{array}$$

by the method of Section 3, assuming that

$$\int_0^a [g(x) - h(x)]\,dx = 0\,.$$

(ii) At what point in your calculation was the above assumption necessary?

5. The general Neumann problem for Laplace's equation for the rectangle,

$$\begin{array}{lllll} \text{D.E.} & u_{xx} + u_{yy} = 0\,, & 0 < x < a\,, & 0 < y < b\,, \\ \text{B.C.} & u_y(x,0) = g(x)\,, & 0 < x < a\,, & \\ & u_y(x,b) = h(x)\,, & 0 < x < a\,, & \\ & u_x(0,y) = f(y)\,, & & 0 < y < b\,, \\ & u_x(a,y) = k(y)\,, & & 0 < y < b\,, \end{array}$$

has a solution provided

$$\int_0^a (g - h)\,dx + \int_0^b (f - k)\,dy = 0\,.$$

Even when this condition is satisfied, the auxiliary assumption of Problem 4 may fail to hold. Show, however, that for a suitably chosen constant α, the problem whose solution is $v = u - \alpha(x^2 - y^2)$ satisfies both conditions, and hence find u.

6. (i) Let D be the rectangle $0 < x < a$, $0 < y < b$. Find a formula for the solution of the Neumann problem for Poisson's equation in D,

$$\text{D.E.} \qquad \Delta u = -q(x,y) \qquad \text{in } D\,,$$

$$\text{B.C.} \qquad \frac{\partial u}{\partial \nu} = 0 \qquad \text{on } C\,,$$

by assuming a series solution

$$u = \sum c_{m,n}\varphi_{m,n}(x,y),$$

where $\varphi_{m,n}(x,y)$ are the eigenfunctions of

$$\text{D.E.} \qquad \Delta\varphi + \lambda\varphi = 0 \qquad \text{in } D,$$

$$\text{B.C.} \qquad \frac{\partial\varphi}{\partial\nu} = 0 \qquad \text{on } C,$$

and using the method of Problem 3.

(ii) At what point in your calculation was the compatibility condition

$$\iint_D q\,dx\,dy = 0$$

used?

11.4. *DIRICHLET'S PROBLEM FOR A CIRCULAR ANNULUS*

We wish to solve the Dirichlet problem for Laplace's equation in the domain D lying between two concentric circles C_1 and C_2, that is, the problem

$$\begin{array}{llll} & \text{D.E.} & \Delta u = 0 & \text{in } D, \\ (11.4.1) & \text{B.C.} & u = g & \text{on } C_1, \\ & & u = f & \text{on } C_2. \end{array}$$

In order to use the method of eigenfunction expansions to solve this problem, it is essential to introduce a polar coordinate system with origin at the common center of the circles. The reason for this will appear as our discussion proceeds. Thus, our problem is to find a function $u = u(r,\theta)$, which satisfies the conditions of problem (11.4.1), and our first step will be to express these conditions in terms of the polar coordinate variables.

The polar coordinates (r,θ) and the rectangular coordinates (x,y) of a point are related by the familiar formulas

$$x = r\cos\theta, \quad y = r\sin\theta, \tag{11.4.2}$$

where we assume $r \geq 0$, but θ can have any real value. From these formulas we find, by a straightforward although tedious calculation, the expression for the Laplacian Δu in terms of derivatives of u with respect to r and θ,

(11.4.3)
$$\Delta u = u_{rr} + \frac{1}{r} u_r + \frac{1}{r^2} u_{\theta\theta}$$
$$= \frac{1}{r} \frac{\partial}{\partial r}\left(r \frac{\partial u}{\partial r}\right) + \frac{1}{r^2} \frac{\partial^2 u}{\partial \theta^2}.$$

We say a *point function* is defined on the annulus if to each point P of the annulus there corresponds a definite value $u = u(P)$. Any point function can be expressed as a function of the polar coordinate variables (r,θ). Not every function of the variables (r,θ), however, has a unique value at each point of the annulus, since the obvious **periodicity condition** $u(r,\theta + 2\pi) = u(r,\theta)$ must be satisfied. A function which does not satisfy the periodicity condition may have different values for the coordinate pairs (r,θ), $(r,\theta \pm 2\pi)$, . . . , $(r,\theta \pm 2n\pi)$, . . . , all of which correspond to the same point. A function which does satisfy the periodicity condition has the same value at all the above coordinate pairs, and thus determines a point function on the annulus.

Let the radii of the two circles of problem (11.4.1) be a and b where $0 < a < b$. Then the solution $u(r,\theta)$ we seek must be a point function, and hence must satisfy the periodicity condition

(11.4.4)
$$u(r,\theta + 2\pi) = u(r,\theta) , \qquad a < r < b , \quad -\infty < \theta < \infty .$$

This condition is a linear and *homogeneous* auxiliary condition. We shall see that the condition plays a role, in the process of finding the solution $u(r,\theta)$, similar to the role of a boundary condition. For this reason we shall include the condition as one of the boundary conditions of the problem, although it does not express a condition holding only on the boundary of the annulus. Thus, the polar coordinate form of problem (11.4.1) is

$$\begin{array}{lll} \text{D.E.} & \frac{1}{r}(ru_r)_r + \frac{1}{r^2} u_{\theta\theta} = 0 , & a < r < b , \\ & & -\infty < \theta < \infty , \\ \text{B.C.} & u(a,\theta) = g(\theta) , & -\infty < \theta < \infty , \\ & u(b,\theta) = f(\theta) , & -\infty < \theta < \infty , \\ & u(r,\theta + 2\pi) = u(r,\theta) , & a < r < b , \\ & & -\infty < \theta < \infty , \end{array} \qquad (11.4.5)$$

where the given functions $f(\theta)$ and $g(\theta)$ satisfy the periodicity conditions $f(\theta + 2\pi) = f(\theta)$, $g(\theta + 2\pi) = g(\theta)$, and are assumed to be piecewise smooth.

We first seek functions of the form $R(r)H(\theta)$, with variables separated, which satisfy the D.E. and the periodicity condition of problem (11.4.5). For a non-trivial solution this leads to the requirement that $R(r)$ and $H(\theta)$ must satisfy

$$(11.4.6)\quad \begin{array}{lll} \text{D.E.} & H'' + \lambda H = 0\,, & -\infty < \theta < \infty\,, \\ \text{B.C.} & H(\theta + 2\pi) = H(\theta)\,, & -\infty < \theta < \infty\,, \end{array}$$

$$(11.4.7)\quad \text{D.E.} \quad r(rR')' - \lambda R = 0\,, \qquad a < r < b\,.$$

Problem (11.4.6) is like an eigenvalue problem, even though the auxiliary condition is not, strictly speaking, a boundary condition. The differential equation has periodic solutions with period 2π only if λ has one of the values $\lambda_n = n^2$, $n = 0, 1, 2, \ldots$. When $\lambda = \lambda_n$, $n \geq 1$, all solutions are periodic and we set

$$(11.4.8)\quad H_n(\theta) = c_{1,n} \cos n\theta + c_{2,n} \sin n\theta\,, \qquad n = 1, 2, \ldots.$$

When $\lambda = \lambda_0$ the only periodic solutions are constant functions,

$$(11.4.9)\quad H_0(\theta) = c_{1,0}\,.$$

The general solution of (11.4.7) when $\lambda = \lambda_n$ is

$$(11.4.10)\quad R_n(r) = c_{3,n} r^n + c_{4,n} r^{-n}\,, \qquad n = 1, 2, \ldots,$$

$$(11.4.11)\quad R_0(r) = c_{3,0} + c_{4,0} \log r\,.$$

Thus, $R_n(r)H_n(\theta)$, with $n \neq 0$, is a linear combination of the four functions

$$(11.4.12)\quad \begin{array}{ll} r^n \cos n\theta\,, & r^{-n} \cos n\theta\,, \\ r^n \sin n\theta\,, & r^{-n} \sin n\theta\,, \end{array} \qquad n = 1, 2, \ldots,$$

and when $n = 0$ it is a linear combination of the functions

$$(11.4.13)\quad 1\,, \quad \log r\,.$$

We have thus found all functions of the special form $R(r)H(\theta)$ which are periodic with period 2π and which satisfy Laplace's equation in the annulus.

Our aim now is to obtain the solution of (11.4.5) by superposition of these functions. We therefore assume a solution of the form

$$(11.4.14)\quad u(r,\theta) = \tfrac{1}{2}(\alpha_0 + \beta_0 \log r) + \sum_{n=1}^{\infty} [(\alpha_n r^n + \beta_n r^{-n}) \cos n\theta + (\gamma_n r^n + \delta_n r^{-n}) \sin n\theta]\,,$$

where the constants α_n, β_n, γ_n, δ_n must be determined so that the boundary conditions at $r = a$ and $r = b$ are satisfied. When we set $r = b$ in (11.4.14) and replace $u(b,\theta)$ by $f(\theta)$, we see that the resulting series must be the Fourier series of $f(\theta)$; hence,

$$\alpha_n b^n + \beta_n b^{-n} = \frac{1}{\pi} \int_{-\pi}^{\pi} f(\theta) \cos n\theta \, d\theta = A_n , \qquad n = 1, 2, \ldots ,$$

(11.4.15)
$$\gamma_n b^n + \delta_n b^{-n} = \frac{1}{\pi} \int_{-\pi}^{\pi} f(\theta) \sin n\theta \, d\theta = B_n , \qquad n = 1, 2, \ldots ,$$

$$\alpha_0 + \beta_0 \log b = \frac{1}{\pi} \int_{-\pi}^{\pi} f(\theta) \, d\theta = A_0 .$$

Similarly, when $r = a$ the series must become the Fourier series of $g(\theta)$, so that

$$\alpha_n a^n + \beta_n a^{-n} = \frac{1}{\pi} \int_{-\pi}^{\pi} g(\theta) \cos n\theta \, d\theta = C_n , \qquad n = 1, 2, \ldots ,$$

(11.4.16)
$$\gamma_n a^n + \delta_n a^{-n} = \frac{1}{\pi} \int_{-\pi}^{\pi} g(\theta) \sin n\theta \, d\theta = D_n , \qquad n = 1, 2, \ldots ,$$

$$\alpha_0 + \beta_0 \log a = \frac{1}{\pi} \int_{-\pi}^{\pi} g(\theta) \, d\theta = C_0 .$$

Now the pair of equations ($n \geq 1$)

(11.4.17)
$$\alpha_n b^n + \beta_n b^{-n} = A_n , \qquad \alpha_n a^n + \beta_n a^{-n} = C_n ,$$

can be solved to give α_n, β_n in terms of the known constants A_n, C_n. We find

(11.4.18)
$$\alpha_n = \frac{A_n/a^n - C_n/b^n}{(b/a)^n - (a/b)^n} , \qquad \beta_n = \frac{a^n C_n - b^n A_n}{(b/a)^n - (a/b)^n} , \qquad n = 1, 2, \ldots .$$

The other constants γ_n, δ_n, α_0, β_0 can be determined in a similar way. This completes the formal solution of (11.4.5).

It is instructive now to consider the problems of Sections 3 and 4 together. In general, the method of eigenfunction expansions will be successful when the following conditions are satisfied: (i) the variables are separable in the differential equation; (ii) the inhomogeneous auxiliary conditions of the problem are given on lines (surfaces) on which *one* of the variables is constant; (iii) the remaining auxiliary conditions are homogeneous, and lead to homogeneous auxiliary conditions for the remaining factors in a solution with separated variables. In the case of problem (11.3.1) the need to satisfy conditions (ii) and (iii) led us to reduce the solution of this problem to that of problems such as (11.3.2). In the case of problem (11.4.1) the need to satisfy condition (ii) led us to introduce polar coordinates. It was not a priori clear that conditions (i) and (iii) would then be satisfied, but this turned out to be so, and made the solution of the problem possible.

11.5. *DIRICHLET'S PROBLEM FOR A DISC*

Let C be a circle of radius b and D be the domain interior to C. The problem

$$(11.5.1) \qquad \begin{array}{lll} \text{D.E.} & \Delta u = 0 & \text{in } D\,, \\ \text{B.C.} & u = f & \text{on } C\,, \end{array}$$

is the Dirichlet problem for Laplace's equation for D. Here f is assumed to be periodic with period 2π and piecewise continuous. To find the solution we introduce a polar coordinate system with origin at the center of the circle. On the basis of the discussion in Section 4, we might expect that problem (11.5.1) is equivalent to

$$(11.5.2) \qquad \begin{array}{lll} \text{D.E.} & \dfrac{1}{r}(ru_r)_r + \dfrac{1}{r^2}u_{\theta\theta} = 0\,, & \begin{array}{l} 0 < r < b\,, \\ -\infty < \theta < \infty\,, \end{array} \\ \text{B.C.} & u(b,\theta) = f(\theta)\,, & -\infty < \theta < \infty\,, \\ & u(r,\theta + 2\pi) = u(r,\theta)\,, & \begin{array}{l} 0 < r < b\,, \\ -\infty < \theta < \infty\,. \end{array} \end{array}$$

However, when we examine the differential equation we notice that in the expression for Δu,

$$\Delta u = u_{rr} + \frac{1}{r}u_r + \frac{1}{r^2}u_{\theta\theta}\,,$$

the coefficients $1/r$, $1/r^2$ become infinite at the origin, where $r = 0$. Thus, the polar coordinate expression for Δu is meaningless at the origin, and for this reason the D.E. in (11.5.2) is required to hold only for $0 < r < b$, and not at $r = 0$. The D.E. in (11.5.2) could be stated in the form "$\Delta u = 0$ in D except perhaps at the origin," whereas the D.E. in (11.5.1) would read "$\Delta u = 0$ everywhere in D, including the origin." Therefore, the D.E. in (11.5.2) is not completely equivalent to the D.E. in (11.5.1). To express the D.E. of (11.5.1) we need the D.E. of (11.5.2) and, in addition, some auxiliary condition relating to the behavior of u at the origin. It would obviously suffice to require that u and its first- and second-order partial derivatives with respect to the rectangular coordinates should exist and be continuous at the origin. However, we shall find that a much simpler condition is sufficient.

It is not hard to find specific examples which show that problem (11.5.2) is not adequately formulated and is not equivalent to problem (11.5.1). In fact, each of the functions

$$\begin{aligned} v_0(r,\theta) &= \log b - \log r \\ v_n(r,\theta) &= (b^{-n}r^n - b^n r^{-n}) \cos n\theta\,, \qquad n = 1, 2, \ldots, \end{aligned} \tag{11.5.3}$$

satisfies the D.E. and the periodicity condition of problem (11.5.2) and is zero on the boundary $r = b$. Hence, if $u(r,\theta)$ is a solution of (11.5.2), then $u(r,\theta) + v_n(r,\theta)$ is another solution, so (11.5.2) does not have a unique solution. On the other hand, we expect from the results of Section 2 that problem (11.5.1) does have a unique solution. The solution of (11.5.1) will also be a solution of (11.5.2), but (11.5.2) has other—"extraneous"—solutions. We conclude that the two problems are not equivalent.

We shall begin the formal process of solving problem (11.5.2) and hope to find some simple additional auxiliary condition which will exclude all of the "extraneous" solutions. We could proceed by the method of separation of variables as in Section 4. However, part of the discussion in that section is obviously relevant to our present problem, and shows that the appropriate eigenfunction expansion for our problem is the Fourier series representation (in the ordinary sense). Hence, it is more convenient and also more instructive to use the method of variation of parameters. We reason that for fixed r, $0 < r < b$, the solution $u(r,\theta)$ is a periodic function of θ with period 2π and has continuous first- and second-order derivatives. It can therefore be represented by its Fourier series, that is,

$$u(r,\theta) = \tfrac{1}{2}A_0(r) + \sum_{n=1}^{\infty} [A_n(r)\cos n\theta + B_n(r)\sin n\theta]\,, \tag{11.5.4}$$

where

$$\begin{Bmatrix} A_n(r) \\ B_n(r) \end{Bmatrix} = \frac{1}{\pi}\int_{-\pi}^{\pi} u(r,\theta) \begin{Bmatrix} \cos n\theta \\ \sin n\theta \end{Bmatrix} d\theta. \tag{11.5.5}$$

From the formula

$$A_n(r) = \frac{1}{\pi}\int_{-\pi}^{\pi} u(r,\theta)\cos n\theta\, d\theta \tag{11.5.6}$$

we obtain

$$r\frac{dA_n}{dr} = \frac{1}{\pi}\int_{-\pi}^{\pi} r\frac{\partial u}{\partial r}\cos n\theta\, d\theta$$

and

$$\begin{aligned}\frac{1}{r}\frac{d}{dr}\left(r\frac{dA_n}{dr}\right) &= \frac{1}{\pi}\int_{-\pi}^{\pi}\frac{1}{r}\frac{\partial}{\partial r}\left(r\frac{\partial u}{\partial r}\right)\cos n\theta\, d\theta \\ &= \frac{1}{\pi}\int_{-\pi}^{\pi} -\frac{1}{r^2}\frac{\partial^2 u}{\partial\theta^2}\cos n\theta\, d\theta \\ &= -\frac{1}{r^2}\frac{1}{\pi}\int_{-\pi}^{\pi}\frac{\partial^2 u}{\partial\theta^2}\cos n\theta\, d\theta\,,\end{aligned} \tag{11.5.7}$$

where we have used the D.E. of (11.5.2). We apply Green's formula to the last integral, and observe that the boundary terms are zero because both u and u_θ are periodic, the periodicity of u_θ being a consequence of that of u. Hence, we obtain

$$\begin{aligned}\frac{1}{r}\frac{d}{dr}\left(r\frac{dA_n}{dr}\right) &= -\frac{1}{r^2}\frac{1}{\pi}\int_{-\pi}^{\pi} u\frac{d^2}{d\theta^2}\cos n\theta\, d\theta \\ &= \frac{n^2}{r^2}\frac{1}{\pi}\int_{-\pi}^{\pi} u\cos n\theta\, d\theta \\ &= \frac{n^2}{r^2}A_n\,.\end{aligned}$$

Thus, $A_n(r)$ is a solution of

$$r^2\frac{d^2A_n}{dr^2} + r\frac{dA_n}{dr} - n^2A_n = 0\,, \qquad 0 < r < b\,. \tag{11.5.8}$$

Letting $r \to b$ in Equation (11.5.6) it follows from the boundary condition $u(b,\theta) = f(\theta)$ that

$$A_n(b) = \alpha_n\,, \tag{11.5.9}$$

where α_n is a Fourier coefficient of $f(\theta)$,

$$\begin{Bmatrix}\alpha_n\\ \beta_n\end{Bmatrix} = \frac{1}{\pi}\int_{-\pi}^{\pi} f(\theta)\begin{Bmatrix}\cos n\theta\\ \sin n\theta\end{Bmatrix} d\theta\,. \tag{11.5.10}$$

Since (11.5.8) is a second-order differential equation, the single condition (11.5.9) is not sufficient to determine $A_n(r)$ uniquely. An additional condition is needed to determine $A_n(r)$, presumably one which arises from the auxiliary condition which is missing in the boundary value problem (11.5.2). The general solution of (11.5.8) is

$$\begin{aligned}A_n(r) &= c_{1,n}r^n + c_{2,n}r^{-n}\,, \qquad n = 1, 2, \ldots\,, \\ A_0(r) &= c_{1,0} + c_{2,0}\log r\,.\end{aligned} \tag{11.5.11}$$

Since $\log r \to -\infty$ and $r^{-n} \to +\infty$ when $r \to 0+$, we see that for each choice of n, $A_n(r)$ is not continuous at $r = 0$ unless $c_{2,n} = 0$. On the other hand, the solution $u(r,\theta)$ we are seeking must certainly be continuous at the origin. If u_0 is the value of $u(r,\theta)$ at the origin, that is, if

$$\lim_{r\to 0+} u(r,\theta) = u_0 , \tag{11.5.12}$$

then from (11.5.6) it follows that

$$\lim_{r\to 0+} A_n(r) = \frac{u_0}{\pi}\int_{-\pi}^{\pi} \cos n\theta \, d\theta = \begin{cases} 2u_0 & \text{if } n = 0 , \\ 0 & \text{if } n = 1, 2, \ldots . \end{cases} \tag{11.5.13}$$

Thus, $A_n(r)$ has a finite limit as $r \to 0+$. This implies that the constants $c_{2,n}$ in (11.5.11) must be zero, so that

$$A_n(r) = c_n r^n , \qquad n = 0, 1, 2, \ldots , \tag{11.5.14}$$

where we have dropped the now superfluous subscript 1, and set $c_{1,n} = c_n$. From (11.5.9) and (11.5.14) we have $\alpha_n = A_n(b) = c_n b^n$, hence $c_n = \alpha_n / b^n$ and

$$A_n(r) = \alpha_n \frac{r^n}{b^n} , \qquad n = 0, 1, 2, \ldots . \tag{11.5.15}$$

By a similar calculation we find that $B_n(r)$ is a solution of

$$r^2 \frac{d^2 B_n}{dr^2} + r \frac{dB_n}{dr} - n^2 B_n = 0 , \qquad 0 < r < b , \tag{11.5.16}$$

and

$$B_n(b) = \beta_n , \tag{11.5.17}$$

where β_n is defined by (11.5.10). Moreover, from the hypothesis that $u(r,\theta)$ is continuous at the origin, we deduce that

$$\lim_{r\to 0+} B_n(r) \text{ exists and is finite} . \tag{11.5.18}$$

The system (11.5.16), (11.5.17), (11.5.18) has the unique solution

$$B_n(r) = \beta_n \frac{r^n}{b^n} , \qquad n = 1, 2, \ldots . \tag{11.5.19}$$

All of the coefficients $A_n(r)$, $B_n(r)$ in (11.5.4) have now been determined. Substitution of (11.5.15), (11.5.19) in (11.5.4) yields the formal solution of the Dirichlet problem for the disc,

$$u(r,\theta) = \frac{1}{2}\alpha_0 + \sum_{n=1}^{\infty} \left(\frac{r}{b}\right)^n (\alpha_n \cos n\theta + \beta_n \sin n\theta) , \tag{11.5.20}$$

where α_n and β_n are the Fourier coefficients of $f(\theta)$, defined by Equation (11.5.10).

To determine the coefficients $A_n(r)$, $B_n(r)$ in (11.5.4), the only condition we needed beyond the conditions of problem (11.5.2) was the simple condition that $u(r,\theta)$ be continuous at the origin. An examination of the preceding discussion will show that, in fact, this condition can be replaced by the still simpler condition that $u(r,\theta)$ be **bounded** at the origin, that is, that there is a constant U_0 such that $|u(r,\theta)| \leq U_0$ for all sufficiently small r, and all θ. In other words, what we have really solved is the problem

$$\text{(11.5.21)}\quad \begin{array}{lll} \text{D.E.} & \dfrac{1}{r}(ru_r)_r + \dfrac{1}{r^2}u_{\theta\theta} = 0\,, & 0 < r < b\,,\ -\infty < \theta < \infty\,, \\ \text{B.C.} & u(b,\theta) = f(\theta)\,, & -\infty < \theta < \infty\,, \\ & u(r,\theta + 2\pi) = u(r,\theta)\,, & 0 < r < b\,,\ -\infty < \theta < \infty\,, \\ & u \text{ is bounded at the origin.} & \end{array}$$

Our procedure has shown that if problem (11.5.21) has a solution, then it has only one solution, and that solution is given by (11.5.20) and (11.5.10). Since any solution of (11.5.1) will certainly be a solution of (11.5.21), we can also say that *if the problem* (11.5.1) *has a solution it has at most one solution, which must be given by* (11.5.20) *and* (11.5.10).

We list the assumptions employed in proving this last statement. It was assumed that: u and its first- and second-order partial derivatives are defined and continuous in D; u satisfies Laplace's equation in D; $f(\theta)$ is periodic with period 2π and piecewise continuous, and the boundary condition is satisfied in the sense that

$$\text{(11.5.22)}\qquad \lim_{r \to b} \int_{-\pi}^{\pi} u(r,\theta)\psi(\theta)\,d\theta = \int_{-\pi}^{\pi} f(\theta)\psi(\theta)\,d\theta\,,$$

where $\psi(\theta)$ is any piecewise continuous function on $-\pi \leq \theta \leq \pi$.

If $f(\theta)$ is continuous and u *is assumed to be continuous in* $D + C$, then Equation (11.5.22) follows immediately from

$$\text{(11.5.23)}\qquad u(b,\theta) = f(\theta)\,, \qquad -\pi \leq \theta \leq \pi\,.$$

This last statement and equation can be taken as an alternative to the boundary condition of the preceding paragraph.

EXERCISES 11c

1. Evaluate γ_n, δ_n, α_0, β_0 in (11.4.14).

2. Solve

$$\text{D.E.} \quad \frac{1}{r}(ru_r)_r + \frac{1}{r^2}u_{\theta\theta} = 0\,, \qquad a < r < b\,, \quad 0 < \theta < \alpha\,,$$

$$\begin{aligned} \text{B.C.} \quad & u(a,\theta) = 0\,, && && 0 < \theta < \alpha\,, \\ & u(b,\theta) = f(\theta)\,, && && 0 < \theta < \alpha\,, \\ & u(r,0) = 0\,, && a < r < b\,, && \\ & u(r,\alpha) = 0\,, && a < r < b\,. && \end{aligned}$$

3. Solve

$$\text{D.E.} \quad \frac{1}{r}(ru_r)_r + \frac{1}{r^2}u_{\theta\theta} = 0\,, \qquad a < r < b\,, \quad 0 < \theta < \alpha\,,$$

$$\begin{aligned} \text{B.C.} \quad & u(a,\theta) = 0\,, && && 0 < \theta < \alpha\,, \\ & u(b,\theta) = 0\,, && && 0 < \theta < \alpha\,, \\ & u(r,0) = f(r)\,, && a < r < b\,, && \\ & u(r,\alpha) = 0\,, && a < r < b\,. && \end{aligned}$$

4. Show how the solution of the problem

$$\text{D.E.} \quad \frac{1}{r}(ru_r)_r + \frac{1}{r^2}u_{\theta\theta} = 0\,, \qquad a < r < b\,, \quad 0 < \theta < \alpha\,,$$

$$\begin{aligned} \text{B.C.} \quad & u(a,\theta) = g(\theta)\,, && && 0 < \theta < \alpha\,, \\ & u(b,\theta) = f(\theta)\,, && && 0 < \theta < \alpha\,, \\ & u(r,0) = k(r)\,, && a < r < b\,, && \\ & u(r,\alpha) = h(r)\,, && a < r < b\,, && \end{aligned}$$

can be represented as a sum of solutions of problems each of which can be solved by the method of separation of variables and eigenfunction expansion.

5. Show directly from the statement of the problem that every solution of the eigenvalue problem (11.4.6) is a solution of the regular Sturm–Liouville problem

$$\begin{aligned} \text{D.E.} \quad & H'' + \lambda H = 0\,, \qquad -\pi < \theta < \pi\,, \\ \text{B.C.} \quad & H(-\pi) = H(\pi)\,, \quad H'(-\pi) = H'(\pi)\,. \end{aligned}$$

[This is problem (4.10.2) with $L = \pi$. The discussions in Sections 4.10 and 11.4 show that (4.10.2) and (11.4.6) have the same solutions, and hence are equivalent. Because of this connection, the B.C. above are called the **periodic boundary conditions.**]

6. Find a non-trivial solution $u(r,\theta) = R(r)H(\theta)$ of the problem

$$\text{D.E.} \quad \frac{1}{r}(ru_r)_r + \frac{1}{r^2}u_{\theta\theta} = 0\,, \qquad a < r < b\,, \quad -\infty < \theta < \infty\,,$$

$$\begin{aligned} \text{B.C.} \quad & u(a,\theta) = 0\,, \qquad -\infty < \theta < \infty\,, \\ & u(b,\theta) = 0\,, \qquad -\infty < \theta < \infty\,. \end{aligned}$$

[This example shows that if in (11.4.5) the functions $f(\theta)$, $g(\theta)$ are periodic with period 2π, the periodicity condition (or some other condition) is still needed to guarantee uniqueness.]

7. Show that all solutions of the Neumann problem for Laplace's equation in a disc D of radius b,

$$\text{D.E.} \qquad \Delta u = 0 \qquad \text{in } D\,,$$

$$\text{B.C.} \qquad \frac{\partial u}{\partial \nu} = f \qquad \text{on } C\,,$$

are solutions of

$$\text{D.E.} \qquad \frac{1}{r}(ru_r)_r + \frac{1}{r^2}u_{\theta\theta} = 0\,, \qquad 0 < r < b\,, \quad -\infty < \theta < \infty$$

$$\begin{aligned} \text{B.C.} \qquad & u_r(b,\theta) = f(\theta)\,, && -\infty < \theta < \infty\,, \\ & u(r,\theta + 2\pi) = u(r,\theta)\,, \qquad 0 < r < b\,, && -\infty < \theta < \infty\,, \\ & u \text{ is bounded at the origin}\,, \end{aligned}$$

and solve this problem, assuming that the function $f(\theta)$, which is periodic with period 2π, satisfies the compatibility condition

$$\int_0^{2\pi} f(\theta)\, d\theta = 0\,.$$

8. Formulate the Dirichlet problem for Poisson's equation in a circular disc of radius b,

$$\begin{aligned} \text{D.E.} \qquad & \Delta u = -q \qquad \text{in } D\,, \\ \text{B.C.} \qquad & u = 0 \qquad \text{on } C\,, \end{aligned}$$

in polar coordinates, and outline a procedure for solving the problem.

9. (i) By letting $a \to 0+$ in your solution for Problem 2, find a series representation for a solution of

$$\text{D.E.} \qquad \frac{1}{r}(ru_r)_r + \frac{1}{r^2}u_{\theta\theta} = 0\,, \qquad 0 < r < b\,, \quad 0 < \theta < \alpha\,,$$

$$\begin{aligned} \text{B.C.} \qquad & u(r,0) = 0\,, && 0 < r < b\,, \\ & u(r,\alpha) = 0\,, && 0 < r < b\,, \\ & u(b,\theta) = f(\theta)\,, && 0 < \theta < \alpha\,. \end{aligned}$$

(ii) Show that the problem has more than one solution.

(iii) Show that the problem has only one solution continuous at the origin, by solving by variation of parameters or by some other means.

10. (i) By letting $a \to 0+$ in your solution for Problem 3, find an integral representation for a solution of

$$\text{D.E.} \qquad \frac{1}{r}(ru_r)_r + \frac{1}{r^2}u_{\theta\theta} = 0\,, \qquad 0 < r < b\,, \quad 0 < \theta < \alpha\,,$$

$$\begin{aligned} \text{B.C.} \qquad & u(r,0) = f(r)\,, && 0 < r < b\,, \\ & u(r,\alpha) = g(r)\,, && 0 < r < b\,, \\ & u(b,\theta) = 0\,, && 0 < \theta < \alpha\,. \end{aligned}$$

(ii) Does this problem have more than one solution?

(iii) Show that on introducing the new variable $\xi = \log(b/r)$, the D.E. becomes

$$u_{\xi\xi} + u_{\theta\theta} = 0\,, \qquad 0 < \xi < \infty\,, \quad 0 < \theta < \alpha\,.$$

Hence, find a solution by Fourier transform methods.

11.6. *POISSON'S INTEGRAL FORMULA FOR THE DISC*

In Section 5 we found, as the formal solution of the Dirichlet problem for Laplace's equation for the disc, the series

$$u(r,\theta) = \frac{1}{2}\alpha_0 + \sum_{n=1}^{\infty}\left(\frac{r}{b}\right)^n (\alpha_n \cos n\theta + \beta_n \sin n\theta)\,, \tag{11.6.1}$$

where

$$\begin{Bmatrix}\alpha_n\\ \beta_n\end{Bmatrix} = \frac{1}{\pi}\int_{-\pi}^{\pi} f(\varphi)\begin{Bmatrix}\cos n\varphi\\ \sin n\varphi\end{Bmatrix} d\varphi\,. \tag{11.6.2}$$

From this series representation we shall derive an important integral formula for the solution of the problem. It will be assumed that $f(\varphi)$ is defined and continuous for $-\pi \leq \varphi \leq \pi$, or equivalently, for $-\infty < \varphi < \infty$ with period 2π.

From (11.6.2) we have

$$\alpha_n \cos n\theta + \beta_n \sin n\theta = \frac{1}{\pi}\int_{-\pi}^{\pi} f(\varphi)(\cos n\theta \cos n\varphi + \sin n\theta \sin n\varphi)\, d\varphi$$

or

$$\alpha_n \cos n\theta + \beta_n \sin n\theta = \frac{1}{\pi}\int_{-\pi}^{\pi} f(\varphi)\cos n(\theta - \varphi)\, d\varphi\,. \tag{11.6.3}$$

As a first consequence, this yields the inequality

$$|\alpha_n \cos n\theta + \beta_n \sin n\theta| \leq \frac{1}{\pi}\int_{-\pi}^{\pi} |f(\varphi)|\, d\varphi = M\,, \tag{11.6.4}$$

where the constant M does not depend on n. Thus, each term of (11.6.1) is majorized by the corresponding term of the series

$$\frac{1}{2}M + \sum_{n=1}^{\infty}\left(\frac{r}{b}\right)^n M\,. \tag{11.6.5}$$

Since (11.6.5) converges when $0 \leq r < b$, it follows that (11.6.1) also converges when $0 \leq r < b$. Thus, (11.6.1) defines a function $u(r,\theta)$ in the interior of the disc.

Substitution of (11.6.3) in (11.6.1) next leads to

$$(11.6.6) \quad u(r,\theta) = \frac{1}{2\pi}\int_{-\pi}^{\pi} f(\varphi)\, d\varphi + \frac{1}{\pi}\sum_{n=1}^{\infty}\left(\frac{r}{b}\right)^n \int_{-\pi}^{\pi} f(\varphi)\cos n(\theta - \varphi)\, d\varphi\, .$$

To simplify this formula, we first observe that interchange of the order of summation and integration gives

$$(11.6.7) \quad u(r,\theta) = \frac{1}{\pi}\int_{-\pi}^{\pi} f(\varphi)\left[\frac{1}{2} + \sum_{n=1}^{\infty}\left(\frac{r}{b}\right)^n \cos n(\theta - \varphi)\right] d\varphi\, .$$

To justify the interchange of order of summation and integration we show, using the Weierstrass M-test, that for each r, $0 \le r < b$, the series

$$(11.6.8) \quad \frac{1}{2} + \sum_{n=1}^{\infty}\left(\frac{r}{b}\right)^n \cos n(\theta - \varphi)$$

is uniformly convergent on the interval $-\pi \le \varphi \le \pi$. Indeed, this series is majorized by the series

$$\frac{1}{2} + \sum_{n=1}^{\infty}\left(\frac{r}{b}\right)^n ,$$

which converges if $0 \le r < b$. Hence, for any fixed r, $0 \le r < b$, the series (11.6.8) can be multiplied by the piecewise continuous function $f(\varphi)$ and integrated term-by-term. This shows that the right hand members of Equations (11.6.6) and (11.6.7) are equal.

A closed expression for the sum of the series (11.6.8) can now be obtained from (6.6.11). In (6.6.11) we let $a = r/b$, $t = \theta - \varphi$, and subtract $\frac{1}{2}$ from each member. The result is

$$(11.6.9) \quad \frac{1}{2}\frac{b^2 - r^2}{b^2 - 2rb\cos(\theta - \varphi) + r^2} = \frac{1}{2} + \sum_{n=1}^{\infty}\left(\frac{r}{b}\right)^n \cos n(\theta - \varphi)\, .$$

Thus, (11.6.7) is equivalent to

$$(11.6.10) \quad u(r,\theta) = \frac{1}{2\pi}\int_{-\pi}^{\pi} f(\varphi)\frac{b^2 - r^2}{b^2 - 2rb\cos(\theta - \varphi) + r^2}\, d\varphi\, .$$

Since (11.6.10) follows from the series representation of Section 5, which was established on the basis of the assumptions stated at the end of that section, we have proved the following theorem.

If $u(r,\theta)$ *is a point function which is continuous in the closed disc* $0 \le r \le b$, $-\pi \le \theta \le \pi$, *has continuous first- and second-order partial derivatives in the (open) disc* $0 \le r < b$, $-\pi \le \theta \le \pi$, *satisfies the*

D.E. $\Delta u = 0\,, \qquad 0 \leq r < b\,, \quad -\pi \leq \theta \leq \pi$

and

B.C. $u(b,\theta) = f(\theta)\,, \qquad -\pi \leq \theta \leq \pi\,,$

then

$$u(r,\theta) = \frac{1}{2\pi}\int_{-\pi}^{\pi} f(\varphi)\,\frac{b^2 - r^2}{b^2 - 2rb\cos(\theta - \varphi) + r^2}\,d\varphi\,.$$

Formula (11.6.10), which appears in the theorem above, is called **Poisson's integral formula for the disc.** Because it gives the solution of the Dirichlet problem for Laplace's equation in the disc in a relatively simple form, Poisson's formula makes apparent several important properties of the solution which are not evident from the series representation of the solution.

The function

$$\text{(11.6.11)} \qquad \frac{1}{2\pi}\,\frac{b^2 - r^2}{b^2 - 2rb\cos(\theta - \varphi) + r^2}\,,$$

which appears in the integral of (11.6.10), is called **Poisson's kernel.** This designation is sometimes applied to the special case in which $b = 1$,

$$\text{(11.6.12)} \qquad \frac{1}{2\pi}\,\frac{1 - r^2}{1 - 2r\cos(\theta - \varphi) + r^2}\,.$$

The more general function (11.6.11) can be obtained from (11.6.12) by replacing r by r/b.

From certain simple properties of Poisson's kernel we can deduce important properties of the solution of the Dirichlet problem for Laplace's equation in the disc. We shall establish two such simple properties of the Poisson kernel and discuss the related properties of the solutions of the Dirichlet problem. Other significant properties of the Poisson kernel, and consequences of the Poisson integral formula are left for Exercises 11d.

The first property to be established is

$$\text{(11.6.13)} \qquad 1 = \int_{-\pi}^{\pi} \frac{1}{2\pi}\,\frac{b^2 - r^2}{b^2 - 2rb\cos(\theta - \varphi) + r^2}\,d\varphi\,,$$

$$0 \leq r < b\,, \quad -\pi \leq \theta \leq \pi\,.$$

This follows immediately from the theorem above, since the constant function $u(r,\theta) = 1$ satisfies the hypothesis of that theorem with $f(\theta) = 1$.

The second property to which we referred is the inequality

$$\frac{1}{2\pi}\frac{b^2 - r^2}{b^2 - 2rb\cos(\theta - \varphi) + r^2} > 0\,, \qquad (11.6.14)$$
$$0 \leq r < b\,, \quad -\pi \leq \theta \leq \pi\,.$$

Since the numerator of (11.6.14) is clearly positive, it remains to show that the denominator is positive. The similarity between the denominator and the positive function $(b - r)^2$ suggests the comparison of these functions. We have

$$b^2 - 2br\cos(\theta - \varphi) + r^2 - (b - r)^2 = 2br[1 - \cos(\theta - \varphi)]\,,$$

from which we get, using a half-angle formula,

$$b^2 - 2br\cos(\theta - \varphi) + r^2 = (b - r)^2 + 4br\sin^2\left(\frac{\theta - \varphi}{2}\right). \qquad (11.6.15)$$

The inequality (11.6.14) follows obviously from the identity (11.6.15).

The simplest consequence of Poisson's formula is obtained by setting $r = 0$ in (11.6.10). We get

$$u(0,0) = \frac{1}{2\pi}\int_{-\pi}^{\pi} f(\varphi)\,d\varphi\,. \qquad (11.6.16)$$

This equation, which is also an immediate consequence of the series representation of Section 5, will be discussed in detail in the following Section 7.

A much less mathematically obvious proposition is the following. Let M and m be the maximum and minimum values of $u(r,\theta)$ for $r = b$ and $-\pi \leq \varphi \leq \pi$, so that

$$m \leq u(b,\varphi) \leq M\,, \qquad -\pi \leq \varphi \leq \pi\,. \qquad (11.6.17)$$

Then,

$$m \leq u(r,\theta) \leq M\,, \qquad -\pi \leq \theta \leq \pi\,, \quad 0 \leq r \leq b\,. \qquad (11.6.18)$$

To establish (11.6.18) we multiply the inequality (11.6.17) by the Poisson kernel (11.6.11), which we denote for convenience by $P(r,\theta,b,\varphi)$. We get, because of (11.6.14),

$$mP(r,\theta,b,\varphi) \leq u(b,\varphi)P(r,\theta,b,\varphi) \leq MP(r,\theta,b,\varphi)\,. \qquad (11.6.19)$$

Integrating (11.6.19) with respect to φ, we have

$$m\int_{-\pi}^{\pi} P(r,\theta,b,\varphi)\,d\varphi \leq \int_{-\pi}^{\pi} u(b,\varphi)P(r,\theta,b,\varphi)\,d\varphi \qquad (11.6.20)$$
$$\leq M\int_{-\pi}^{\pi} P(r,\theta,b,\varphi)\,d\varphi\,.$$

On setting $f(\varphi) = u(b,\varphi)$ in the relation (11.6.10) and taking (11.6.13) into account, we see that the inequality (11.6.20) is identical with inequality (11.6.18).

The proposition just established is the statement, for the special case in which the domain is a disc, of a theorem for general domains which will be discussed in Section 8.

EXERCISES 11d

1. Find a formula, analogous to (11.6.10), for the solution of Exercises 11c, Problem 9.

2. (i) Use Poisson's integral to show that if u satisfies Laplace's equation and is nonnegative in a disc of radius b, then u satisfies *Harnack's inequality*

$$\frac{b-r}{b+r}\,u(O) \leq u(P) \leq \frac{b+r}{b-r}\,u(O)\,,$$

where O is the center of the disc, P is any point of the disc, and $r = |OP|$.

(ii) Use Harnack's inequality to show that if u satisfies Laplace's equation in the entire plane and is nonnegative, then u is a constant.

(iii) Deduce that if u satisfies Laplace's equation in the entire plane and is bounded, then u is a constant.

3. Show that given arbitrary $\epsilon > 0$, $\sigma > 0$, there is a δ such that

$$0 < \frac{1}{2\pi}\,\frac{b^2 - r^2}{b^2 - 2br\cos(\theta - \varphi) + r^2} \leq \epsilon$$

for all r such that $0 < |b - r| < \delta$ and all θ and φ for which $\sigma \leq |\theta - \varphi| \leq 2\pi - \sigma$.

4. (i) Let $u(r,\theta)$ be the solution of problem (11.5.1) corresponding to the boundary function $f(\theta)$. Show that if f is continuous and nonnegative for $-\pi \leq \theta \leq \pi$, and is positive in some interval $[\alpha,\beta]$, $-\pi \leq \alpha < \beta \leq \pi$, then $u(r,\theta) > 0$ for all r, $0 \leq r < b$, and all θ.

(ii) Show that the solution $u(r,\theta)$ of problem (11.5.1) depends on the totality of boundary values $f(\theta)$ in the following sense. If (r_0,θ_0) is any point in the interior of the disc and (α,β) is any interval however small, $-\pi < \alpha < \beta < \pi$, there is a solution $v(r,\theta)$ of (11.5.1) corresponding to boundary values $g(\theta)$, such that $f(\theta) = g(\theta)$ except for $\alpha < \theta < \beta$, but $u(r_0,\theta_0) \neq v(r_0,\theta_0)$.

11.7. *THE MEAN VALUE THEOREM*

It is convenient and customary to call a function, which is continuous together with its first- and second-order partial derivatives in a domain D and which satisfies Laplace's equation in D, a **harmonic** function in D. Here D and Laplace's equation may be two- or three-dimensional.

Let u be a harmonic function in a plane domain D with boundary C, and suppose also that u is continuous in $D + C$. Let D_b be a disc of radius b with boundary C_b, such that $D_b + C_b$ lies entirely in $D + C$. If we introduce polar coordinates with origin at the center of D_b, we may apply equation (11.6.16), with $f(\varphi) = u(b,\varphi)$, obtaining

$$u(0,0) = \frac{1}{2\pi} \int_{-\pi}^{\pi} u(b,\varphi)\, d\varphi \tag{11.7.1}$$

or

$$u(0,0) = \frac{1}{2b\pi} \int_{-\pi}^{\pi} u(b,\varphi) b\, d\varphi. \tag{11.7.2}$$

Now since the element of arc length on the boundary C_b is $ds = b\, d\varphi$, formula (11.7.2) is the same as

$$u(0,0) = \frac{1}{2\pi b} \int_{C_b} u\, ds\, . \tag{11.7.3}$$

We shall say that if C is any curve of finite length L, and u is a function defined on C, then the *mean value* or average value of u on C is given by

$$\frac{1}{L} \int_{C} u\, ds. \tag{11.7.4}$$

With this terminology the result of the discussion ending with equation (11.7.3) can be stated as follows.

If u is a harmonic function in a plane domain D, with boundary C, and u is continuous in $D + C$, then the value of u at the center of any disc, which lies entirely in $D + C$, is equal to the mean value of u on the boundary of the disc.

This statement is called the **mean value theorem** for harmonic functions in two dimensions. We shall present another proof of this theorem which is independent of any representation of harmonic functions, and which has the virtue that it can be extended immediately to yield the corresponding theorem for harmonic functions in three dimensions and analogous statements for certain other related equations. Most important, our new proof will enable us to demonstrate the very close connection between the mean value property and the property of being harmonic.

The hypothesis of the mean value theorem insures that if D_b is chosen as above, then for any disc D_r, of radius $r < b$, concentric with D_b, Equation (11.2.7) holds,

$$\iint_{D_r} \Delta u \, dx \, dy = \int_{C_r} \frac{\partial u}{\partial \nu} \, ds \, .$$

If we introduce polar coordinates with origin at the center of D_b, this equation becomes

$$\int_0^r \int_0^{2\pi} \Delta u \, \rho \, d\rho \, d\theta = \int_0^{2\pi} \frac{\partial u}{\partial r}(r,\theta) r \, d\theta \, . \tag{11.7.5}$$

If u is harmonic, then the left side of (11.7.5) is zero, and dividing the equation by r, we get

$$0 = \int_0^{2\pi} \frac{\partial u}{\partial r}(r,\theta) \, d\theta \, . \tag{11.7.6}$$

Now set

$$I(r) = \int_0^{2\pi} u(r,\theta) \, d\theta \, . \tag{11.7.7}$$

Then $I(r)$ is smooth for $0 \leq r \leq b$, and from (11.7.6),

$$0 = \frac{dI(r)}{dr}, \qquad 0 < r < b \, . \tag{11.7.8}$$

Thus, $I(r)$ is constant for $0 \leq r \leq b$, so that, in particular,

$$I(0) = I(r) = I(b) \, . \tag{11.7.9}$$

To calculate $I(0)$ we use the continuity of I and u, and the mean value theorem of the calculus,

$$\begin{aligned} I(0) &= \lim_{r \to 0} I(r) = \lim_{r \to 0} \int_0^{2\pi} u(r,\theta) \, d\theta \\ &= 2\pi \lim_{\substack{r \to 0 \\ 0 < \tilde{\theta} < 2\pi.}} u(r,\tilde{\theta}) = 2\pi u(0,0) \, . \end{aligned} \tag{11.7.10}$$

From (11.7.7), (11.7.9), and (11.7.10) follows (11.7.2), and the mean value theorem is proved.

By reversing the preceding argument we can prove the following.

Let u be continuous together with its first- and second-order partial derivatives in a domain D, and suppose u has the property stated in the conclusion of the mean value theorem. Then u is harmonic in D.

To prove this we calculate Δu, obtaining some continuous function, say p,

$$\Delta u = p \, , \tag{11.7.11}$$

and show that $p = 0$ at every point in D. Since every point of D is the center of some disc which, with its boundary, lies in D, it is clearly sufficient to adopt the notation of the preceding discussion, and to show that $p = 0$ at the center of D_b. Substituting (11.7.11) and (11.7.7) in (11.7.5), we get

$$\int_0^r \int_0^{2\pi} p(\rho,\theta)\rho \, d\rho \, d\theta = r\frac{dI(r)}{dr}. \tag{11.7.12}$$

The mean value property implies that $I(r)$ is constant for $0 \leq r \leq b$, $I(r) = 2\pi u(0,0)$, so that the right side of (11.7.12) is zero. Thus,

$$\int_0^r \int_0^{2\pi} p(\rho,\theta)\rho \, d\rho \, d\theta = 0\,, \qquad 0 \leq r \leq b\,. \tag{11.7.13}$$

Now suppose p is not 0 for $r = 0$, but is, say, positive. Then for sufficiently small r the integrand of (11.7.13), and hence the integral of that equation, will be positive. This contradiction proves the assertion.

This characterization of harmonic functions in terms of one of their seemingly simpler properties is quite remarkable. Even more striking is the proposition, proved in many advanced textbooks on partial differential equations, that the conclusion holds without the assumption of the existence and continuity of the first- and second-order partial derivatives of the function.

EXERCISES 11e

1. The mean value of a function u on a plane domain D of area A is defined by

$$\frac{1}{A}\iint_D u \, dx \, dy\,.$$

Use the mean value theorem for harmonic functions in the plane to show that the value of a harmonic function at the center of a disc is equal to its mean value on the disc. (This is sometimes called the *interior* mean value theorem.)

2. If Δ is the two-dimensional Laplace operator, a solution of the differential equation

$$\Delta\Delta w = \Delta(\Delta w) = 0$$

is called a (two-dimensional) biharmonic function. Use the result of Problem 1 and Equation (11.7.5) to prove a "mean value theorem" for biharmonic functions.

3. State and prove a mean value theorem for harmonic functions in three dimensions.

4. (i) Let Δ be the two-dimensional Laplace operator and let u be a solution of

$$\Delta u + \gamma u = 0$$

in a domain D which contains the disc D_b of radius b and its boundary C_b. Define $I(r)$ by (11.7.7) and show that I satisfies the ordinary differential equation

$$rI'' + I' + \gamma r I = 0 .$$

(ii) Given that the only solutions of the ordinary differential equation of part (i) which are continuous at $r = 0$ are constant multiples of a function $J_0(r\sqrt{\gamma})$ (a non-elementary function which we shall study in the next chapter), and that $J_0(0) = 1$, derive a mean value formula for the partial differential equation of part (i).

5. (i) Find the general solution of the ordinary differential equation

$$r^2 I'' + 2rI' + \gamma r^2 I = 0$$

by use of the substitution

$$I(r) = \frac{\Phi(r)}{r} .$$

(ii) Let Δ be the three-dimensional Laplace operator and let u be a solution of

$$\Delta u + \gamma u = 0$$

in a domain D which contains the spherical surface S_b of radius b, and its interior. Prove the mean value formula

$$u_0 \frac{\sin b\sqrt{\gamma}}{b\sqrt{\gamma}} = \frac{1}{4\pi b^2} \iint_{S_b} u \, dS ,$$

where u_0 is the value of u at the center of S_b.

11.8. *MAXIMUM PRINCIPLES*

Consider a source-free heat conducting body of finite extent and suppose that a fixed temperature distribution is maintained on the boundary. Then the temperature in the interior of the body will approach an equilibrium temperature distribution. From physical considerations it is clear that at no point in the interior of the body will the equilibrium temperature be greater than the temperature at *every* point on the boundary. For otherwise there would be a tendency for heat to flow from the hotter interior point to the cooler boundary points, without any compensating flow of heat from the boundary to the interior point. Thus the temperature at the interior point would decrease, in contradiction to our assumption of equilibrium. We conclude that at no point in the interior will the equilibrium temperature be greater than the maximum temperature on the boundary. For a similar reason it is clear that at no point in the interior will the equilibrium temperature be less than the minimum temperature on the boundary.

The conclusions of these intuitive arguments can be formulated as propositions about the solutions of certain partial differential equations. If the medium is uniform, then the equilibrium temperature distribution satisfies Laplace's equation. Conversely, we believe that any solution of Laplace's equation can be interpreted as an equilibrium temperature distribution. Hence the discussion indicates that any solution of $\Delta u = 0$ will attain its maximum and minimum values on the boundary of the domain in which it is defined.

The physical argument applies also to the case of a nonuniform medium whose conductivity κ varies from point to point. Thus, in the two-dimensional case, the argument indicates that a solution of

$$\frac{\partial}{\partial x}\left[\kappa(x,y)\,\frac{\partial u}{\partial x}\right] + \frac{\partial}{\partial y}\left[\kappa(x,y)\,\frac{\partial u}{\partial y}\right] = 0 \tag{11.8.1}$$

will attain its maximum and minimum values on the boundary of its domain of definition.

If there are heat sources present in the medium, it is still possible to draw certain similar conclusions. For example, in Poisson's equation,

$$\Delta u = -q(x,y)\,, \tag{11.8.2}$$

if $q(x,y)$ is positive throughout the domain under consideration, then we interpret the solution as an equilibrium temperature distribution when heat is being produced at a positive rate within the medium. Under such circumstances the maximum temperature may perhaps be attained in the interior, but we expect the minimum temperature to occur on the boundary. Similarly, if $q(x,y)$ were negative, we would expect the maximum temperature to occur on the boundary.

There are other areas in which intuitive arguments about temperature distributions indicate the validity of propositions about the maxima and minima of solutions of partial differential equations. We will formulate more precise statements of some of these propositions and then establish them by mathematical arguments.

Our first proposition is called the **maximum principle for harmonic functions.** This proposition, which we state below, is often called the "weak form" of the maximum principle. The reason for this qualification will be indicated at the end of the section.

Let D be a finite domain with boundary C and let u be a function continuous in $D + C$ and harmonic in D. Then the maximum value of u is achieved on the boundary.

The maximum principle implies an analogous minimum principle. We apply the maximum principle to the harmonic function $v = -u$, which achieves its maximum value at exactly the same points at which u achieves its minimum value. Since the maximum value of v is achieved on the boundary, it follows that *the minimum value of u is achieved on the boundary.*

Before we prove the maximum principle we will show, as one application, how it can be used to establish a *uniqueness theorem* for the Dirichlet problem for Laplace's equation in a finite domain with continuous boundary data. Consider the problem

$$(11.8.3) \qquad \begin{array}{lll} \text{D.E.} & \Delta u = 0 & \text{in } D\,, \\ \text{B.C.} & u = f & \text{on } C\,, \end{array}$$

where D is a finite domain with boundary C and f is a continuous function defined on C. By a solution of (11.8.3) we mean a function which has continuous first- and second-order partial derivatives and satisfies the D.E. in D, is continuous in $D + C$, and satisfies the B.C. in the sense of equality. We wish to show that there is at most one solution of this problem. The difference of two such solutions is a function w continuous in $D + C$, harmonic in D, and satisfying $w = 0$ on C. By the maximum principle, w achieves its maximum value on C, so the maximum value of w is zero and hence $w \leq 0$ in D. Similarly, w achieves its minimum value on C, so that $w \geq 0$ in D. Therefore $w = 0$ at all points of D, and the solution of (11.8.3) is unique.

We can also use the maximum principle to show that the solution of Dirichlet's problem *depends continuously on the boundary data.* This means, roughly speaking, that a small change in the boundary data results in only small changes of the solution at interior points. We formulate this precisely, as follows. *Let D be a finite domain with boundary C, and let f_1, f_2 be continuous functions defined on C. Let u_1 and u_2 be the solutions of $\Delta u = 0$ in D such that on C, $u_1 = f_1$, $u_2 = f_2$. If $|f_1 - f_2| < \epsilon$ at all points of C, then $|u_1 - u_2| < \epsilon$ at all points of D.* To prove this, we note that $v = u_1 - u_2$ is the solution of the problem $\Delta v = 0$ in D, $v = f_1 - f_2$ on C. By the maximum principle, the maximum M and minimum m of v are achieved on the boundary. Since

$$-\epsilon < f_1 - f_2 < \epsilon \qquad \text{on } C\,,$$

we have

$$-\epsilon < m \leq M < \epsilon$$

so that, at any interior point,

$$-\epsilon < m \leq v \leq M < \epsilon$$

and hence $|v| < \epsilon$ in D, which was to be proved.

This result has immediate and important physical interpretations. For example, in terms of equilibrium heat conduction, it implies that a small error in the measurement of the temperature on the boundary results in a small error in the theoretically calculated values of the temperature in the interior. Since physical measurement inevitably involves some error, it is clear that if some statement such as that above did not hold, the comparison of theory and experiment would not be possible.

We turn now to the proof of the maximum principle for harmonic functions of two variables. We first recall from elementary calculus that if a real valued function is continuous on a finite closed interval, then there is at least one point of the closed interval where the function achieves its maximum value. Similarly, a continuous function of several variables defined on a finite closed region achieves its maximum value at one or more points of the closed region. Therefore, if the function u is harmonic in a finite region D and continuous in $D + C$, there is a point Q in $D + C$ such that $u(Q) = M$, where the constant M is the maximum value of u in $D + C$. If Q is on the boundary of D, then u achieves its maximum value on the boundary. If Q is not on the boundary, we still have to find a point Q_1 on the boundary such that $u(Q_1) = M$. Now if Q is not on the boundary, then there are discs with centers at Q contained in D. Let D_1 be the largest disc with this property, so that the radius of the disc is the shortest distance from Q to the boundary of D (see Fig. 11.1). The boundary C_1 of D_1 is contained in $D + C$, and on C_1 there is at least one point Q_1 of the boundary C. We will show that $u = M$ at every point of C_1, and hence, $u(Q_1) = M$, as follows. We apply the mean value theorem to the harmonic function $v = u - M$ to obtain

$$v(Q) = \frac{1}{L} \int_{C_1} v\, ds\,, \tag{11.8.4}$$

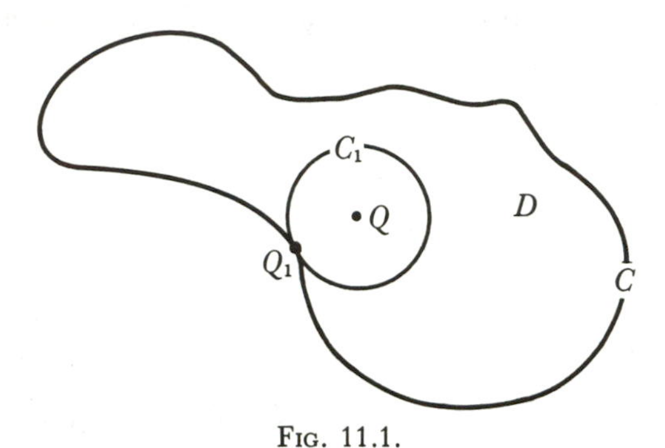

FIG. 11.1.

where L is the length of C_1. From $u \leq M$ on C_1, it follows that $v \leq 0$ on C_1, and hence,

$$\int_{C_1} v\, ds \leq 0 .$$

Moreover, if v were negative at a point of the circle C_1, then, because v is continuous, v would be negative on an entire arc of the circle and we would have

$$\int_{C_1} v\, ds < 0 . \tag{11.8.5}$$

But (11.8.4) and (11.8.5) imply that $v(Q) < 0$, which is impossible since $v(Q) = u(Q) - M = 0$. Thus, the assumption that v is negative at any point on C_1 leads to a contradiction. Therefore, $v = 0$ at all points of C_1 and $u(Q_1) = M$. This proves the maximum principle.

For the continuation of our discussion of maximum and minimum values, the reader should recall from calculus the definition of local maximum and minimum points of a function of one variable and the conditions, in terms of the derivatives of the function, which must be satisfied at such points. A function $u(x,y)$ defined in a domain D is said to have a local maximum at a point (x_0,y_0) if $u(x,y) \leq u(x_0,y_0)$ for all points (x,y) in some disc with center (x_0,y_0) and positive radius δ. If this is the case, then the function $u(x,y_0)$ of x has a local maximum at $x = x_0$ and the function $u(x_0,y)$ of y has a local maximum at $y = y_0$. If u has continuous first- and second-order partial derivatives, we deduce from these statements about functions of one variable that

$$\begin{gathered} u_x(x_0,y_0) = u_y(x_0,y_0) = 0 , \\ u_{xx}(x_0,y_0) \leq 0 , \quad u_{yy}(x_0,y_0) \leq 0 . \end{gathered} \tag{11.8.6}$$

Similarly, if u has a local minimum at (x_0,y_0),

$$\begin{aligned} u_x(x_0,y_0) &= u_y(x_0,y_0) = 0\,, \\ u_{xx}(x_0,y_0) \geq 0\,, &\quad u_{yy}(x_0,y_0) \geq 0\,. \end{aligned} \tag{11.8.7}$$

Equations (11.8.6) and (11.8.7) imply, in particular, that $\Delta u \leq 0$ at a local maximum and $\Delta u \geq 0$ at a local minimum. This has the following consequences for solutions of Poisson's equation:

A solution of $\Delta u = -q(x,y)$ cannot have a local maximum in a domain in which q is negative and cannot have a local minimum in a domain in which q is positive.

The proof of the first statement merely requires the observation that the statements $\Delta u \leq 0$ at a local maximum, and $\Delta u = -q(x,y) > 0$ are in contradiction. The second statement is established similarly. Notice that because of the first pairs of Equations (11.8.6) and (11.8.7), both the statements above and their proof hold for the more general equations

$$\Delta u + a(x,y)u_x + b(x,y)u_y = -q(x,y)\,,$$

$$\frac{\partial}{\partial x}[\kappa(x,y)u_x] + \frac{\partial}{\partial y}[\kappa(x,y)u_y] = -q(x,y)\,,$$

where $\kappa(x,y) > 0$ has first-order partial derivatives.

Statements about local maxima and minima immediately yield statements about (absolute) maximum and minimum values in domains. Suppose, for example, that u satisfies $\Delta u = -q(x,y)$, with $q(x,y)$ negative, in a domain D with boundary C, and that u is continuous in $D + C$. Then, because of the last condition, u attains its maximum value at some point of $D + C$. If the maximum value were attained at a point of D, then it would be a local maximum. Since we have shown that u cannot have a local maximum in D, the maximum value must be attained on the boundary C, in agreement with our earlier intuitive discussion.

There is a statement about local maxima and minima of harmonic functions, from which the theorem we first proved in this section follows as in the preceding paragraph. This statement, which we present below, is usually called the maximum principle for harmonic functions, and our earlier statement the "weak form" of the principle.

A function which is harmonic in a domain does not have a local maximum or minimum in the domain unless it is a constant function.

We shall not prove this theorem, since the weak form of the maximum principle for harmonic functions was proved independently, and is all that is required for our applications.

EXERCISES 11f

1. Give an intuitive argument based on consideration of equilibrium temperature distributions to show that when $\gamma < 0$ no solution of

$$\Delta u + \gamma u = 0$$

can have a positive value at a local maximum in the interior of its domain of definition.

2. (i) Let D be a finite domain in which $a(x,y)$, $b(x,y)$, and $c(x,y)$ are continuous and assume that

$$c(x,y) < 0 \qquad \text{in } D\,.$$

Let u be a solution of

$$\Delta u + a(x,y)u_x + b(x,y)u_y + c(x,y)u = 0 \qquad \text{in } D\,.$$

Show that u cannot have a positive value at a local maximum in D, and cannot have a negative value at a local minimum in D.

(ii) Use the conclusion of part (i) to show that under the hypothesis of part (i) and the additional assumption that u is continuous in $D + C$, the problem

$$\begin{array}{lrl} \text{D.E.} & \Delta u + a(x,y)u_x + b(x,y)u_y + c(x,y)u = 0 & \text{in } D\,, \\ \text{B.C.} & u = f & \text{on } C\,, \end{array}$$

has at most one solution.

3. The (weak) maximum principle for Laplace's equation can be inferred from the corresponding statement for Poisson's equation as follows (D is a domain with boundary C). Let u be a solution of $\Delta u = 0$ in D which is continuous in $D + C$. If u does not attain its maximum on C, then it does at a point (x_0, y_0) of D. For sufficiently small positive ϵ the function $v = u + \epsilon(x - x_0)^2$ also does not attain its maximum on C. This is a contradiction of the maximum principle for Poisson's equation. Complete the details of this argument.

4. Let D be the domain $0 < x < L$, $0 < t < T_0$, let C be the boundary of D, and let Γ be the part of C comprised of the three segments $t = 0$, $0 \le x \le L$; $x = 0$, $0 \le t \le T_0$; $x = L$, $0 \le t \le T_0$. The following assertion is called the **maximum principle for the heat equation**: *if u is continuous in $D + C$ and satisfies $u_t - u_{xx} = 0$ in D, then the maximum of u in $D + C$ is attained on Γ*. Give a physical argument supporting the assertion. From the maximum principle deduce a corresponding minimum principle.

5. Let $u(x,t)$ be continuous for $0 \le x \le L$, $0 \le t$, and satisfy

$$u_t = u_{xx} + q(x,t)\,, \qquad 0 < x < L\,, \quad 0 < t\,.$$

Let D, C, and Γ be defined as in Problem 4. Show that *if $q(x,t) < 0$ in $D + C$, then the maximum of u is attained on Γ, and if $q(x,t) > 0$ in $D + C$,* *then the minimum of u is attained on* Γ. [Recall that if a function $\varphi(t)$ is continuous together with its first derivative in the interval $0 \leq t \leq T$, and the maximum value of φ in this interval is $\varphi(T)$, then $\varphi'(T) \geq 0$.]

6. (i) Let $u(x,t)$ be continuous for $0 \leq x \leq L$, $0 \leq t$ and satisfy

$$u_t = u_{xx}\,, \qquad 0 < x < L\,, \quad 0 < t\,.$$

Use the result of Problem 5 to prove the assertion of Problem 4, taking Problem 3 as a model.

(ii) State and prove a uniqueness theorem for the initial-boundary value problem

$$\begin{array}{lll} \text{D.E.} & u_t = u_{xx}\,, & 0 < x < L\,, \quad 0 < t\,, \\ \text{B.C.} & u(0,t) = 0\,, & \\ & u(L,t) = 0\,, & 0 < t\,, \\ \text{I.C.} & u(x,0) = f(x)\,. & \end{array}$$

Include in your statement all assumptions about u and f that are required.

11.9. *BOUNDARY VALUE PROBLEMS IN INFINITE DOMAINS*

In this section we consider boundary value problems for Laplace's equation and related equations in quarter planes, half planes, semi-infinite strips, and infinite strips. These problems in infinite domains can be viewed as limiting cases of the problems in rectangular domains considered in Section 3. This view provides a physical interpretation of the problems we consider and can serve as a guide to their solution. We shall present a solution procedure by means of examples.

Before considering these examples, we remark on a necessary preliminary reduction. We may take the sides of the domain in a problem to be parallel to the coordinate axes. If not all sides of the domain are parallel to *one* of the coordinate axes, then, as in Section 3, we can reduce the solution of the problem to the solution of problems in which the boundary conditions are *homogeneous* on all sides parallel to one of the coordinate axes. Accordingly, we need only consider problems in which either all sides of the domain are parallel to one of the coordinates axes, or, if this is not so, the boundary conditions are homogeneous on all sides parallel to one of the coordinate axes.

The solution method we shall employ in general is the transform method of Chapter 9. The reader will recall from Chapter 9 that the use of this method requires assumptions about the data and the solution which, though reasonable, are not a priori necessary and on more careful analysis may prove to be excessively stringent or unnecessary. We shall consider first a problem in which the possibly artificial restrictions involved in the use of the transform method do not arise, namely the following Dirichlet problem for Laplace's equation in a semi-infinite strip,

$$(11.9.1)\qquad \begin{array}{lll} \text{D.E.} & u_{xx} + u_{yy} = 0\,, \quad 0 < x < \pi\,, & 0 < y < \infty\,, \\ \text{B.C.} & u(0,y) = 0\,, & 0 < y < \infty\,, \\ & u(\pi,y) = 0\,, & 0 < y < \infty\,, \\ & u(x,0) = g(x)\,, \quad 0 < x < \pi\,. & \end{array}$$

This problem can be viewed as the limiting case of problem (11.3.2), with $a = \pi$, when $b \to \infty$. For (11.3.2) we found that the appropriate eigenfunction expansion is the Fourier sine series with respect to x. We solve (11.9.1) using this representation and the method of variation of parameters.

If $u(x,y)$ is the solution of (11.9.1), then for each fixed y

$$(11.9.2)\qquad u(x,y) = \sum_{n=1}^{\infty} B_n(y) \sin nx\,,$$

where

$$(11.9.3)\qquad B_n(y) = \frac{2}{\pi} \int_0^{\pi} u(x,y) \sin nx\, dx\,.$$

Differentiating (11.9.3) twice with respect to y, and using the D.E. of (11.9.1) we obtain

$$(11.9.4)\qquad \frac{d^2}{dy^2} B_n(y) = -\frac{2}{\pi} \int_0^{\pi} u_{xx}(x,y) \sin nx\, dx\,.$$

Applying Green's formula to (11.9.4) and using the fact that $\sin nx$ and $u(x,y)$ vanish for $x = 0$ and $x = \pi$, we get the ordinary differential equation for $B_n(y)$,

$$(11.9.5)\qquad \frac{d^2}{dy^2} B_n(y) - n^2 B_n(y) = 0\,, \qquad 0 < y < \infty\,.$$

Setting $x = 0$ in (11.9.3) we have

$$(11.9.6)\qquad B_n(0) = b_n = \frac{2}{\pi} \int_0^{\pi} g(x) \sin nx\, dx\,,$$

as one auxiliary condition for (11.9.5). However, (11.9.5) is a second-order differential equation and two auxiliary conditions are needed to specify a solution. The general solution of (11.9.5) is a linear combination of e^{-ny} and e^{ny}. A physically plausible condition is that $u(x,y)$ be bounded for $0 < x < \pi$, $0 < y$. If we impose this requirement, it follows immediately from (11.9.3) that B_n must satisfy

$$B_n(y) \text{ is bounded as } y \to \infty \,. \tag{11.9.7}$$

This condition will be satisfied if and only if the coefficient of e^{ny} in the representation of $B_n(y)$ is zero. Thus, the solution of (11.9.1) is found to be

$$u(x,y) = \sum_{n=1}^{\infty} b_n e^{-ny} \sin nx \,, \tag{11.9.8}$$

with b_n given by (11.9.6).

The auxiliary condition that $u(x,y)$ be bounded (or some other condition) must be adjoined to the conditions stated in (11.9.1) for the problem to be well determined. That this is so and that the condition is not just an artificial one associated with our method of solution can also be seen directly. The function

$$\psi(x,y) = (e^{y} - e^{-y}) \sin x$$

is clearly a solution of Laplace's equation, which is zero on the entire boundary of the semi-infinite strip of (11.9.1). Thus, if u is any solution of (11.9.1), $u + \psi$ is also a solution, so that the problem does not have a unique solution.

Now we consider a problem, the method of solution of which is typical for the problems of this section,

$$\begin{array}{llll} & \text{D.E.} & u_{xx} + u_{yy} = 0 \,, & -\infty < x < \infty \,, \\ & & & 0 < y < \infty \,, \\ (11.9.9) & \text{B.C.} & u(x,0) = f(x) \,, & -\infty < x < \infty \,, \\ & & u(x,y) \text{ bounded} \,, & -\infty < x < \infty \,, \\ & & & 0 < y < \infty \,. \end{array}$$

This is Dirichlet's problem for Laplace's equation in a half plane. Interpretation of this problem as a limiting case of a problem in a rectangle, or comparison of the problem with those discussed in Chapter 9, suggests that we undertake to find the solution using the Fourier transform with respect to x.

Proceeding formally, we set

$$U(\lambda,y) = \int_{-\infty}^{\infty} u(x,y) e^{i\lambda x} \, dx \,. \tag{11.9.10}$$

Differentiating (11.9.10) twice with respect to y, and substituting the D.E. of (11.9.9), we have

$$\frac{\partial^2 U}{\partial y^2} = -\int_{-\infty}^{\infty} u_{xx} e^{i\lambda x}\,dx\,. \tag{11.9.11}$$

Applying Green's formula to (11.9.11), assuming that

$$\lim_{x\to\pm\infty} u(x,y) = \lim_{x\to\pm\infty} u_x(x,y) = 0\,, \qquad 0 < y < \infty\,, \tag{11.9.12}$$

we get the ordinary differential equation for U as a function of y,

$$\frac{\partial^2 U}{\partial y^2} - \lambda^2 U = 0\,, \qquad 0 < y < \infty\,, \quad -\infty < \lambda < \infty\,. \tag{11.9.13}$$

Setting $y = 0$ in (11.9.10) and using the B.C. of (11.9.9), we get one auxiliary condition for (11.9.13),

$$U(\lambda,0) = F(\lambda) = \int_{-\infty}^{\infty} f(x)e^{i\lambda x}\,dx\,, \qquad -\infty < \lambda < \infty\,. \tag{11.9.14}$$

The second auxiliary condition, corresponding to the requirement in problem (11.9.9) that u be bounded, we take to be

$$U(\lambda,y) \text{ is bounded as } y \to \infty\,, \qquad -\infty < \lambda < \infty\,, \tag{11.9.15}$$

in analogy with (11.9.7). The solution of (11.9.13) that satisfies the auxiliary conditions (11.9.14), (11.9.15) is

$$U(\lambda,y) = F(\lambda)e^{-|\lambda|y}\,, \qquad -\infty < \lambda < \infty\,. \tag{11.9.16}$$

The solution of problem (11.9.9), obtained by substituting (11.9.16) and (11.9.14) in the inversion formula for the Fourier transform, namely,

$$u(x,y) = \frac{1}{2\pi}\int_{-\infty}^{\infty} U(\lambda,y)e^{-i\lambda x}\,d\lambda\,, \tag{11.9.17}$$

is

$$u(x,y) = \frac{1}{2\pi}\int_{-\infty}^{\infty}\left(\int_{-\infty}^{\infty} f(\xi)e^{i\lambda\xi}\,d\xi\right)e^{-|\lambda|y}e^{-i\lambda x}\,d\lambda\,. \tag{11.9.18}$$

We can, however, obtain another, simpler, representation of the solution. By definition, $F(\lambda)$ is the Fourier transform of $f(x)$ and, according to Exercises 9a, Problem 5(vii), $e^{-|\lambda|y}$ is the Fourier transform of

$$\frac{1}{\pi}\frac{y}{x^2 + y^2}\,.$$

Hence, by the convolution theorem, the solution $u(x,y)$ whose Fourier transform $U(\lambda,y)$ is the product (11.9.16) must be given by

$$u(x,y) = \frac{1}{\pi}\int_{-\infty}^{\infty} \frac{yf(\xi)}{(x-\xi)^2 + y^2}\,d\xi\,. \tag{11.9.19}$$

Formula (11.9.19) is called **Poisson's integral formula for the half plane.**

We leave as exercises for the reader the solution of boundary value problems in other domains cited in the first paragraph of this section. There are two essential features of the general solution procedure. The first is the choice of a transform, from those given in Chapter 9, that is appropriate to the geometry of the domain and to the type of the homogeneous boundary condition. The second is the use of boundedness of the transform as an auxiliary condition when the transform is not otherwise completely specified. Boundary value problems for corresponding domains in space can be solved similarly, using the transforms given in Sections 10.8 and 10.9.

EXERCISES 11g

1. Solve

D.E.	$u_{xx} + u_{yy} = 0\,,$	$0 < x < \infty\,,$	$0 < y < \infty\,,$
B.C.	$u_x(0,y) = 0\,,$		$0 < y < \infty\,,$
	$u(x,0) = f(x)\,,$	$0 < x < \infty\,,$	
	$u(x,y)$ bounded	$0 < x < \infty\,,$	$0 < y < \infty\,.$

2. Solve

D.E.	$u_{xx} + u_{yy} = 0\,,$	$0 < x < \infty\,,$	$0 < y < \infty\,,$
B.C.	$u(0,y) = g(y)\,,$		$0 < y < \infty\,,$
	$u(x,0) = 0\,,$	$0 < x < \infty\,,$	
	$u(x,y)$ bounded ,	$0 < x < \infty\,,$	$0 < y < \infty\,.$

3. Solve

D.E.	$u_{xx} + u_{yy} = 0\,,$	$0 < x < a\,,$	$0 < y < \infty\,,$
B.C.	$u(0,y) = f(y)\,,$		$0 < y < \infty\,,$
	$u(a,y) = k(y)\,,$		$0 < y < \infty\,,$
	$u(x,0) = 0\,,$	$0 < x < a\,,$	
	$u(x,y)$ bounded ,	$0 < x < a\,,$	$0 < y < \infty\,.$

4. Solve

D.E.	$u_{xx} + u_{yy} = 0\,,$	$0 < x < a\,,$	$0 < y < \infty\,,$
B.C.	$u(0,y) = 0\,,$		$0 < y < \infty\,,$
	$u(a,y) = f(y)\,,$		$0 < y < \infty\,,$
	$u_y(x,0) = 0\,,$	$0 < x < a\,,$	
	$u(x,y)$ bounded ,	$0 < x < a\,,$	$0 < y < \infty\,.$

5. Solve

D.E.	$u_{xx} + u_{yy} = 0\,,$	$0 < x < a\,,$	$-\infty < y < \infty\,,$
B.C.	$u(0,y) = 0\,,$		$-\infty < y < \infty\,,$
	$u(a,y) = f(y)\,,$		$-\infty < y < \infty\,,$
	$u(x,y)$ bounded ,	$0 < x < a\,,$	$-\infty < y < \infty\,.$

6. Solve

$$\text{D.E.}\quad u_{xx} + u_{yy} + u_{zz} = 0\,, \qquad -\infty < x < \infty\,,\quad -\infty < y < \infty\,,\quad 0 < z < \infty\,,$$

$$\text{B.C.}\quad u(x,y,0) = f(x,y)\,, \qquad -\infty < x < \infty\,,\quad -\infty < y < \infty\,,$$

$$u(x,y,z) \text{ bounded}\,, \qquad -\infty < x < \infty\,,\quad -\infty < y < \infty\,,\quad 0 < z < \infty\,.$$

7. Solve

$$\text{D.E.}\quad u_{xx} + u_{yy} + u_{zz} = 0\,, \qquad 0 < x < \infty\,,\quad -\infty < y < \infty\,,\quad 0 < z < \infty\,,$$

$$\text{B.C.}\quad u_x(0,y,z) = 0\,, \qquad -\infty < y < \infty\,,\quad 0 < z < \infty\,,$$

$$u(x,y,0) = f(x,y)\,, \qquad 0 < x < \infty\,,\quad -\infty < y < \infty\,,$$

$$u(x,y,z) \text{ bounded}\,, \qquad 0 < x < \infty\,,\quad -\infty < y < \infty\,,\quad 0 < z < \infty\,.$$

8. Solve the boundary value problem for an infinite sector (r, θ polar coordinates),

$$\text{D.E.}\quad \Delta u = \frac{1}{r}\,(ru_r)_r + \frac{1}{r^2}\,u_{\theta\theta} = 0\,, \qquad 0 < r < \infty\,,\quad 0 < \theta < \alpha\,,$$

$$\text{B.C.}\quad u(0,r) = 0\,, \qquad 0 < r < \infty\,,$$

$$u(\alpha,r) = f(r)\,, \qquad 0 < r < \infty\,,$$

$$u(r,\theta) \text{ bounded}\,, \qquad 0 \le r < \infty\,,\quad 0 < \theta < \alpha\,,$$

by first introducing the new variable $\xi = \log(1/r)$.

9. (i) Show that for $-\infty < x < \infty\,,\ 0 < y < \infty\,,$

$$\frac{1}{\pi}\int_{-\infty}^{\infty} \frac{y}{(x-\xi)^2 + y^2}\,d\xi = 1\,.$$

(ii) Show that if $f(x)$ is bounded on $(-\infty,\infty)$, explicitly $|f(x)| \le M$ for $-\infty < x < \infty$, then the integral in (11.9.19) is convergent, and the function $u(x,y)$ defined by the integral satisfies $|u(x,y)| \le M$ for $-\infty < x < \infty$, $0 < y < \infty$.

(iii) Use the theorem of Weierstrass proved in Section 9.6 to show that if $f(x)$ is bounded on $(-\infty,\infty)$ and $u(x,y)$ is given by (11.9.19), then

$$\lim_{y\to 0+} u(x,y) = f(x)$$

for each value of x at which f is continuous.

10. Let u be given in terms of f by (11.9.19). Show that for any point (x_0, y_0), where $y_0 > 0$, and for any finite interval $a \le x \le b$, the value of u at (x_0, y_0) cannot be determined from the values of f only in the interval $a \le x \le b$. (Use the method of Exercises 9f, Problem 5.)

11. (i) Show that the solution obtained in Problem 6 can be transformed to

$$u(x,y,z) = \int_{-\infty}^{\infty}\int_{-\infty}^{\infty} p(x-\xi, y-\eta, z)f(\xi,\eta)\, d\xi\, d\eta ,$$

where $p(x, y, z)$ is the function

$$\frac{1}{(2\pi)^2}\int_{-\infty}^{\infty}\int_{-\infty}^{\infty} \exp\left[-z\sqrt{\lambda^2+\mu^2}\right] \exp\left[-i(\lambda x + \mu y)\right] d\lambda\, d\mu .$$

(ii) By introducing polar coordinates $x = r\cos\theta, y = r\sin\theta$ in the (x, y) plane, and also polar coordinates $\lambda = \rho\cos\varphi$, $\mu = \rho\sin\varphi$ as new integration variables, show that

$$p(x,y,z) = \frac{-1}{(2\pi)^2}\frac{\partial}{\partial z}F(z,r) ,$$

where

$$F(z,r) = \int_{-\pi}^{\pi} \frac{d\varphi}{z + ir\cos\varphi} .$$

(iii) Use the substitution $t = \tan(\theta/2)$ (or any other method) to evaluate the last integral of part (ii), obtaining

$$F(z,r) = \frac{2\pi}{\sqrt{r^2+z^2}}$$

and hence, finally,

$$p(x,y,z) = \frac{z}{2\pi(x^2+y^2+z^2)^{3/2}} .$$

12. The problem

D.E.	$\Delta\Delta u = 0$	in D,
B.C.	$u = f$,	
	$\dfrac{\partial u}{\partial \nu} = g$	on C,

is called the first boundary value problem for the biharmonic equation in D. Find a series solution for this problem when D is the unit disc $x^2 + y^2 < 1$.

13. Find a solution of the first boundary value problem for the biharmonic equation in the half plane $-\infty < x < \infty$, $0 < y < \infty$, assuming that the solution and its partial derivatives have appropriate behavior at infinity.

14. Find a solution of

D.E.	$\Delta\Delta u = 0$	in D,
B.C.	$u = f$,	
	$\Delta u = g$	on C,

when D is the unit disc $x^2 + y^2 < 1$.

15. Find a bounded solution of

D.E. $u_{xx} + u_{xy} + u_{yy} = 0\,, \qquad -\infty < x < \infty\,, \quad 0 < y < \infty\,,$
B.C. $u(x,0) = f(x)\,, \qquad -\infty < x < \infty\,,$

where $f(x)$ is bounded and absolutely integrable on $(-\infty,\infty)$. Express your answer in the form

$$u(x,y) = \int_{-\infty}^{\infty} k(x - \xi, y) f(\xi)\, d\xi\,.$$

12

Special problems involving Bessel functions

12.1. PRELIMINARY DISCUSSION OF PROBLEMS IN A DISC

The problems in the plane that we have solved explicitly, with the exception of those discussed in Sections 11.4 and 11.5, have been set in domains bounded by coordinate lines of a rectangular coordinate system. In Sections 11.4 and 11.5, we solved boundary value problems for Laplace's equation in domains bounded by coordinate lines of a polar coordinate system. Our first objective in this chapter is the solution of initial-boundary value problems for the heat and wave equations in these latter domains.

Consider the initial-boundary value problem (10.1.13) for the heat equation when the domain D is a disc of radius b. Following some of the discussion in Sections 11.4 and 11.5, we introduce a polar coordinate system with origin at the center of the disc and formulate problem (10.1.13) as follows,

$$\text{D.E.} \quad u_t = k\left[\frac{1}{r}\frac{\partial}{\partial r}(ru_r) + \frac{1}{r^2}u_{\theta\theta}\right], \quad 0 < r < b\,,\ -\infty < \theta < \infty\,,\ 0 < t\,,$$

$$(12.1.1) \quad \text{B.C.} \quad \begin{cases} u(b,\theta,t) = 0\,, & -\infty < \theta < \infty\,,\ 0 < t\,, \\ u(r,\theta + 2\pi,t) = u(r,\theta,t)\,, & 0 < r < b\,,\ -\infty < \theta < \infty\,,\ 0 < t\,, \\ u \text{ is bounded at the origin}\,, & 0 < t\,, \end{cases}$$

$$\text{I.C.} \quad u(r,\theta,0) = f(r,\theta)\,, \quad 0 < r < b\,,\ -\infty < \theta < \infty\,.$$

Here f is supposed to be a point function, so that it satisfies the condition $f(r,\theta + 2\pi) = f(r,\theta)$, $-\infty < \theta < \infty$. In the formulation (12.1.1), the D.E., the first B.C., and the I.C. are the straightforward expressions in polar coordinates of the corresponding conditions of the problem (10.1.13). The second B.C., the *periodicity condition*, expresses the requirement that the solution u be a point function. We know from the discussion in Section 11.5 that some additional condition is needed to insure that u and its relevant partial derivatives exist, are continuous, and satisfy the D.E. of (10.1.13) at the origin. We will see that the third B.C. of (12.1.1), the *boundedness condition*, is sufficient for this purpose.

For the solution of problem (12.1.1) we shall use the method employed in Chapters 4 and 10 for the solution of similar problems. The principal feature of this method is the reduction of the solution of the problem to the solution of eigenvalue problems for ordinary differential equations. In applying this method to problem (12.1.1) we shall encounter ordinary differential equations that we have not studied before. Thus, although no new technique is involved, there is an essentially new element in the process of solution of (12.1.1).

To present this essentially new element as clearly as possible, with a minimum of technical detail, we consider first the special case of (12.1.1) in which f and u are independent of θ, that is, they are functions of r and t alone. In this case the problem and its solution are said to be **circularly symmetric.** Stated physically, we assume that the temperature is constant on each circle concentric with the disc, so that heat flows only along the radii of the disc. Such a heat flow is called a **plane radial flow.**

For the special case of radial flow, problem (12.1.1) becomes

$$\begin{array}{llll} & \text{D.E.} & u_t = k\dfrac{1}{r}\dfrac{\partial}{\partial r}(ru_r)\,, & 0 < r < b\,,\ 0 < t\,, \\ (12.1.2) & \text{B.C.} & \begin{cases} u(b,t) = 0\,, \\ u \text{ is bounded as } r \to 0+\,, \end{cases} & \begin{array}{l} 0 < t\,, \\ 0 < t\,, \end{array} \\ & \text{I.C.} & u(r,0) = f(r)\,, & 0 < r < b\,. \end{array}$$

To solve problem (12.1.2) we use the method of Chapter 4. If $\varphi(r)\psi(t)$ is a function which satisfies the D.E. and B.C. of (12.1.2), then by substitution and separation of variables we find that φ and ψ must satisfy

$$(12.1.3) \quad \text{D.E.} \quad \psi' + \lambda k \psi = 0\,, \qquad 0 < t\,,$$

$$\text{(12.1.4)}\qquad \begin{aligned} &\text{D.E.} \quad \frac{1}{r}\frac{d}{dr}(r\varphi') + \lambda\varphi = 0\,, \qquad 0 < r < b\,,\\ &\text{B.C.} \quad \varphi \text{ is bounded as } r \to 0+\,, \qquad \varphi(b) = 0\,. \end{aligned}$$

Problem (12.1.4) is an eigenvalue problem for φ. The D.E. in (12.1.4) is, however, one that we have not considered previously. Before proceeding with our discussion of the eigenvalue problem (12.1.4) and the solution of (12.1.2) we must determine the solutions of this D.E. and some of their properties. We do this in the next section.

12.2. *BESSEL FUNCTIONS OF ORDER ZERO*

The ordinary differential equation

$$\text{(12.2.1)}\qquad \frac{1}{r}\frac{d}{dr}(r\varphi') + \lambda\varphi = \varphi'' + \frac{1}{r}\varphi' + \lambda\varphi = 0$$

is an equation which cannot be solved in closed form in terms of elementary functions. Since it is a linear homogeneous equation of second order, its general solution is a linear combination of any two linearly independent solutions. A procedure for determining infinite series representations of such solutions can be found in most textbooks on elementary ordinary differential equations under the heading "*method of Frobenius*" or "*series solution of equations with regular singular points.*" We shall not employ this procedure here. Instead, we shall find one solution of (12.2.1) by using a special device and the relation of (12.2.1) to partial differential equations which led us to consider the equation in Section 1. A second, linearly independent solution will be discussed briefly later in this section.

To find a first solution of (12.2.1) we recall that if $u = v\psi(t)$ is any solution of

$$u_t = k\Delta u\,,$$

then v satisfies

$$\Delta v + \lambda v = 0\,;$$

that is, in rectangular coordinates,

$$\text{(12.2.2)}\qquad v_{xx} + v_{yy} + \lambda v = 0\,,$$

and in polar coordinates,

$$\text{(12.2.3)}\qquad \begin{aligned} &v_{rr} + \frac{1}{r}v_r + \frac{1}{r^2}v_{\theta\theta} + \lambda v = 0\,,\\ &v(r,\theta + 2\pi) = v(r,\theta)\,,\\ &v \text{ is bounded at the origin}\,. \end{aligned}$$

Equation (12.2.1) is the special case of (12.2.3) in which v is independent of θ.

Now let $v(r,\theta)$ be any solution of (12.2.3) and integrate the differential equation of (12.2.3) with respect to θ, from $\theta = -\pi$ to $\theta = \pi$. We get

(12.2.4) $$\int_{-\pi}^{\pi} v_{rr}\, d\theta + \frac{1}{r}\int_{-\pi}^{\pi} v_r\, d\theta + \frac{1}{r^2}\int_{-\pi}^{\pi} v_{\theta\theta}\, d\theta + \lambda \int_{-\pi}^{\pi} v\, d\theta = 0\,.$$

Since v, and hence v_θ, is periodic in θ with period 2π, we have

$$\int_{-\pi}^{\pi} v_{\theta\theta}\, d\theta = v_\theta(r,\pi) - v_\theta(r,-\pi) = 0\,.$$

It follows, if we interchange the order of differentiation with respect to r and integration with respect to θ in (12.2.4), that

(12.2.5) $$\varphi(r) = \int_{-\pi}^{\pi} v(r,\theta)\, d\theta$$

is a solution of (12.2.1). Thus, every solution of (12.2.3) yields a solution of (12.2.1).

To find solutions of (12.2.3) we first find solutions of (12.2.2) using the method of exponential solutions of Section 1.5. The function $v = \exp(\alpha x + \beta y)$ will be a solution of (12.2.2) for any constants α, β satisfying

(12.2.6) $$\alpha^2 + \beta^2 + \lambda = 0\,.$$

Choosing $\alpha = 0$, $\beta = i\sqrt{\lambda}$ as a convenient solution of (12.2.6), we get the solution

(12.2.7) $$v = e^{iy\sqrt{\lambda}}$$

of (12.2.2), and the introduction of polar coordinates into (12.2.7) yields a solution of (12.2.3),

(12.2.8) $$v = \exp\left[ir\sqrt{\lambda}\sin\theta\right]\,.$$

Finally, integrating (12.2.8) with respect to θ, from $\theta = -\pi$ to $\theta = \pi$, and multiplying by $1/2\pi$, we obtain a solution of (12.2.1),

(12.2.9) $$J_0(r\sqrt{\lambda}) = \frac{1}{2\pi}\int_{-\pi}^{\pi} \exp\left[ir\sqrt{\lambda}\sin\theta\right] d\theta\,.$$

The introduction of the factor $1/2\pi$ is clearly permissible, and is made to satisfy the convention that $J_0(0) = 1$.

The integral in (12.2.9) can be written in various equivalent forms by means of simple transformations. Using Euler's formula (1.4.4) and employing the odd and even character, respectively, of the sine and cosine, we get

$$J_0(r\sqrt{\lambda}) = \frac{1}{\pi}\int_0^{\pi} \cos\,(r\sqrt{\lambda}\sin\theta)\,d\theta\,. \tag{12.2.10}$$

Making the change of variable $\omega = \theta - \pi/2$ in (12.2.10), we obtain

$$J_0(r\sqrt{\lambda}) = \frac{2}{\pi}\int_0^{\pi/2} \cos\,(r\sqrt{\lambda}\cos\omega)\,d\omega\,. \tag{12.2.11}$$

Finally, the substitution $t = \cos\omega$ in (12.2.11) yields

$$J_0(r\sqrt{\lambda}) = \frac{2}{\pi}\int_0^{1} \frac{\cos\,(rt\sqrt{\lambda})}{\sqrt{1-t^2}}\,dt\,. \tag{12.2.12}$$

When $\lambda = 1$, Equation (12.2.1) becomes, if we write ψ in place of φ,

$$\frac{1}{r}\frac{d}{dr}(r\psi') + \psi = \psi'' + \frac{1}{r}\psi' + \psi = 0\,, \tag{12.2.13}$$

which is called **Bessel's differential equation of order zero.** Setting $\lambda = 1$ in any of the *integral representations* (12.2.9) to (12.2.12) we obtain a solution $J_0(r)$ of this differential equation. The function $J_0(r)$ is called the **Bessel function of order zero of the first kind.** The statement that $J_0(r)$ is a solution of (12.2.13) can be verified directly by differentiation and substitution of one of its integral representations in (12.2.13), without reference to the derivation of the representation. For this purpose, Equation (12.2.12) is most convenient.

The most frequently encountered **integral representation** of $J_0(r)$ is (12.2.10), with $\lambda = 1$, that is,

$$J_0(r) = \frac{1}{\pi}\int_0^{\pi} \cos\,(r\sin\theta)\,d\theta\,. \tag{12.2.14}$$

We obtain a power series expansion of $J_0(r)$ by substituting the power series expansion of the cosine in (12.2.14) and integrating term-by-term. This yields

$$J_0(r) = \frac{1}{\pi}\sum_{j=0}^{\infty}\frac{(-1)^j r^{2j}}{(2j)!}\int_0^{\pi}\sin^{2j}\theta\,d\theta\,. \tag{12.2.15}$$

Substituting in (12.2.15) the value

$$\int_0^{\pi}\sin^{2j}\theta\,d\theta = \pi\,\frac{(2j)!}{2^{2j}(j!)^2}\,,$$

obtained from a table of definite integrals, we have the **series representation** for $J_0(r)$

$$J_0(r) = \sum_{j=0}^{\infty} \frac{(-1)^j}{(j!)^2 2^{2j}} r^{2j} \tag{12.2.16}$$
$$= 1 - \frac{1}{(1!)^2}\left(\frac{r}{2}\right)^2 + \frac{1}{(2!)^2}\left(\frac{r}{2}\right)^4 - \cdots .$$

Since the power series expansion of the cosine is convergent for all values of its argument, the integration (12.2.15) is permissible for all values of r, and the *power series* (12.2.16) *converges to* $J_0(r)$ *for all values of* r. We infer, in particular, that $J_0(r)$ *is continuous and has continuous derivatives of all orders, for all values of* r. We note also that from

$$J_0(r\sqrt{\lambda}) = \sum_{j=0}^{\infty} \frac{(-1)^j}{(j!)^2 2^{2j}} \lambda^j r^{2j} \tag{12.2.17}$$
$$= 1 - \lambda\left(\frac{r}{2}\right)^2 + \frac{1}{(2!)^2}\lambda^2\left(\frac{r}{2}\right)^4 - \cdots ,$$

it follows that $J_0(r\sqrt{\lambda})$ has continuous derivatives of all orders with respect to λ, for all values of λ.

From the first few terms of the power series for $J_0(r)$ we get a good approximation to $J_0(r)$ for small values of r. A different kind of representation which furnishes a good approximation to $J_0(r)$ for large values of r can be obtained as follows. In (12.2.12), with $\lambda = 1$, we introduce a new variable of integration τ, by setting $rt = r - \tau$, that is, $\tau = r(1 - t)$. We get,

$$J_0(r) = \frac{2}{\pi}\int_0^r \frac{\cos(r - \tau)}{\sqrt{2r\tau}\,\sqrt{1 - \tau/2r}}\,d\tau , \tag{12.2.18}$$

or, after a simple calculation,

$$J_0(r) = \frac{2}{\pi\sqrt{r}}\,[P(r)\cos r + Q(r)\sin r] , \tag{12.2.19}$$

where

$$P(r) = \int_0^r \frac{\cos\tau}{\sqrt{2\tau}\,\sqrt{1 - \tau/2r}}\,d\tau , \tag{12.2.20}$$
$$Q(r) = \int_0 \frac{\sin\tau}{\sqrt{2\tau}\,\sqrt{1 - \tau/2r}}\,d\tau .$$

Now, it is plausible, and correct although difficult to prove, that

$$\lim_{r\to\infty} P(r) = \frac{1}{\sqrt{2}}\int_0^{\infty} \frac{\cos\tau}{\sqrt{\tau}}\,d\tau = \frac{1}{\sqrt{2}}\sqrt{\frac{\pi}{2}}, \tag{12.2.21}$$
$$\lim_{r\to\infty} Q(r) = \frac{1}{\sqrt{2}}\int_0^{\infty} \frac{\sin\tau}{\sqrt{\tau}}\,d\tau = \frac{1}{\sqrt{2}}\sqrt{\frac{\pi}{2}},$$

where the values of the integrals on the right of (12.2.21) have been taken from a table of definite integrals. Equations (12.2.21) may be rewritten as

$$
\begin{aligned}
P(r) &= \frac{1}{\sqrt{2}}\sqrt{\frac{\pi}{2}} + \delta_1(r)\,, \qquad \lim_{r\to\infty} \delta_1(r) = 0\,,\\
Q(r) &= \frac{1}{\sqrt{2}}\sqrt{\frac{\pi}{2}} + \delta_2(r)\,, \qquad \lim_{r\to\infty} \delta_2(r) = 0\,.
\end{aligned} \tag{12.2.22}
$$

Substituting (12.2.22) in (12.2.19) and setting

$$\delta_1(r)\cos r + \delta_2(r)\sin r = \sqrt{\frac{\pi}{2}}\,\delta(r)\,,$$

we get

$$J_0(r) = \sqrt{\frac{2}{\pi r}}\left[\cos\left(r - \frac{\pi}{4}\right) + \delta(r)\right], \qquad \lim_{r\to\infty} \delta(r) = 0\,. \tag{12.2.23}$$

Equation (12.2.23) is called an **asymptotic representation** of $J_0(r)$ as $r \longrightarrow \infty$.

We see from (12.2.23) that for large values of r, $J_0(r)$ behaves approximately like

$$\sqrt{\frac{2}{\pi r}}\cos\left(r - \frac{\pi}{4}\right).$$

In particular, we infer from (12.2.23) that for sufficiently large integers m, $J_0(r)$ has a zero in each of the successive intervals $(m\pi + \pi/4, [m+1]\pi + \pi/4)$. At the endpoints of such intervals, $\cos(r - \pi/4)$ takes the values $+1$ and -1, successively. Hence, if m is sufficiently large so that $|\delta(r)| \leq \frac{1}{2}$ for r in these intervals, $J_0(r)$ must be successively positive and negative at the endpoints of the intervals, and so must have a zero in each interval. Thus, $J_0(r)$ has infinitely many positive zeros. A more careful analysis (see Exercises 12a, Problem 12) shows that these zeros form an increasing infinite sequence which diverges to ∞. We arrange the zeros of J_0 in order of increasing magnitude and denote them by β_m, $m = 1, 2, \ldots$, so that,

$$
\begin{aligned}
&J_0(\beta_m) = 0\,, \qquad m = 1, 2, \ldots\,,\\
&0 < \beta_1 < \beta_2 < \cdots\,.
\end{aligned} \tag{12.2.24}
$$

Using the power series (12.2.16) and the asymptotic representation (12.2.23) we can sketch the graph of $J_0(r)$. Because of the importance of Bessel functions in many applications their values have been calculated and tabulated in as much detail as those of the trigonometric functions. Tables of values of $J_0(r)$, in particular, can be found even in quite elementary collections of mathematical tables. The graph of $J_0(r)$ based on these tabulated values is given in Figure 12.1.

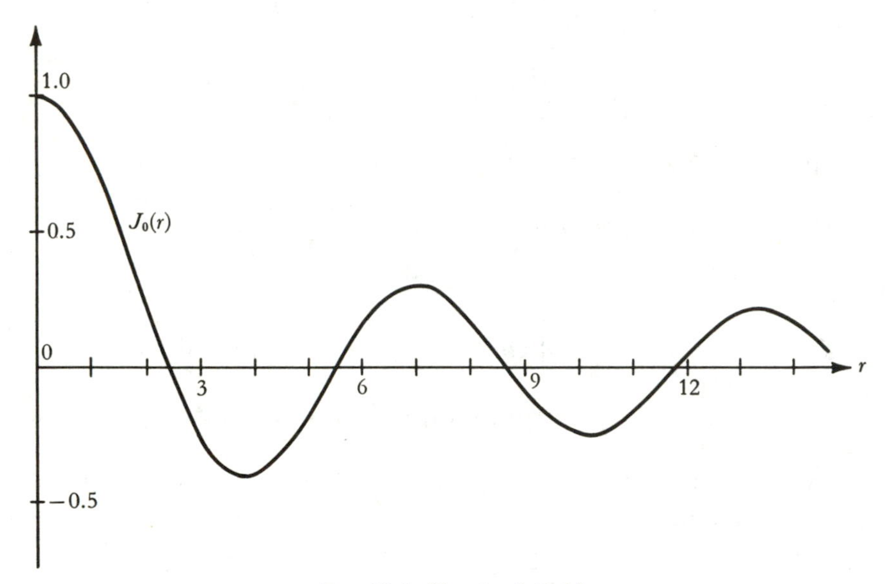

FIG. 12.1. Graph of $J_0\ (r)$

We have seen in a special case that a solution $\varphi(r)$ of Equation (12.2.1) is given by $\varphi(r) = \psi(r\sqrt{\lambda})$, where ψ is a solution of Equation (12.2.13), Bessel's equation of order zero. It is easy to verify directly that this is a general relation, that is $\varphi(r)$ is a solution of (12.2.1) if and only if $\varphi(r) = \psi(r\sqrt{\lambda})$ where ψ is a solution of Equation (12.2.13). We may accordingly restrict our attention to Equation (12.2.13) in the remainder of this section.

Any real-value solution of Bessel's equation of order zero which is linearly independent of $J_0(r)$ is called a **Bessel function of order zero of the second kind.** If $Z_0(r)$ is any Bessel function of the second kind then the general solution of Bessel's equation (12.2.13) is

$$\psi(r) = AJ_0(r) + BZ_0(r)\,, \tag{12.2.25}$$

where A and B are arbitrary constants. Since the only functions which are not linearly independent of J_0 are constant multiples of J_0, it is clear that every function (12.2.25) with $B \neq 0$ is a function of the second kind. Thus, there are many functions of the second kind, but any two such functions differ only by a factor and an added multiple of J_0. The principal property of functions of the second kind is the following: *Every Bessel function of order zero of the second kind becomes infinite as* $r \to 0$, *and conversely, every solution of Bessel's equation of order zero which becomes infinite as* $r \to 0$ *is a function of the second kind.* The converse is a clear consequence of the definition of linear independence, since $J_0(r) \to 1$ as $r \to 0$. To prove the direct statement it is sufficient to produce one function of the second kind and show that it has the stated property. Since every function of the second kind is obtained from one such function by multiplying by a non-zero constant and adding a constant multiple of J_0, the general statement will follow immediately. We shall find a solution of Bessel's equation of order zero which is linearly independent of J_0 using the *method of reduction of order*. We seek a solution $\psi(r)$ of (12.2.13) of the form

$$\psi(r) = \chi(r)J_0(r)\,, \tag{12.2.26}$$

where χ is a function to be determined. Substituting (12.2.26) in the differential equation (12.2.13) we obtain the first order differential equation for χ',

$$\chi''J_0 + \left(2J_0' + \frac{J_0}{r}\right)\chi' = 0\,, \tag{12.2.27}$$

a solution of which is

$$\chi'(r) = \frac{1}{rJ_0^2(r)}\,. \tag{12.2.28}$$

(We neglect arbitrary constants since we are seeking a particular solution.) Integration of (12.2.28) yields $\chi(r)$. To determine the behavior of $\chi(r)$ as $r \to 0$, we observe that since $J_0(r)$ can be expanded in a power series and $J_0(0) \neq 0$, $1/J_0^2(r)$ can be expanded in a power series. The first few terms of the expansion can be determined using Taylor's theorem. We have

$$\chi'(r) = \frac{1}{r}\left(1 + \frac{1}{2}r^2 + \frac{5}{32}r^4 + \cdots\right), \tag{12.2.29}$$

so that, on integration,

$$\chi(r) = \log r + \frac{r^2}{4} + \frac{5}{128}r^4 + \cdots = \log r + \tilde{\chi}(r)\,. \tag{12.2.30}$$

Thus, we have a solution

$$Z_0(r) = \chi(r)J_0(r) = [\log r + \tilde{\chi}(r)]J_0(r) \tag{12.2.31}$$

of Bessel's equation of order zero, and since $\lim_{r\to 0} J_0(r) = 1$ and

$\lim_{r\to 0} \tilde{\chi}(r) = 0$, it is clear that $\lim_{r\to 0+} Z_0(r) = -\infty$. This proves our statement.

It is desirable to have a "standard" function of the second kind for the representation of the general solution of Bessel's equation of order zero. The function (12.2.31) is inconvenient for this purpose. There are several such standard functions, the most widely used of which is due to Weber and is usually denoted by $Y_0(r)$. This function is given by

$$Y_0(r) = \frac{2}{\pi} J_0(r) \log r + \sum_{k=0}^{\infty} c_k r^{2k} , \tag{12.2.32}$$

where the coefficients c_k, $k = 0, 1, 2, \ldots$, are definite non-zero constants whose values are of no importance for our discussions. For further details we refer the reader to books on the special functions of mathematical physics, with the caution that, in addition to the existence of several standard functions of the second kind, there are variations in the notation for these functions which are possible sources of confusion. The general solution of Bessel's equation of order zero, following our choice of a function of the second kind, is given by

$$\psi(r) = AJ_0(r) + BY_0(r) , \tag{12.2.33}$$

where A and B are arbitrary constants.

EXERCISES 12a

1. Verify that solutions of (12.2.6) are given by

$$\alpha = i\sqrt{\lambda} \sin \gamma , \quad \beta = i\sqrt{\lambda} \cos \gamma ,$$

where γ is any real number. Show that for any choice of γ

$$J_0(r\sqrt{\lambda}) = \frac{1}{2\pi} \int_{-\pi}^{\pi} \exp (\alpha r \cos \theta + \beta r \sin \theta) \, d\theta .$$

2. Use (12.2.12), with $\lambda = 1$, to show that

$$J_0''(r) + J_0(r) = -\frac{1}{r} J_0'(r) ,$$

and thus verify that $J_0(r)$ is a solution of Bessel's equation of order zero.

3. Given that $J_0(r)$ has a power series expansion, determine the series directly by the *method of undetermined coefficients*. Show directly that the series is convergent for all values of r.

4. Use (12.2.12) and the inversion formula for the Fourier transform to find the Fourier transform of $J_0(ax)$ (a some positive constant).

5. Show that if $\varphi(r)$ is a solution of the differential equation

$$\frac{d^2\varphi}{dr^2} + \frac{1}{r}\frac{d\varphi}{dr} + (\sqrt{\lambda}\,\gamma r^{\gamma-1})^2\varphi = 0$$

(λ, γ constants), and new variables are introduced by $\rho = r^\gamma\sqrt{\lambda}$, $\varphi(r) = \psi(\rho)$, then $\psi(\rho)$ is a solution of Bessel's equation of order zero

$$\frac{d^2\psi}{d\rho^2} + \frac{1}{\rho}\frac{d\psi}{d\rho} + \psi = 0\,.$$

Conclude that the general solution of the given equation is

$$\varphi(r) = AJ_0(r^\gamma\sqrt{\lambda}) + BY_0(r^\gamma\sqrt{\lambda})\,,$$

where A and B are arbitrary constants.

6. Show that given any differential equation of second order,

$$\varphi'' + \alpha(r)\varphi' + \beta(r)\varphi = 0\,,$$

a function $\mu(r)$ can be chosen so that $\psi(r)$ defined by

$$\psi(r)\mu(r) = \varphi(r)$$

satisfies a differential equation

$$\psi'' + N(r)\psi = 0\,,$$

in which the first-order derivative is absent. (The latter differential equation is sometimes called the **normal form** of the given differential equation.) Find explicit formulas for $\mu(r)$ and $N(r)$ in terms of $\alpha(r)$ and $\beta(r)$.

7. Apply the transformation to normal form found in Problem 6 to Bessel's equation of order zero. What does your result suggest about the behavior for large values of r of *any* solution $\varphi(r)$ of Bessel's equation of order zero?

8. The differential equation

$$\psi'' + \frac{1}{r}\psi' - \psi = 0$$

is called the **modified Bessel equation of order zero.** Show that the function

$$I_0(r) = J_0(ir)$$

is a real-valued solution of this equation which is bounded as $r \to 0$, and is the only (except for constant factors) such function. $I_0(r)$ is called the **modified Bessel function of order zero of the first kind.**

9. Consider the double Fourier transform formulas

$$F(\lambda,\mu) = \int_{-\infty}^{\infty}\int_{-\infty}^{\infty} f(x,y)e^{i(\lambda x+\mu y)}\,dx\,dy\,,$$

$$f(x,y) = \frac{1}{(2\pi)^2}\int_{-\infty}^{\infty}\int_{-\infty}^{\infty} F(\lambda,\mu)e^{-i(\lambda x+\mu y)}\,d\lambda\,d\mu\,,$$

for the special case when f is a *plane radial function;* that is,

$$f(x,y) = g(r) ,$$

where

$$r = \sqrt{x^2 + y^2} .$$

Show that in this case $F(\lambda,\mu)$ is also a radial function; that is, $F(\lambda,\mu) = 2\pi G(\rho)$ where $\rho = \sqrt{\lambda^2 + \mu^2}$. More explicitly show that

$$G(\rho) = \int_0^\infty g(r) J_0(\rho r) r \, dr ,$$

$$g(r) = \int_0^\infty G(\rho) J_0(\rho r) \rho \, d\rho .$$

The function $G(\rho)$ defined by the first of the latter pair of formulas above is called the **Hankel transform** of $g(r)$. The second formula of the pair is the **inversion formula for the Hankel transform.**

10. The **Gamma function,** $\Gamma(\mu)$, is defined for $\mu > 0$ by

$$\Gamma(\mu) = \int_0^\infty e^{-t} t^{\mu-1} \, dt .$$

(i) Show by integration by parts that

$$\Gamma(\mu + 1) = \mu \Gamma(\mu) .$$

(ii) Use the equation of part (i) and $\Gamma(1) = 1$ to show that if n is a positive integer

$$\Gamma(n + 1) = n!$$

11. Apply the transformation to normal form found in Problem 6 to the differential equation

$$\frac{1}{r^2} \frac{d}{dr} (r^2 \varphi') + \lambda \varphi = 0 ,$$

and thus find the general solution of the given equation.

12. Let $\psi(r) = \sqrt{r}\, J_0(r)$ and let r_0, r_1, $0 < r_0 < r_1$ be two successive zeros of $\psi(r)$. Then $\psi(r_0) = 0$, $\psi'(r_0) = c \neq 0$. Let $M^2 = 1 + 1/(4\beta^2)$, where β is the smallest positive zero of $J_0(r)$, and let $w(r)$ be the solution of

$$w'' + M^2 w = 0 , \qquad 0 < r < \infty ,$$
$$w(r_0) = 0 , \qquad w'(r_0) = c .$$

(i) Use this differential equation and the result of Problem 7 to find dW/dr where W is the Wronskian

$$W = \psi' w - \psi w' .$$

Hence show that

$$\psi'(r_1) w(r_1) = \int_{r_0}^{r_1} \left[M^2 - \left(1 + \frac{1}{4r^2} \right) \right] \psi(r) w(r) \, dr .$$

(ii) Let r_2 be the first zero of $w(r)$ to the right of r_0. Show that the assumption $r_2 \geq r_1$ leads to a contradiction, by considering the sign of $\psi'(r_1)w(r_1)$ and the sign of the integral in the identity of part (i).

(iii) Conclude that

$$r_1 - r_0 \geq \frac{2\pi\beta}{\sqrt{1 + 4\beta^2}} .$$

Thus, any two successive zeros of $J_0(r)$ are separated by at least this minimal distance.

12.3. *EIGENVALUE PROBLEMS AND EIGENFUNCTION EXPANSIONS ASSOCIATED WITH BESSEL'S EQUATION OF ORDER ZERO*

The eigenvalue problem (12.1.4) can now be solved easily. The general solution of the differential equation of (12.1.4) is

$$\varphi(r) = AJ_0(r\sqrt{\lambda}) + BY_0(r\sqrt{\lambda}) . \tag{12.3.1}$$

The boundary condition of (12.1.4) at $r = 0$,

$$\varphi \text{ is bounded as } r \to 0_+ \tag{12.3.2}$$

requires that $B = 0$ in (12.3.1), since

$$\lim_{r\to 0} J_0(r\sqrt{\lambda}) = 1 , \quad \lim_{r\to 0^+} Y_0(r\sqrt{\lambda}) = -\infty .$$

It is important to emphasize that since the boundary condition (12.3.2) implies that $B = 0$ in (12.3.1), (12.3.2) implies that much more stringent boundary conditions at $r = 0$ are satisfied. Indeed, $\varphi(r)$ and any number of its derivatives are continuous in any interval $0 \leq r \leq b$ including the point $r = 0$.

Since φ must be a non-trivial function, we must have $A \neq 0$ in Equation (12.3.1) and the boundary condition of (12.1.4) at $r = b$,

$$0 = \varphi(b) = AJ_0(b\sqrt{\lambda}) , \tag{12.3.3}$$

requires that $b\sqrt{\lambda}$ be a zero of J_0. In fact, we obtain all eigenvalues of (12.1.4) by requiring that $b\sqrt{\lambda}$ be one of the positive real zeros of J_0 specified in (12.2.24). We shall see that J_0 has no non-real zeros (Exercises 12b, Problem 2), and since J_0 is an even function, negative real zeros furnish the same eigenvalues and eigenfunctions as the positive real zeros. Thus we determine an infinite sequence of eigenvalues

$$\lambda_m = \left(\frac{\beta_m}{b}\right)^2, \qquad m = 1, 2, \ldots, \tag{12.3.4}$$

and corresponding eigenfunctions

$$(12.3.5) \quad \varphi_m(r) = J_0(r\sqrt{\lambda_m}) = J_0\left(\frac{\beta_m}{b} r\right), \qquad m = 1, 2, \ldots,$$

of the problem (12.1.4). Notice that each eigenvalue has multiplicity one; that is, with each eigenvalue there is associated only one linearly independent eigenfunction. This property depends only on the conclusion that $B = 0$ in (12.3.1), and hence only on the boundary condition at $r = 0$.

We now seek a formula for the expansion of a given function $f(r)$ in a series of the eigenfunctions (12.3.5), as we did for similar problems in Chapter 4. In fact, the eigenvalue problem (12.1.4) is a special case of the problem

$$(12.3.6) \quad \begin{array}{lll} \text{D.E.} & B\varphi = \lambda\varphi\,, & 0 < r < b\,, \\ \text{B.C.} & \varphi \text{ is bounded as } r \to 0+\,, & c_1\varphi'(b) + c_2\varphi(b) = 0\,, \end{array}$$

where

$$(12.3.7) \quad B\varphi = -\frac{1}{r}\frac{d}{dr}(r\varphi')\,,$$

and c_1, c_2 are constants, not both zero. The eigenvalue problem specified by (12.3.6) and (12.3.7) is a *singular Sturm–Liouville problem*, as defined in Section 4.11. It is singular because the Sturm–Liouville differential operator (12.3.7) is singular at $r = 0$. This singularity of (12.3.7) accounts for the unusual boundary condition at $r = 0$ in (12.3.6). Although the Sturm–Liouville problem defined by (12.3.6)–(12.3.7) is singular, its eigenfunctions have the properties established in Section 4.11 for the eigenfunctions of regular Sturm–Liouville problems. This is not obvious, but we establish it below.

Let B be the operator defined by (12.3.7) and let $f(r)$ and $g(r)$ be continuous together with their first and second derivatives in the interval $0 < r \leq b$. Then it follows from (4.11.3) and (4.11.10), or by direct calculation, that for all $a > 0$,

$$(12.3.8) \quad \int_a^b [(Bf)g - f(Bg)]r\,dr = -[r(f'g - fg')]_a^b\,.$$

Suppose further that f and g satisfy

$$(12.3.9) \quad \begin{array}{ll} \lim\limits_{r\to 0+} f(r) < \infty\,, & \lim\limits_{r\to 0+} rf'(r) = 0\,, \\ \lim\limits_{r\to 0+} g(r) < \infty\,, & \lim\limits_{r\to 0+} rg'(r) = 0\,. \end{array}$$

Then, letting $a \to 0$ in (12.3.8), we get

$$\int_0^b [(Bf)g - f(Bg)]r\,dr = b[f(b)g'(b) - f'(b)g(b)]\,. \tag{12.3.10}$$

This is *Green's formula* for the operator B.

Now, let $\varphi_m(r)$ and $\varphi_n(r)$ be eigenfunctions of the problem (12.3.6)–(12.3.7) belonging to different eigenvalues λ_m, λ_n. As we observed at the beginning of this section, it follows from the D.E. and B.C. at $r = 0$ of (12.3.6)–(12.3.7) that any eigenfunction of the problem is continuous together with its first and second derivatives in the interval $0 \leq r \leq b$, in particular, at $r = 0$. Thus, the hypotheses under which Green's formula (12.3.10) holds are certainly satisfied by $\varphi_m(r)$ and $\varphi_n(r)$. Applying (12.3.10) to φ_n and φ_m and using the D.E. of (12.3.6), we obtain

$$(\lambda_m - \lambda_n)\int_0^b \varphi_m(r)\varphi_n(r)r\,dr = b[\varphi_m(b)\varphi_n'(b) - \varphi_m'(b)\varphi_n(b)]\,. \tag{12.3.11}$$

An easy calculation shows that the right side of (12.3.11) vanishes in consequence of the B.C. at $r = b$ of (12.3.6). Thus, we conclude that

$$\int_0^b \varphi_m(r)\varphi_n(r)r\,dr = 0\,, \qquad m \neq n\,, \tag{12.3.12}$$

that is, *the eigenfunctions of the problem* (12.3.6)–(12.3.7) *form an orthogonal system on the interval* $[0,b]$ *relative to the weight function* r.

We restrict our attention now to the special case of (12.3.6)–(12.3.7) in which $c_1 = 0$, $c_2 = 1$. This corresponds to the problem (12.1.4) with which we began this discussion. We have seen that the eigenvalues of this problem form an increasing infinite sequence, each eigenvalue being of multiplicity one, and we have selected one eigenfunction corresponding to each eigenvalue. Suppose that a given piecewise continuous function $f(r)$ can be represented on the interval $(0,b)$ by an infinite series of the eigenfunctions (12.3.5),

$$f(r) = \sum_{m=1}^{\infty} A_m J_0(r\sqrt{\lambda_m})\,, \qquad \lambda_m = (\beta_m/b)^2\,. \tag{12.3.13}$$

Then, by a familiar argument, it follows formally from the orthogonality statement above that the coefficients A_m must be given by

$$A_m = \frac{\int_0^b f(r)J_0(r\sqrt{\lambda_m})r\,dr}{\int_0^b [J_0(r\sqrt{\lambda_m})]^2 r\,dr}\,. \tag{12.3.14}$$

It is desirable and possible to calculate the *normalizing constants*,

$$\int_0^b [J_0(r\sqrt{\lambda_m})]^2 r\,dr\,, \tag{12.3.15}$$

appearing in (12.3.14). To do this we use the procedure introduced in Section 4.7. For any values λ, μ (not necessarily eigenvalues of any problem) we have, if B is the operator (12.3.7),

$$BJ_0(r\sqrt{\lambda}) = \lambda J_0(r\sqrt{\lambda}) , \quad BJ_0(r\sqrt{\mu}) = \mu J_0(r\sqrt{\mu}) . \tag{12.3.16}$$

We set $g = J_0(r\sqrt{\lambda})$ and $f = J_0(r\sqrt{\mu})$ in Green's formula (12.3.10), and use (12.3.16) to obtain

$$\begin{aligned}(\mu - \lambda) \int_0^b J_0(r\sqrt{\mu})J_0(r\sqrt{\lambda})r\,dr \\ = b[J_0(b\sqrt{\mu})\sqrt{\lambda}\,J_0'(b\sqrt{\lambda}) - \sqrt{\mu}\,J_0'(b\sqrt{\mu})J_0(b\sqrt{\lambda})] .\end{aligned} \tag{12.3.17}$$

Now we set $\lambda = \lambda_m$, where λ_m is an eigenvalue given by (12.3.4), and suppose that $\mu \neq \lambda_m$. Then since $J_0(b\sqrt{\lambda_m}) = 0$, Equation (12.3.17) becomes, on division by $\mu - \lambda_m$,

$$\int_0^b J_0(r\sqrt{\mu})J_0(r\sqrt{\lambda_m})r\,dr = \frac{b\sqrt{\lambda_m}\,J_0'(b\sqrt{\lambda_m})J_0(b\sqrt{\mu})}{\mu - \lambda_m} . \tag{12.3.18}$$

Finally, we take the limit of both sides of (12.3.18) as $\mu \to \lambda_m$, calculating the value of the limit of the right side with the aid of l'Hospital's rule. We get

$$\int_0^b [J_0(r\sqrt{\lambda_m})]^2 r\,dr = \frac{b^2}{2}[J_0'(b\sqrt{\lambda_m})]^2 , \tag{12.3.19}$$

as the value of the mth normalizing constant.

If we substitute (12.3.19) in (12.3.14), and also replace λ_m by the value given by (12.3.4), we get

$$A_m = \frac{2}{b^2[J_0'(\beta_m)]^2} \int_0^b f(r)J_0\left(\frac{\beta_m r}{b}\right) r\,dr . \tag{12.3.20}$$

According to the definitions of Section 6.9, (12.3.20) defines the mth *Fourier coefficient* of $f(r)$ relative to the system $\{J_0(\beta_m r/b)\}$ of functions orthogonal on $[0,b]$ with respect to the weight function r. The infinite series

$$f(r) \sim \sum_{m=1}^{\infty} A_m J_0(\beta_m r/b) \tag{12.3.21}$$

formed with these coefficients is the *Fourier series* of $f(r)$ relative to the system $\{J_0(\beta_m r/b)\}$. In Section 6.9 we stated a *pointwise convergence* theorem for the Fourier series of a function relative to the eigenfunctions of a regular Sturm–Liouville problem. A corresponding statement holds for the Fourier series (12.3.21) relative to the eigenfunctions of the problem (12.1.4). Although the proof of the theorem is beyond the scope of this book, we state the theorem below.

Theorem. *Let $f(r)$ be piecewise smooth on the interval* $[0,b]$, *and let* $\beta_1 < \beta_2 < \cdots < \beta_m < \cdots$ *be the successive positive zeros of* $J_0(r)$. *Then for each value of r in the interval* $0 < r < b$ *the series* (12.3.21), *with coefficients given by* (12.3.20), *converges and if r is a point of continuity of f,*

$$f(r) = \sum_{m=1}^{\infty} A_m J_0(\beta_m r/b) .$$

If r is a point of discontinuity of f, the sum of the series (12.3.21) *is*

$$\frac{f(r+) + f(r-)}{2} .$$

For other special cases of the eigenvalue problem (12.3.6)–(12.3.7) the formal calculations of the eigenvalues, eigenfunctions, and normalizing constants, and of the Fourier coefficients and Fourier series of a given function relative to the eigenfunctions of the problem, are analogous to those given above for the special case corresponding to the problem (12.1.4). The details are left to Exercises 12b. A pointwise convergence theorem of the same form as that given above for the eigenfunction expansion (12.3.20)–(12.3.21) holds also for the eigenfunction expansions calculated in these problems.

12.4. *PROBLEMS WITH ROTATIONAL SYMMETRY*

We are now prepared to complete the solution of the initial-boundary value problem (12.1.2). We found in Section 1 that the solutions with separated variables of the D.E. and B.C. of this problem are determined by the eigenvalue problem (12.1.4) and the D.E. (12.1.3). The solution of (12.1.4) is given by (12.3.4) and (12.3.5). If we set $\lambda = \lambda_m$ in (12.1.3) and solve the equation, we obtain the sequence

(12.4.1) $$\varphi_m(r)\psi_m(t) = J_0(\beta_m r/b) \exp\left[-(\beta_m/b)^2 kt\right] , \quad m = 1, 2, 3, \ldots ,$$

of functions satisfying the D.E. and B.C. of (12.1.2). Superposition of these functions yields

(12.4.2) $$u(r,t) = \sum_{m=1}^{\infty} A_m J_0(\beta_m r/b) \exp\left[-(\beta_m/b)^2 kt\right] ,$$

and (12.4.2) will be the formal solution of (12.1.2) if the constants A_m can be chosen so that the I.C. of (12.1.2) is satisfied. This requires that

$$f(r) = u(r,0) = \sum_{m=1}^{\infty} A_m J_0(\beta_m r/b) , \qquad 0 < r < b , \tag{12.4.3}$$

and, according to the theorem stated at the end of Section 3, (12.4.3) will hold if $f(r)$ is piecewise smooth and

$$A_m = \frac{2}{b^2[J_0'(\beta_m)]^2} \int_0^b f(r) J_0\left(\frac{\beta_m r}{b}\right) r\, dr . \tag{12.4.4}$$

Thus (12.4.2) and (12.4.4) furnish the formal solution of the initial-boundary value problem (12.1.2).

There are a number of other *initial-boundary value problems* in the plane whose solutions can be determined in terms of Bessel functions of order zero. Problems of radial heat flow in a disc when the boundary condition is a homogeneous condition of the second or third kind can be solved in the same way as problem (12.1.2). When the displacement function of a circular membrane satisfies a linear homogeneous boundary condition and the initial displacement and velocity are circularly symmetric, the displacement function will be circularly symmetric for all time. The problem of determining such a circularly symmetric vibration can be solved in the same way as the corresponding heat conduction problem, with only the differences mentioned at the beginning of Section 8.4. The solutions of all these problems can be determined in terms of the Bessel function J_0. (See Exercises 12b, Problems 6 and 7.)

Problems of radial heat flow in a heat conducting annulus and of circularly symmetric vibration of an annular membrane can be solved in a similar fashion. The solutions of such problems can be expressed in terms of the Bessel functions J_0 and Y_0. (See Exercises 12b, Problem 13.)

When the data of an *initial value problem* for the heat equation or wave equation in the plane are circularly symmetric, the solution has the same property, and can be determined in terms of J_0 using the Hankel transform defined in Exercises 12a, Problem 9. (See Exercises 12b, Problem 8.)

There is an important class of *boundary value problems* whose solutions can be expressed in terms of Bessel functions of order zero. Consider the problem of determining the equilibrium temperature distribution in a right circular cylinder composed of a uniform and isotropic heat conducting substance, when the surface of the cylinder is maintained at prescribed temperature. Let the radius of the cylinder be b and let its height be l. We take the base of the cylinder to lie in the (x,y)-plane and its axis to be coincident with the z-axis, and introduce cylindrical coordinates

$$x = r\cos\theta\,, \quad y = r\sin\theta\,, \quad z = z\,. \tag{12.4.5}$$

Then the equilibrium temperature distribution in the cylinder, $u(r,\theta,z)$, is the solution of the problem

$$\begin{array}{lll}
\text{D.E.} & \dfrac{1}{r}(ru_r)_r + \dfrac{1}{r^2}u_{\theta\theta} + u_{zz} = 0\,, & 0 < r < b\,,\ -\infty < \theta < \infty\,,\ 0 < z < l\,, \\
\text{B.C.} & u(r,\theta,0) = f(r,\theta)\,, & \\
 & u(r,\theta,l) = g(r,\theta)\,, & 0 < r < b\,,\ -\infty < \theta < \infty\,, \\
 & u(b,\theta,z) = h(\theta,z)\,, & -\infty < \theta < \infty\,,\ 0 < z < l\,, \\
 & u(r,\theta + 2\pi,z) = u(r,\theta,z)\,, & 0 < r < b\,,\ -\infty < \theta < \infty\,,\ 0 < z < l\,, \\
 & u \text{ is bounded as } r \to 0+\,, & -\infty < \theta < \infty\,,\ 0 < z < l\,.
\end{array} \tag{12.4.6}$$

The differential operator in (12.4.6) is the three-dimensional *Laplace operator in cylindrical coordinates.*

Now suppose that all the functions appearing in problem (12.4.6) are independent of θ, so that (12.4.6) becomes

$$\begin{array}{lll}
\text{D.E.} & \dfrac{1}{r}(ru_r)_r + u_{zz} = 0\,, & 0 < r < b\,,\ 0 < z < l\,, \\
\text{B.C.} & u(r,0) = f(r)\,, & 0 < r < b\,, \\
 & u(r,l) = g(r)\,, & 0 < r < b\,, \\
 & u(b,z) = h(z)\,, & 0 < z < l\,, \\
 & u \text{ is bounded as } r \to 0+\,, & 0 < z < l\,.
\end{array} \tag{12.4.7}$$

We say that problem (12.4.7) and its solution are **axially symmetric.** The solution of (12.4.7), and of related problems in which one or more of the B.C. are replaced by conditions of the second or third kind, can be effected using the methods employed in Chapter 11 for the solution of boundary value problems. The solutions of these problems can be expressed in terms of J_0, or the modified Bessel function

I_0, defined in Exercises 12a, Problem 8, or both. (See Exercises 12b, Problems 9, 10 and 11.)

It is of interest to consider here the spatial analogue of problem (12.1.2), the problem of radial heat flow in a sphere, because of the peculiar fact that this and other problems with the same D.E. can be solved in terms of elementary functions. Let the radius of the sphere be b. A **spherically symmetric** temperature distribution in the sphere is a function $u(r,t)$ which depends only on t and on

$$r = \sqrt{x^2 + y^2 + z^2}\,. \tag{12.4.8}$$

In the case of spherical symmetry the three-dimensional Laplace operator in spherical coordinates becomes

$$\Delta u = \frac{1}{r^2}(r^2 u_r)_r\,. \tag{12.4.9}$$

If the surface of the sphere is maintained at zero temperature and the initial temperature is $f(r)$, then the radial heat flow in the sphere will be described by the solution $u(r,t)$ of the problem

$$\begin{array}{llll} & \text{D.E.} & u_t = k\dfrac{1}{r^2}(r^2u_r)_r\,, & 0 < r < b\,,\ 0 < t\,, \\ (12.4.10) & \text{B.C.} & u(b,t) = 0\,, & 0 < t\,, \\ & & u \text{ is bounded as } r \to 0+\,, & 0 < t\,, \\ & \text{I.C.} & u(r,0) = f(r)\,, & 0 < r < b\,. \end{array}$$

This problem and related problems can be solved with the aid of the result of Exercises 12a, Problem 11. (See Exercises 12b, Problems 12 and 13.)

We mention finally that there are a number of problems, unrelated to those discussed above, whose solutions can be expressed in terms of Bessel functions of order zero. In the process of solution of such problems one is often led, not to Bessel's equation, but to an equation related to Bessel's equation by a simple change of variable like that in Exercises 12a, Problem 5. (See Exercises 12b, Problem 15.)

EXERCISES 12b

1. Assuming that the eigenfunctions of problem (12.3.6)–(12.3.7) satisfy the conditions

$$\lim_{r\to 0+} \varphi(r) < \infty\,, \quad \lim_{r\to 0+} r\varphi'(r) = 0\,,$$

show that all eigenvalues of the problem are real if c_1 and c_2 are real.

2. (i) Establish Green's first formula for the operator B defined by (12.3.7).

(ii) Making the assumptions stated in Problem 1, show that if $c_1c_2 \geq 0$ then eigenvalues of the problem (12.3.6)–(12.3.7) are real and nonnegative. Infer, in particular, that J_0 has only real zeros.

3. For the eigenvalue problem

$$\text{D.E.} \qquad \frac{1}{r}\frac{d}{dr}(r\varphi') + \lambda\varphi = 0\,, \qquad 0 < r < b\,,$$

$$\text{B.C.} \qquad \varphi(r) \text{ is bounded as } r \to 0+\,, \qquad \varphi'(b) = 0\,,$$

assume that the eigenvalues form an increasing sequence of nonnegative real numbers diverging to ∞ and do the following:

(i) Find an equation whose roots are the eigenvalues of the problem, and for each eigenvalue find a corresponding eigenfunction. [Is $\lambda = 0$ an eigenvalue?]

(ii) Find formulas for the Fourier coefficients and Fourier series of a given function $f(r)$ relative to the system of eigenfunctions determined in part (i).

(iii) Calculate the normalizing constants of the eigenfunctions determined in part (i).

4. Repeat Problem 3 with the boundary condition at b replaced by

$$\varphi'(b) + h\varphi(b) = 0\,, \qquad h > 0\,.$$

5. Show that the eigenvalues of Problem 3 form an increasing sequence of nonnegative real numbers which diverges to ∞, as follows:

(i) Verify that the eigenvalues are real and nonnegative with the aid of part (ii) of Problem 2.

(ii) Use the uniqueness theorem for the initial value problem for ordinary differential equations to show that $J_0(r)$ and $J_0'(r)$ have no common zero.

(iii) Show that between each pair of successive zeros of $J_0(r)$ there is at least one zero of $J_0'(r)$.

(iv) Use the differential equation satisfied by $J_0(r)$ to show that in the interval between successive zeros of $J_0(r)$, $rJ_0'(r)$ is monotone, so that $J_0'(r)$ has at most one zero in each such interval.

6. Formulate and solve the problem of radial heat flow in a disc of radius b, when the circumference of the disc is insulated, and the initial temperature is

$$f(r) = \begin{cases} 1 & 0 \leq r < \dfrac{b}{2}\,, \\[2ex] 0 & \dfrac{b}{2} \leq r < b\,. \end{cases}$$

7. A circular membrane of radius b and tension–surface density ratio c^2, has its circumference fixed in a horizontal plane, but is otherwise free to move under the influence of gravity. Initially the membrane is at rest in the plane of its boundary. When the membrane is released at time zero it executes transverse circularly symmetric vibrations.

(i) Formulate the problem which determines the displacement function, $u(r,t)$, of the membrane.

(ii) Find the equilibrium displacement of the membrane.

(iii) Solve the problem formulated in part (i).

8. Use the Hankel transform and inversion formula to find the solution, $u(r,t)$, of the circularly symmetric initial value problem for the wave equation,

$$\text{D.E.} \qquad u_{tt} = c^2 \frac{1}{r}(ru_r)_r\,, \qquad 0 < r < \infty\,, \quad 0 < t\,,$$

$$\text{I.C.} \qquad u(r,0) = f(r)\,,$$
$$u_t(r,0) = g(r)\,, \qquad 0 < r < \infty\,.$$

Employ the method of variation of parameters, with the aid of Green's formula (12.3.10), and make any assumptions about f, g, and u that seem necessary.

9. Solve problem (12.4.7) with $h(z) = 0$, $g(r) = 0$, and $f(r)$ any piecewise smooth function.

10. Solve problem (12.4.7) with $f(r) = g(r) = 0$ and $h(z)$ any piecewise smooth function.

11. Solve problem (12.4.7) with $g(r) = 0$, and with $f(r)$ and $h(z)$ any piecewise smooth functions.

12. Solve problem (12.4.10).

13. Formulate and solve the problem of radial heat flow in a spherical shell of inner radius a and outer radius b when the spherical boundaries of the shell are maintained at temperature zero, and the initial temperature distribution is $f(r)$, $a < r < b$.

14. (i) Formulate the problem of radial heat flow in an annulus of inner radius a and outer radius b when the circular boundaries of the annulus are maintained at temperature zero, and the initial temperature distribution is $f(r)$, $a < r < b$.

(ii) Find the eigenvalue problem satisfied by $\varphi(r)$ when $\varphi(r)\psi(t)$ satisfies the D.E. and B.C. of part (i). Verify that this problem is a regular Sturm–Liouville problem so that the results of Sections 4.11 and 6.9 can be applied to it.

(iii) Find an equation whose roots are the eigenvalues of the problem of part (ii), and for each eigenvalue find a corresponding eigenfunction.

(iv) Solve the problem of part (i).

15. The partial differential equation satisfied by the horizontal component, u, of the displacement of an oscillating chain was found in Exercises 8a, Problem 6. If we make the change of variable $r = L - s$, in effect measuring the length of the chain from its free end instead of from its fixed end, this differential equation becomes

$$\text{D.E.} \qquad u_{tt} = g(ru_r)_r\,, \qquad 0 < r < L\,, \quad 0 < t\,.$$

The function $u(r,t)$ also satisfies the

$$\text{B.C.} \qquad u \text{ is bounded as } r \to 0+\,, \quad u(L,t) = 0\,, \qquad 0 < t\,.$$

Find $u(r,t)$ if the horizontal components of the displacement and velocity of the chain at time zero are given by the

$$\text{I.C.} \qquad u(r,0) = f(r)\,, \quad u_t(r,0) = g(r)\,, \qquad 0 < r < L\,.$$

12.5. *BESSEL FUNCTIONS OF ANY REAL ORDER*

The linear homogeneous second-order ordinary differential equation

$$(12.5.1) \quad \frac{1}{r}\frac{d}{dr}(r\psi') + \left(1 - \frac{\nu^2}{r^2}\right)\psi = \psi'' + \frac{1}{r}\psi' + \left(1 - \frac{\nu^2}{r^2}\right)\psi = 0\,,$$

where ν is any nonnegative real number, is called **Bessel's equation of order** ν. In the process of solution of problem (12.1.1), problem (12.4.6), and related problems, we shall encounter the differential equation

$$(12.5.2) \qquad \frac{1}{r}\frac{d}{dr}(r\varphi') + \left(\lambda - \frac{\nu^2}{r^2}\right)\varphi = 0\,.$$

It is easy to verify that every solution of (12.5.2) has the form

$$(12.5.3) \qquad \varphi(r) = \psi(r\sqrt{\lambda})\,,$$

where ψ is a solution of Bessel's equation (12.5.1).

In the special and important case of Bessel's equation of *integer order*, that is, of order $\nu = n$, $n = 0, 1, 2, \ldots$, the solutions of the equation have representations and other properties quite analogous to, and including, those established in Section 2 for functions of order zero. We consider this case of integral order first.

The function (12.2.8) which we used to get an integral representation of J_0 is, with respect to θ, a periodic function of period 2π with continuous first and second order derivatives. Accordingly (12.2.8), with $\lambda = 1$, can be expanded in a uniformly convergent Fourier series which, for convenience, we take to be in complex form. We have

$$\exp[ir\sin\theta] = \sum_{n=-\infty}^{\infty} J_n(r)e^{in\theta}\,, \tag{12.5.4}$$

where

$$J_n(r) = \frac{1}{2\pi}\int_{-\pi}^{\pi} \exp[ir\sin\theta]\exp[-in\theta]\,d\theta\,, \qquad n = 0, \pm1, \pm2, \ldots. \tag{12.5.5}$$

The functions (12.5.5), with $n \geq 0$, are solutions of (12.5.1) with $\nu = n$, that is: $J_n(r)$ is a solution of Bessel's equation of order n. To show this we multiply Equation (12.2.3), which is satisfied by (12.2.8), by $e^{-in\theta}$ and then integrate the equation with respect to θ from $\theta = -\pi$ to $\theta = \pi$. From Green's formula for the operator $d^2/d\theta^2$, we get

$$\int_{-\pi}^{\pi} v_{\theta\theta}e^{-in\theta}\,d\theta = -n^2\int_{-\pi}^{\pi} ve^{-in\theta}\,d\theta\,, \tag{12.5.6}$$

since $v(r,\theta)$ and $e^{-in\theta}$ are periodic functions of θ with period 2π. The demonstration is concluded just as it was in Section 2 for the case $n = 0$. In the preceding argument we made no use of the assumption $n \geq 0$, so clearly $J_{-n}(r)$ is also a solution of Bessel's equation of order n. However, as we shall show below

$$J_{-n}(r) = (-1)^n J_n(r)\,, \tag{12.5.7}$$

so that we have determined only one linearly independent solution of Bessel's equation for each integer order.

The integral (12.5.5) is easily transformed into

$$J_n(r) = \frac{1}{\pi}\int_0^{\pi} \cos(r\sin\theta - n\theta)\,d\theta\,. \tag{12.5.8}$$

The function $J_n(r)$ is called **Bessel's function of order n of the first kind.** Equation (12.5.8) is an *integral representation* of $J_n(r)$.

If we set $t = e^{i\theta}$ in (12.5.4) we obtain

$$\exp\left[\frac{r}{2}\left(t - \frac{1}{t}\right)\right] = \sum_{-\infty}^{\infty} J_n(r)t^n\,. \tag{12.5.9}$$

The function on the left of (12.5.9) is called the **generating function** of the sequence $\{J_n(r)\}$ of Bessel functions. The phrase "generating function of a sequence" is generally employed to denote a function whose expansion in a series of integer powers of a variable has the terms of the sequence as coefficients. We shall deduce important properties of the Bessel functions of integer order of the first kind from the relation (12.5.9).

We observe first that

$$\exp\left[\frac{r}{2}\left(t - \frac{1}{t}\right)\right] = \exp\left[\frac{r}{2}\,t\right] \cdot \exp\left[-\frac{r}{2}\,t^{-1}\right]$$

$$(12.5.10) \qquad = \left[\sum_{j=0}^{\infty} \frac{1}{j!}\left(\frac{r}{2}\right)^{j} t^{j}\right]\left[\sum_{k=0}^{\infty} \frac{1}{k!}\,(-1)^{k}\left(\frac{r}{2}\right)^{k} t^{-k}\right]$$

$$= \sum_{j=0}^{\infty}\sum_{k=0}^{\infty} \frac{1}{j!}\,\frac{(-1)^{k}}{k!}\left(\frac{r}{2}\right)^{j+k} t^{j-k}\,.$$

The coefficient of t^n, $n \geq 0$, in the last series of (12.5.10) is the sum of those terms for which $j - k = n$ and $k \geq 0$, that is,

$$(12.5.11) \qquad \sum_{k=0}^{\infty} \frac{(-1)^{k}}{(k+n)!k!}\left(\frac{r}{2}\right)^{n+2k}.$$

Since the coefficient of t^n in (12.5.9) is $J_n(r)$, we have the *series representation*

$$(12.5.12) \qquad J_n(r) = \sum_{k=0}^{\infty} \frac{(-1)^{k}}{(k+n)!k!}\left(\frac{r}{2}\right)^{n+2k}$$

$$= \left(\frac{r}{2}\right)^{n} \sum_{k=0}^{\infty} \frac{(-1)^{k}}{(k+n)!k!}\left(\frac{r}{2}\right)^{2k}.$$

The corresponding calculation of the coefficient of t^{-n}, $n > 0$, in (12.5.10) is performed by summing those terms for which $j - k = -n$ and $j \geq 0$, and establishes (12.5.7). It is clear that the series (12.5.12) converges for all values of r, so that $J_n(r)$ is continuous and has continuous derivatives of all orders for all values of r, including $r = 0$. Further, for $n > 0$, $J_n(0) = 0$.

If we differentiate Equation (12.5.9) with respect to r, we get

$$(12.5.13) \qquad \frac{1}{2}\left(t - \frac{1}{t}\right)\exp\left[\frac{r}{2}\left(t - \frac{1}{t}\right)\right] = \sum_{n=-\infty}^{\infty} J_n'(r)t^n\,.$$

Substitution of (12.5.9) in the left side of (12.5.13) yields

$$(12.5.14) \qquad \sum_{n=-\infty}^{\infty} J_n(r)t^{n+1} - \sum_{n=-\infty}^{\infty} J_n(r)t^{n-1} = \sum_{n=-\infty}^{\infty} 2J_n'(r)t^n\,,$$

which becomes, on replacement of the summation index n by $n - 1$ in the first series, and by $n + 1$ in the second series on the left side,

$$(12.5.15) \qquad \sum_{n=-\infty}^{\infty} [J_{n-1}(r) - J_{n+1}(r)]t^n = \sum_{n=-\infty}^{\infty} 2J_n'(r)t^n\,.$$

Equating coefficients of corresponding powers of t in the two series of the identity (12.5.15) we obtain the **recurrence relation**

(12.5.16) $$2J_n'(r) = J_{n-1}(r) - J_{n+1}(r)\,.$$

Similarly, beginning with differentiation of (12.5.9) with respect to t, we obtain successively

$$\frac{r}{2}\left(1+\frac{1}{t^2}\right)\exp\left[\frac{r}{2}\left(t-\frac{1}{t}\right)\right] = \sum_{n=-\infty}^{\infty} nJ_n(r)t^{n-1}\,,$$

$$\sum_{n=-\infty}^{\infty} J_n(r)t^n + \sum_{n=-\infty}^{\infty} J_n(r)t^{n-2} = \sum_{n=-\infty}^{\infty} \frac{2n}{r} J_n(r)t^{n-1}\,,$$

$$\sum_{n=-\infty}^{\infty} [J_{n-1}(r) + J_{n+1}(r)]t^{n-1} = \sum_{n=-\infty}^{\infty} \frac{2n}{r} J_n(r)t^{n-1}\,,$$

and thus a *second recurrence relation*

(12.5.17) $$\frac{2n}{r} J_n(r) = J_{n-1}(r) + J_{n+1}(r)\,.$$

By subtraction and addition of (12.5.16) and (12.5.17) followed by multiplication by $r/2$, we get the equivalent pair of recurrence relations

(12.5.18) $$nJ_n(r) - rJ_n'(r) = rJ_{n+1}(r)\,,$$

(12.5.19) $$nJ_n(r) + rJ_n'(r) = rJ_{n-1}(r)\,.$$

We emphasize that in each of the preceding recurrence relations, what is important is the arithmetic connection between the indices of the functions appearing in the equation, and any change of indices which preserves this connection is permissible. Thus, for example, if we replace n by $n - 1$ in (12.5.18) we obtain the equivalent relation

(12.5.20) $$(n-1)J_{n-1}(r) - rJ_{n-1}'(r) = rJ_n(r)\,.$$

We shall employ the recurrence relations proved above to establish further properties of the functions J_n later in this section and in Exercises 12c. Our present interest is in the significance of these relations for other solutions of Bessel's equation. The fact that J_{n+1} can be calculated in terms of J_n from (12.5.18) suggests the following statement. *If Z_n is a solution of Bessel's equation of order n, and if Z_{n+1} is defined by*

(12.5.21) $$Z_{n+1}(r) = \frac{n}{r} Z_n(r) - Z_n'(r)\,,$$

then Z_{n+1} is a solution of Bessel's equation of order $n + 1$. This statement can be established by a straightforward calculation, which is left to Exercises 12c, Problem 5.

We use the statement just made to define (Weber's) **Bessel functions of order n of the second kind,** denoted by $Y_n(r)$, recursively in terms of $Y_0(r)$; that is, we define

$$
\begin{aligned}
Y_1(r) &= -Y_0'(r)\,, \\
Y_2(r) &= \frac{1}{r}\,Y_1(r) - Y_1'(r)\,, \\
&\;\vdots \\
Y_{n+1}(r) &= \frac{n}{r}\,Y_n(r) - Y_n'(r)\,.
\end{aligned}
\tag{12.5.22}
$$

From this definition and from the definition of $Y_0(r)$ in (12.2.32) follows the principal property

$$\lim_{r\to 0+} Y_n(r) = -\infty\,, \qquad n > 0\,. \tag{12.5.23}$$

(See Exercises 12c, Problem 6.) Equation (12.5.23) and

$$\lim_{r\to 0} J_n(r) = 0\,, \qquad n > 0\,,$$

imply the linear independence of J_n and Y_n. Thus *the general solution of Bessel's equation of integer order n* ($n = 0, 1, 2, \ldots$) is given by

$$\psi = AJ_n(r) + BY_n(r)\,, \tag{12.5.24}$$

where A and B are arbitrary constants. We remark, although we shall not use the fact, that the functions $Y_n(r)$ which, by definition, satisfy a recurrence relation of the form (12.5.18) also satisfy a recurrence relation of the form (12.5.19).

None of the discussion above holds for Bessel's equation of non-integral order. The most direct and the simplest approach in this case is that via *the method of Frobenius* from which it is found that, if $\nu > 0$ is not an integer, Bessel's equation (12.5.1) has solutions of the form

$$r^{\nu} \sum_{k=0}^{\infty} a_k r^{2k} \tag{12.5.25}$$

and

$$r^{-\nu} \sum_{k=0}^{\infty} b_k r^{2k}\,, \tag{12.5.26}$$

where a_0 and b_0 may have any non-zero values, which when chosen determine the values of the remaining coefficients in the respective power series of (12.5.25), (12.5.26). For suitable choices of a_0 and b_0

the functions defined by (12.5.25) and (12.5.26) are denoted by $J_\nu(r)$ and $J_{-\nu}(r)$, respectively, and are called **Bessel functions of the first kind of order ν and of order $-\nu$**. The appropriate choice of the value of a_0 is

$$a_0 = \frac{1}{2^\nu \Gamma(\nu + 1)},$$

where Γ denotes the Gamma function which was defined and briefly discussed in Exercises 12a, Problem 10. With this choice (12.5.25) becomes

$$J_\nu(r) = \sum_{k=0}^{\infty} \frac{(-1)^k}{k!\Gamma(\nu + k + 1)} \left(\frac{r}{2}\right)^{2k+\nu} \tag{12.5.27}$$

which is identical with (12.5.12) when $\nu = n$, n a positive integer. We mention, without further discussion, that the definition of the Gamma function can be extended to negative non-integer real values, and with this extended definition (12.5.27) is meaningful and defines $J_{-\nu}$ when ν is replaced by $-\nu$.

Those properties of J_ν and $J_{-\nu}$ that can be inferred immediately from their series representations and are important for our purposes, are the following. For any *positive non-integral real number* ν,

$$\lim_{r \to 0+} J_\nu(r) = 0, \quad \lim_{r \to 0+} rJ'_\nu(r) = 0, \tag{12.5.28}$$

$$\lim_{r \to 0+} |J_{-\nu}(r)| = \infty; \tag{12.5.29}$$

$J_\nu(r)$ and $J_{-\nu}(r)$ are *linearly independent;* the *general solution of Bessel's equation of order* ν is

$$\psi(r) = AJ_\nu(r) + BJ_{-\nu}(r), \tag{12.5.30}$$

where A and B are arbitrary constants. The functions J_ν also satisfy the *recurrence relations* (12.5.18) and (12.5.19) with n replaced by ν. The verification of this statement, which is left to Exercises 12c, Problem 8, utilizes the property of the Gamma function

$$\Gamma(\mu + 1) = \mu\Gamma(\mu), \tag{12.5.31}$$

that was established for nonnegative μ in Exercises 12a, Problem 10.

From now on our main interest will be in *Bessel functions of the first kind of any nonnegative real order*. We shall denote these functions by $J_\nu(r)$, with the understanding that ν may have any nonnegative real value including the integer values 0, 1, 2, ... For these functions we have the *asymptotic representations* as $r \to \infty$

(12.5.32) $$J_\nu(r) = \sqrt{\frac{2}{\pi r}}\left[\cos\left(r - \frac{\nu\pi}{2} - \frac{\pi}{4}\right) + \delta_\nu(r)\right], \quad \lim_{r\to\infty} \delta_\nu(r) = 0 .$$

The graph of $J_\nu(r)$ can be sketched with the aid of (12.5.32) and (12.5.27). The proof of (12.5.32) is beyond the scope of this book.

For each $\nu \geq 0$, the zeros of $J_\nu(r)$ are all real, and the positive zeros form an infinite increasing sequence diverging to ∞. We shall denote the *positive zeros* of J_ν arranged in order of magnitude by $\beta_{\nu,m}$, so that

(12.5.33) $$J_\nu(\beta_{\nu,m}) = 0 , \qquad m = 1, 2, 3, \ldots ,$$
$$0 < \beta_{\nu,1} < \beta_{\nu,2} < \cdots < \beta_{\nu,m} < \cdots .$$

Since

(12.5.34) $$J_\nu(r) = r^\nu P_\nu(r) ,$$

where $P_\nu(r)$ is an even power series which converges for all values of r, the only zeros of J_ν other than $\beta_{\nu,m}$, $m = 1, 2, \ldots$, are $-\beta_{\nu,m}$, $m = 1, 2, \ldots$, and 0 (if $\nu \neq 0$). These statements about the zeros of J_ν are established by arguments identical with those employed in the discussion of the zeros of J_0 in Section 2. (See also Exercises 12c, Problem 12.)

Eigenvalue problems for the differential equation (12.5.2), with given $\nu \geq 0$, can be discussed with results exactly corresponding to, and including, those obtained in Section 4 for the special case $\nu = 0$. That this is so for the formal results is almost evident from the earlier statements of this section and the observation that Green's formula for the operator

(12.5.35) $$-\frac{1}{r}\frac{d}{dr}(r\varphi') + \frac{\nu^2}{r^2}\varphi$$

is identical to that for (12.3.7), namely (12.3.10).

We state these results explicitly for the eigenvalue problem

(12.5.36)
D.E. $$\frac{1}{r}\frac{d}{dr}(r\varphi') + \left(\lambda - \frac{\nu^2}{r^2}\right)\varphi = 0 , \qquad 0 < r < b ,$$
B.C. φ is bounded as $r \to 0+$, $\varphi(b) = 0$,

where $\nu \geq 0$ is fixed. The *eigenvalues* of (12.5.36) are given by

(12.5.37) $$\lambda_{\nu,m} = \left(\frac{\beta_{\nu,m}}{b}\right)^2, \qquad m = 1, 2, \ldots ,$$

and to each eigenvalue there corresponds one (linearly independent) *eigenfunction*, namely,

$$\varphi_{\nu,m}(r) = J_\nu(r\sqrt{\lambda_{\nu,m}}) = J_\nu\left(\beta_{\nu,m}\frac{r}{b}\right), \qquad m = 1, 2, \ldots . \tag{12.5.38}$$

If $f(r)$ is piecewise continuous on the interval $[0,b]$, the *Fourier coefficients* of $f(r)$ relative to the system (12.5.38) of eigenfunctions are defined by

$$C_{\nu,m} = \frac{2}{b^2[J'_\nu(\beta_{\nu,m})]^2}\int_0^b f(r)J_\nu\left(\beta_{\nu,m}\frac{r}{b}\right)r\,dr\,, \qquad m = 1, 2, \ldots, \tag{12.5.39}$$

and the *Fourier series* of $f(r)$ relative to the system is given by

$$f(r) \sim \sum_{m=1}^{\infty} C_{\nu,m}J_\nu\left(\beta_{\nu,m}\frac{r}{b}\right). \tag{12.5.40}$$

We can make an improvement in the form of (12.5.39) by observing that from (12.5.33) and the recurrence relation (12.5.18), with n replaced by ν, we have

$$-J'_\nu(\beta_{\nu,m}) = J_{\nu+1}(\beta_{\nu,m})\,. \tag{12.5.41}$$

Upon substitution of (12.5.41), (12.5.39) becomes

$$C_{\nu,m} = \frac{2}{b^2[J_{\nu+1}(\beta_{\nu,m})]^2}\int_0^b f(r)J_\nu\left(\beta_{\nu,m}\frac{r}{b}\right)r\,dr\,. \tag{12.5.42}$$

Finally, we state, without proof, the

Theorem. *Let $f(r)$ be piecewise smooth on the interval $[0,b]$. Then for each value of r in $0 < r < b$, the series (12.5.40) with coefficients given by Equation (12.5.42) converges and*

$$\frac{f(r+) + f(r-)}{2} = \sum_{m=1}^{\infty} C_{\nu,m}J_\nu\left(\beta_{\nu,m}\frac{r}{b}\right), \qquad 0 < r < b\,.$$

In particular, if f is continuous at r the sum of the series is $f(r)$.

The particular eigenfunction expansions cited in the preceding theorem are usually called **Fourier–Bessel series of order ν.** Other eigenfunction expansions associated with the differential equation (12.5.2) in an interval $0 < r < b$ are frequently called **Dini series.**

12.6. *PROBLEMS IN POLAR AND CYLINDRICAL COORDINATES*

The general problem (12.1.1) of heat conduction in a disc, which we solved in Section 4 for a special case, is a problem in two space variables. Accordingly, our approach to the general problem will be similar to that employed in Section 10.3.

We begin by seeking solutions of the D.E. and B.C. of (12.1.1) having the form

$$u(r,\theta,t) = \varphi(r,\theta)\psi(t)\,. \tag{12.6.1}$$

In order that this function be non-trivial, the factors must be non-trivial solutions of

$$\text{D.E.} \qquad \psi' + \lambda k\psi = 0\,, \qquad 0 < t\,, \tag{12.6.2}$$

and

$$\begin{aligned} &\text{D.E.} \qquad \varphi_{rr} + \frac{1}{r}\varphi_r + \frac{1}{r^2}\varphi_{\theta\theta} + \lambda\varphi = 0\,, && 0 < r < b\,,\ -\infty < \theta < \infty\,, \\ &\text{B.C.} \qquad \varphi(b,\theta) = 0\,, && -\infty < \theta < \infty\,, \\ &\qquad\qquad \varphi(r,\theta + 2\pi) = \varphi(r,\theta)\,, && 0 < r < b\,,\ -\infty < \theta < \infty\,, \\ &\qquad\qquad \varphi \text{ is bounded at the origin}\,, \end{aligned} \tag{12.6.3}$$

where λ is a constant. Our task is to find the values of λ for which (12.6.3) has a non-trivial solution, and for those values of λ to determine the corresponding non-trivial solutions. Thus, we have to solve an eigenvalue problem for a partial differential equation.

We undertake to solve the eigenvalue problem (12.6.3) by further separation of variables, that is: we seek eigenfunctions of the form

$$\varphi(r,\theta) = R(r)H(\theta)\,. \tag{12.6.4}$$

The factors $R(r)$, $H(\theta)$ must satisfy

$$\begin{aligned} &\text{D.E.} \qquad H'' + \mu H = 0\,, && -\infty < \theta < \infty\,, \\ &\text{B.C.} \qquad H(\theta + 2\pi) = H(\theta)\,, && -\infty < \theta < \infty\,, \end{aligned} \tag{12.6.5}$$

and

$$\begin{aligned} &\text{D.E.} \qquad R'' + \frac{1}{r}R' + \left(\lambda - \frac{\mu^2}{r^2}\right)R = 0\,, \qquad 0 < r < b\,, \\ &\text{B.C.} \qquad R(r) \text{ is bounded as } r \to 0+\,, \quad R(b) = 0\,, \end{aligned} \tag{12.6.6}$$

where μ is a second constant to be determined. If (12.6.4) is to be non-trivial, then both factors $R(r)$ and $H(\theta)$ must be non-trivial. Thus μ and H must be eigenvalues and eigenfunctions, respectively, of the eigenvalue problem (12.6.5). The problem (12.6.5) was solved in Section 11.4. The eigenvalues are given by

$$\mu_n = n^2\,, \qquad n = 0, 1, 2, \ldots\,. \tag{12.6.7}$$

The corresponding eigenfunctions are, for $n \neq 0$, linear combinations of the independent eigenfunctions

$$H_{n,1}(\theta) = \cos n\theta , \quad H_{n,2}(\theta) = \sin n\theta , \qquad n = 1, 2, \ldots , \tag{12.6.8}$$

and, for $n = 0$,

$$H_0(\theta) = 1 . \tag{12.6.9}$$

Substituting (12.6.7) in (12.6.6) we obtain a sequence of eigenvalue problems of the form (12.5.36) with $\nu = n$, $n = 0, 1, 2, \ldots$. For each value of n, according to the statements of Section 5, the eigenvalues of (12.6.6) are

$$\lambda_{n,m} = \left(\frac{\beta_{n,m}}{b}\right)^2 , \qquad m = 1, 2, \ldots , \tag{12.6.10}$$

and the corresponding eigenfunctions are

$$R_{n,m}(r) = J_n\left(\beta_{n,m}\frac{r}{b}\right) , \qquad m = 1, 2, \ldots . \tag{12.6.11}$$

Thus, we have found eigenvalues of the problem (12.6.3) which form a double sequence,

$$\lambda_{n,m} = \left(\frac{\beta_{n,m}}{b}\right)^2 , \qquad n = 0, 1, 2, \ldots ; \quad m = 1, 2, \ldots . \tag{12.6.12}$$

Each eigenvalue $\lambda_{n,m}$ with $n \neq 0$ has multiplicity two, and the corresponding eigenfunctions $\varphi_{n,m}(r,\theta)$ are linear combinations of $R_{n,m}(r)H_{n,1}(\theta)$ and $R_{n,m}H_{n,2}(\theta)$; that is,

$$\varphi_{n,m}(r,\theta) = J_n\left(\beta_{n,m}\frac{r}{b}\right) [A_{n,m} \cos n\theta + B_{n,m} \sin n\theta] , \tag{12.6.13}$$

$$n = 1, 2, \ldots ; \quad m = 1, 2, \ldots ,$$

where $A_{m,n}$ and $B_{m,n}$ are any constants. The eigenvalues $\lambda_{0,m}$ have multiplicity one and corresponding eigenfunctions

$$\varphi_{0,m}(r,\theta) = \frac{A_{0,m}}{2} J_0\left(\beta_{0,m}\frac{r}{b}\right) , \qquad m = 1, 2, \ldots , \tag{12.6.14}$$

where $A_{0,m}/2$ are any constants.

The solution of (12.6.2) with $\lambda = \lambda_{n,m}$, is

$$\psi_{n,m}(t) = \exp[-\lambda_{n,m}kt] , \tag{12.6.15}$$

so that, referring to (12.6.1), the function $u(r,\theta,t)$ defined by

$$\sum_{n=0}^{\infty}\sum_{m=1}^{\infty} \psi_{n,m}(t)\varphi_{n,m}(r,\theta)$$

$$(12.6.16)\qquad = \sum_{m=1}^{\infty} \frac{A_{0,m}}{2} J_0\left(\beta_{0,m}\frac{r}{b}\right) e^{-\lambda_{0,m}kt} + \sum_{n=1}^{\infty}\sum_{m=1}^{\infty} e^{-\lambda_{n,m}kt} J_n\left(\beta_{n,m}\frac{r}{b}\right)[A_{n,m}\cos n\theta + B_{n,m}\sin n\theta]$$

will be a formal solution of problem (12.1.1), provided the constants $A_{n,m}$, $B_{n,m}$ can be chosen so that the I.C. of the problem is satisfied. This last condition requires that

$$(12.6.17)\qquad f(r,\theta) = \sum_{m=1}^{\infty} \frac{A_{0,m}}{2} J_0\left(\beta_{0,m}\frac{r}{b}\right) + \sum_{n=1}^{\infty}\sum_{m=1}^{\infty} J_n\left(\beta_{n,m}\frac{r}{b}\right)[A_{n,m}\cos n\theta + B_{n,m}\sin n\theta],$$

where the right side is obtained by setting $t = 0$ in (12.6.16).

To conclude our discussion of the formal solution of problem (12.1.1), we shall show that for suitable functions $f(r,\theta)$ the coefficients $A_{n,m}$, $B_{n,m}$ can be determined so that Equation (12.6.17) is satisfied if the second series in the equation is regarded as an iterated series. We will also find formulas for these coefficients. The method we shall use is similar to that employed in Section 10.5 to show the convergence of the iterated series of eigenfunctions considered there.

Let $f(r,\theta)$ be continuous and have continuous first partial derivatives for $0 \leq r \leq b$, $-\infty < \theta < \infty$, and suppose that, as a function of θ, f is periodic with period 2π. Then, for each fixed value of r in $[0,b]$, $f(r,\theta)$ can be expanded in a convergent Fourier series

$$(12.6.18)\qquad f(r,\theta) = \frac{\tilde{A}_0(r)}{2} + \sum_{n=1}^{\infty} \tilde{A}_n(r)\cos n\theta + \tilde{B}_n(r)\sin n\theta, \qquad -\infty < \theta < \infty,$$

where

$$(12.6.19)\qquad \begin{aligned} \tilde{A}_n(r) &= \frac{1}{\pi}\int_{-\pi}^{\pi} f(r,\theta)\cos n\theta\, d\theta, \\ \tilde{B}_n(r) &= \frac{1}{\pi}\int_{-\pi}^{\pi} f(r,\theta)\sin n\theta\, d\theta, \end{aligned} \qquad n = 0, 1, 2, \ldots.$$

The functions $\tilde{A}_n(r)$, $\tilde{B}_n(r)$ defined by (12.6.19) are smooth functions of r on $[0,b]$. Hence, according to the theorem stated at the end of Section 5, for each value of n we have

$$
\begin{aligned}
\tilde{A}_n(r) &= \sum_{m=1}^{\infty} A_{n,m} J_n\left(\beta_{n,m}\frac{r}{b}\right), \\
&\qquad\qquad\qquad\qquad 0 < r < b, \\
\tilde{B}_n(r) &= \sum_{m=1}^{\infty} B_{n,m} J_n\left(\beta_{n,m}\frac{r}{b}\right),
\end{aligned}
\tag{12.6.20}
$$

where the constants $A_{n,m}$, $B_{n,m}$ are given by

$$
\begin{aligned}
A_{n,m} &= \frac{2}{b^2[J_{n+1}(\beta_{n,m})]^2}\int_0^b \tilde{A}_n(r) J_n\left(\beta_{n,m}\frac{r}{b}\right) r\,dr, \\
&\qquad\qquad\qquad\qquad m = 1, 2, \ldots, \\
B_{n,m} &= \frac{2}{b^2[J_{n+1}(\beta_{n,m})]^2}\int_0^b \tilde{B}_n(r) J_n\left(\beta_{n,m}\frac{r}{b}\right) r\,dr.
\end{aligned}
\tag{12.6.21}
$$

On substitution of (12.6.20) in (12.6.18) we obtain

$$
\begin{aligned}
f(r,\theta) &= \sum_{m=1}^{\infty} \frac{A_{0,m}}{2} J_0\left(\frac{\beta_{0,m}r}{b}\right) \\
&\quad + \sum_{n=1}^{\infty}\sum_{m=1}^{\infty} J_n\left(\beta_{n,m}\frac{r}{b}\right)[A_{n,m}\cos n\theta + B_{n,m}\sin n\theta],
\end{aligned}
\tag{12.6.22}
$$

where the first term on the right is a convergent series and the second term is a convergent iterated series for $0 < r < b$, $-\infty < \theta < \infty$. The coefficients $A_{n,m}$, $B_{n,m}$ can be expressed directly in terms of $f(r,\theta)$ by substitution of (12.6.19) in (12.6.21). We obtain

$$
\begin{aligned}
A_{n,m} &= \frac{2}{\pi b^2[J_{n+1}(\beta_{n,m})]^2}\int_{-\pi}^{\pi}\int_0^b f(r,\theta) J_n\left(\beta_{n,m}\frac{r}{b}\right)\cos n\theta\, r\,dr\,d\theta, \\
&\qquad\qquad n = 0, 1, 2, \ldots; \quad m = 1, 2, \ldots, \\
B_{n,m} &= \frac{2}{\pi b^2[J_{n+1}(\beta_{n,m})]^2}\int_{-\pi}^{\pi}\int_0^b f(r,\theta) J_n\left(\beta_{n,m}\frac{r}{b}\right)\sin n\theta\, r\,dr\,d\theta.
\end{aligned}
\tag{12.6.23}
$$

The method of solution of problem (12.1.1) described above can be applied, with obvious modifications, to other initial-boundary value problems in two-dimensional domains bounded by coordinate curves of a polar coordinate system, and to initial-boundary value problems and boundary value problems in three-dimensional domains bounded by coordinate surfaces of a cylindrical coordinate system. Some examples are given in Exercises 12c.

EXERCISES 12c

1. Verify the relation (12.5.3) between solutions of (12.5.2) and (12.5.1).

2. Use the recurrence relations (12.5.18) and (12.5.19) to show that:

(i) $J_1(r) = -J_0'(r)$;
(ii) $J_{-1}(r) = -J_1(r)$;
(iii) $J_{-2}(r) = J_2(r)$;
(iv) $J_{-n}(r) = (-1)^n J_n(r)$, by mathematical induction.

3. Show that the *recurrence relations*

$$\frac{d}{dr}[r^{-n}J_n(r)] = -r^{-n}J_{n+1}(r)\,,$$

$$\frac{d}{dr}[r^{n}J_n(r)] = r^{n}J_{n-1}(r)\,,$$

are equivalent, respectively, to (12.5.18) and (12.5.19).

4. (i) Use the result of Problem 3 to show that

$$\int_0^s r^{n+1}J_n(r)\,dr = s^{n+1}J_{n+1}(s)\,.$$

(ii) Apply the theorem stated at the end of Section 3 to the function $f(r) = 1$, $0 \le r \le 1$, and use the result of part (i) to show that, if $\beta_1 < \beta_2 < \cdots < \beta_m < \cdots$ are the positive zeros of J_0,

$$1 = 2\sum_{m=1}^{\infty}\frac{J_0(\beta_m r)}{\beta_m J_1(\beta_m)}\,, \qquad 0 < r < 1\,.$$

5. Let $Z_n(r)$ be a solution of

$$Z_n'' + \frac{1}{r}Z_n + \left(1 - \frac{n^2}{r^2}\right)Z_n = 0$$

and define $Z_{n+1}(r)$ by

$$Z_{n+1} = \frac{n}{r}Z_n - Z_n'\,.$$

Using the equations stated above, establish successively the following equations:

$$rZ_{n+1}' = (n+1)Z_n' + \left(r - \frac{n^2+n}{r}\right)Z_n\,,$$

$$\frac{d}{dr}(rZ_{n+1}') = \left[r - \frac{(n+1)^2}{r}\right]Z_n' - n\left[1 - \frac{(n+1)^2}{r^2}\right]Z_n\,,$$

$$Z_{n+1}'' + \frac{1}{r}Z_{n+1}' + \left[1 - \frac{(n+1)^2}{r^2}\right]Z_{n+1} = 0\,.$$

6. Use the definition (12.5.22) and the series representations (12.2.32) and (12.2.16) to calculate series representations of $Y_1(r)$ and $Y_2(r)$. Show that

$$\lim_{r\to 0+} Y_1(r) = \lim_{r\to 0+} Y_2(r) = -\infty \, .$$

More explicitly, show that

$$Y_1(r) = -\frac{2}{\pi r} + \frac{\epsilon_1(r)}{r}, \qquad \lim_{r\to 0+} \epsilon_1(r) = 0 \, ,$$

$$Y_2(r) = -\frac{4}{\pi r^2} + \frac{\epsilon_2(r)}{r^2}, \qquad \lim_{r\to 0+} \epsilon_2(r) = 0 \, .$$

7. Given that, if $\nu > 0$ is not an integer, Bessel's equation (12.5.11) has two solutions of the form

$$\psi(r) = r^\gamma \sum_{j=0}^{\infty} c_j r^{2j} \, ,$$

use the *method of undetermined coefficients* to calculate the two values of γ and corresponding sequences c_0 (arbitrary), c_1, c_2, ...

8. Show that the functions $J_\nu(r)$ defined by (12.5.27) satisfy the recurrence relations

$$\frac{d}{dr}\,[r^{-\nu} J_\nu(r)] = -r^{-\nu} J_{\nu+1}(r) \, ,$$

$$\frac{d}{dr}\,[r^{\nu} J_\nu(r)] = r^{\nu} J_{\nu-1}(r) \, ,$$

and hence, as in Problem 3, that the functions J_ν satisfy the relations (12.5.18) and (12.5.19) with n replaced by ν.

9. Show that if a system of functions Z_ν satisfies the recurrence relations

$$\nu Z_\nu(r) - r Z'_\nu(r) = r Z_{\nu+1}(r) \, ,$$
$$\nu Z_\nu(r) + r Z'_\nu(r) = r Z_{\nu-1}(r) \, ,$$

then each function Z_ν is a solution of Bessel's equation of order ν. Conclude from this statement and the result of Problem 8 that Bessel's functions of the first kind of order ν are solutions of Bessel's equation of order ν.

10. Use (12.5.18) to show that J_n and J_{n+1} have no common zeros, except possibly at $r = 0$.

11. Use the result of Problem 3 to show that between each pair of successive positive zeros of $J_n(r)$ there is *exactly* one zero of $J_{n+1}(r)$, as follows.
   (i) Show that between each pair of successive positive zeros of $J_n(r)$ there is at least one zero of $J_{n+1}(r)$.
   (ii) Show that between each pair of successive zeros of $J_n(r)$ the function $r^{n+1} J_{n+1}(r)$ is monotone, so that $J_{n+1}(r)$ has no more than one zero between two successive zeros of $J_n(r)$.

12. Given that the positive zeros of $J_0(r)$ form an increasing sequence diverging to ∞, use the results of Problems 10 and 11 to show that the positive zeros of $J_n(r)$, $n = 1, 2, \ldots$ have the same property, and further that, if $\{\beta_{n,m}\}$ denotes the positive zeros of $J_n(r)$ arranged in order of increasing magnitude,

$$0 < \beta_{n,1} < \beta_{n+1,1} < \beta_{n,2} < \beta_{n+1,2} < \beta_{n,3} < \cdots .$$

13. For the eigenvalue problem ($\nu > 0$)

$$\text{D.E.} \qquad \frac{1}{r}\frac{d}{dr}(r\varphi') + \left(\lambda - \frac{\nu^2}{r^2}\right)\varphi = 0\,, \qquad 0 < r < b\,,$$

$$\text{B.C.} \qquad \varphi(r) \text{ is bounded as } r \to 0+\,, \quad \varphi'(b) = 0\,,$$

assume that the eigenvalues form an increasing sequence of positive numbers diverging to ∞ and do the following.

(i) Find an equation whose roots are the eigenvalues of the problem and for each eigenvalue find a corresponding eigenfunction.

(ii) Find formulas for the Fourier coefficients and Fourier series of a piecewise continuous function $f(r)$ relative to the eigenfunctions determined in part (i).

(iii) Calculate the normalizing constants of the eigenfunctions determined in part (i).

14. Solve the boundary value problem (12.4.6) with $h(\theta,z) = 0$, $g(r,\theta) = 0$, and $f(r,\theta)$ any smooth function of r and θ which is periodic in θ with period 2π.

15. Formulate and solve the problem of vibration of a *semi-circular* membrane of radius b with boundary fixed in a plane when the initial displacement and initial velocity are any given functions f and g satisfying appropriate conditions.

16. Formulate and solve the problem of heat conduction in a flat plate in the shape of a *circular sector* of radius b and angular opening ω when the straight edges of the plate are insulated, the circular edge is maintained at temperature zero, and the initial temperature is any function f satisfying suitable conditions.

17. Repeat Problem 16 when the entire boundary of the plate is insulated.

18. Formulate and solve the problem of *axially symmetric* heat conduction in a cylinder of radius b and height l when the lateral surface of the cylinder is maintained at temperature zero, the top and bottom of the cylinder are insulated, and the initial temperature is a smooth function $f(r,z)$, $0 \le r \le b$, $0 \le z \le l$.

19. Show that the general solution of

$$r^2\varphi'' + (2\delta - 2\gamma n + 1)r\varphi' + [\gamma^2\lambda r^{2\gamma} + \delta(\delta - 2\gamma n)]\varphi = 0$$

is

$$\varphi(r) = r^{\gamma n-\delta}[AJ_n(r^\gamma\sqrt{\lambda}) + BY_n(r^\gamma\sqrt{\lambda})]\,,$$

where A and B are arbitrary constants.

20. Using the relation

$$e^{i(a+b)\sin\theta} = e^{ia\sin\theta}\cdot e^{ib\sin\theta}\,,$$

derive the *addition formula*

$$J_n(a+b) = \sum_{n=-\infty}^{\infty} J_m(a)J_{n-m}(b)$$

for Bessel functions of the first kind of integer order.

21. The function e^{-t^2+2tx} is a generating function for the *Hermite polynomials* $H_n(x)$, $n = 0, 1, 2, \ldots$, that is,

$$e^{-t^2+2tx} = \sum_{n=0}^{\infty} \frac{H_n(x)}{n!}\,t^n\,.$$

(i) Calculate $H_n(x)$, distinguishing the cases $n = 2m$ (even) and $n = 2m + 1$ (odd).

(ii) Find two recurrence relations satisfied by the Hermite polynomials.

(iii) Use the result of part (ii) to show that $H_n(x)$ satisfies the differential equation

$$H_n'' - 2xH_n' + 2nH_n = 0\,.$$

(iv) Find an integral representation of $H_n(x)$.

(v) The Hermite polynomials form an orthogonal system on $-\infty < x < \infty$ with respect to a certain weight function. Find the weight function and prove the orthogonality relation.

Suggestions for further reading

The following is a brief list of books suggested for collateral or supplementary reading, grouped under the heading of the principal subject to which they are relevant. Within each group the choice of material, level of presentation, and background required of the reader vary appreciably. In each case we have tried to include some relatively elementary books and not to include those which seemed to be too advanced for the purposes of this list. References to more advanced books can be found in some of the books listed below.

PARTIAL DIFFERENTIAL EQUATIONS

Epstein, B., *Partial Differential Equations*, McGraw-Hill, New York, 1962.

Garabedian, P. R., *Partial Differential Equations*, Wiley, New York, 1964.

Petrovsky, I. G., *Lectures on Partial Differential Equations*, Interscience, New York, 1954.

Sobolev, S. L., *Partial Differential Equations of Mathematical Physics*, Addison-Wesley, Reading, Mass., 1964.

Weinberger, H. F., *A First Course in Partial Differential Equations*, Blaisdell, New York, 1965.

FOURIER SERIES AND FOURIER INTEGRALS

Bateman Manuscript Project, *Tables of Integral Transforms*, Vol. 1, McGraw-Hill, New York, 1954.

Seeley, R., *An Introduction to Fourier Series and Integrals*, Benjamin, New York, 1966.

Sneddon, I. N., *Fourier Transforms*, McGraw-Hill, New York, 1951.

Tolstov, G. P., *Fourier Series*, Prentice-Hall, Englewood Cliffs, New Jersey, 1962.

EIGENVALUE PROBLEMS AND EIGENFUNCTION EXPANSIONS

Birkhoff, G. and G. Rota, *Ordinary Differential Equations*, Ginn, Boston, 1962.

Coddington, E. A. and N. Levinson, *Theory of Ordinary Differential Equations*, McGraw-Hill, New York, 1955.

Courant, R. and D. Hilbert, *Methods of Mathematical Physics*, Vol. I, Interscience, New York, 1953.

Ince, E. L., *Ordinary Differential Equations*, Dover, New York, 1956.

Miller, K. S., *Linear Differential Equations in the Real Domain*, Norton, New York, 1963.

Sagan, H., *Boundary and Eigenvalue Problems in Mathematical Physics*, Wiley, New York, 1962.

MATHEMATICAL PHYSICS

Carslaw, H. S. and J. C. Jaeger, *Conduction of Heat in Solids*, 2nd ed., Oxford, 1959.

Morse, P. M., *Vibration and Sound*, McGraw-Hill, New York, 1948.

Sternberg, W. J. and T. L. Smith, *The Theory of Potential and Spherical Harmonics*, 2nd ed., University of Toronto, Toronto, 1952.

SPECIAL FUNCTIONS OF MATHEMATICAL PHYSICS

Jahnke, E. and F. Emde, *Tables of Functions*, Dover, New York, 1965.

Lebedev, N. N., *Special Functions and their Applications*, Prentice-Hall, Englewood Cliffs, New Jersey, 1965.

Sneddon, I. N., *Special Functions of Mathematical Physics and Chemistry*, 2nd ed., Interscience, New York, 1961.

Answers and hints for selected problems

EXERCISES 1a

2. $u(x,y) = f(y)e^x + g(y)e^{-x}$, f and g arbitrary.
5. $u(x,y) = f(x)g(y)$, f and g arbitrary.
7. $u(x,y) = f(x^2 + y^2)$, f arbitrary.
9. (a) $yu_x + xu_y = 0$,
   (b) $y^2u_{xy} - xyu_{yy} + xu_y = 0$.

EXERCISES 1b

1. linear homogeneous: (c), (g)
   linear inhomogeneous: (b), (d)
   non-linear: (a), (e), (f), (h).
2. $u = (\frac{1}{2} + c)u_2 + (\frac{1}{2} - c)u_1$, c arbitrary.

EXERCISES 1c

1. (b) $U = \exp\left[\dfrac{1 \pm \sqrt{1 - 4\lambda}}{2}y\right] \sin x\sqrt{\lambda}$,

   $V = \exp\left[\dfrac{1 \pm \sqrt{1 - 4\lambda}}{2}y\right] \cos x\sqrt{\lambda}$, λ any constant.

   (c) If $u = X(x)Y(y)$, then

$$-\frac{X''}{X} = \frac{Y''}{Y'' + Y}.$$

   (e) $u(x,y,z;\lambda,\mu) = \exp\left[-\lambda x + \mu y + \dfrac{1 + \lambda\mu}{\mu - \lambda}z\right]$, λ and μ any constants.
3. Use the principle of superposition.
5. (b) $u(x,y) = [A + Be^{\lambda x}] \exp(\lambda \tan^{-1} y)$, λ, A, and B any constants.

EXERCISES 2a

2. $u(x,y) = -ye^{x+2y}$.

5. $u_0(x) = Ce^{2x}$, where C is any constant. $u(x,y) = \varphi(y)e^{2x}$, where φ is any function that satisfies the condition $\varphi(0) = C$.

8. $P(s,t) = \exp\left[\dfrac{s + 7t}{4}\right]$.

EXERCISES 2b

1. (a) $u_{xx} + u_{yy} - u = 0$,
   (b) $u_{xx} + u_{yy} + u = 0$,
   (c) $u_{xx} + u_{yy} + u = 0$,
   (d) $u_{xx} - u_{yy} + u = 0$,
   (e) $u_{xx} + u = 0$.

3. If $u_{xx} - u_{yy} = 0$, then $u(x,y) = F(x + y) + G(x - y)$, F and G arbitrary.

5. Let $a(x,y)$ be any coefficient. By hypothesis, for any (x_0,y_0), $a(x,y) = a(x + x_0, y + y_0)$. Setting $x = y = 0$, we get $a(0,0) = a(x_0,y_0)$, which implies, since (x_0,y_0) is arbitrary, that $a(x,y)$ is a constant function.

EXERCISES 3a

2. B.C. $\quad u(0,t) = 0, \quad u(L,t) = \dfrac{1}{L}\displaystyle\int_0^L u(x,t)\,dx \qquad 0 < t.$

5. $u_t = k\left[\dfrac{1}{r}(ru_r)_r + \dfrac{1}{r^2}u_{\theta\theta}\right]$.

6. $\dfrac{\partial u}{\partial t} = \dfrac{k}{r^2}\dfrac{\partial}{\partial r}\left(r^2\dfrac{\partial u}{\partial r}\right)$.

EXERCISES 3b

2. FL/κ.

5. (ii)

$$u(x) = \begin{cases} \dfrac{\kappa_0\kappa_2 U}{L_0\kappa_1\kappa_2 + \kappa_0 L_1\kappa_2 + \kappa_0\kappa_1 L_2}\,x & 0 < x < L_1 \\[2ex] \dfrac{\kappa_0\kappa_1 U}{L_0\kappa_1\kappa_2 + \kappa_0 L_1\kappa_2 + \kappa_0\kappa_1 L_2}\,[x - (L_1 + L_2)] + U, & L_1 < x < L_1 + L_2. \end{cases}$$

7. $\dfrac{A_1(A_2 + A_3)U}{A_1A_2 + A_1A_3 + A_2A_3}, \dfrac{A_1A_2U}{A_1A_2 + A_1A_3 + A_2A_3}.$

11.

$$u(x) = \begin{cases} \dfrac{Q}{10k}[3x] & 0 \le x \le 3 \\[2ex] \dfrac{Q}{10k}[-5(x - 3)^2 + 3(x - 3) + 9] & 3 \le x \le 4 \\[2ex] \dfrac{Q}{10k}[-7(x - 4) + 7] & 4 \le x \le 5. \end{cases}$$

12. $\kappa A u_x(0,t) - C u_t(0,t) = 0.$

13. D.E. $\quad u_t = k u_{xx} - V u_x \quad 0 < x < L, \quad 0 < t.$

EXERCISES 4a

1. $\varphi_1(x) = e^{-2x}\left(\cos x\sqrt{\lambda - 4} + \dfrac{2 \sin x\sqrt{\lambda - 4}}{\sqrt{\lambda - 4}}\right), \varphi_2(x) = e^{-2x}\dfrac{\sin x\sqrt{\lambda - 4}}{\sqrt{\lambda - 4}}.$

2. (a) $\varphi_1(x) = \cos(\sqrt{\lambda}\log x), \varphi_2(x) = \dfrac{\sin(\sqrt{\lambda}\log x)}{\sqrt{\lambda}}.$

4. The roots of $m^3 - \lambda = 0$ are $m = s$, $m = \left(-\dfrac{1}{2} + i\dfrac{\sqrt{3}}{2}\right)s$, $m = \left(-\dfrac{1}{2} - i\dfrac{\sqrt{3}}{2}\right)s$, where $s = \lambda^{1/3}$.

$$\varphi_3(x) = \frac{1}{3s^2}\left[\exp(xs) - \exp\left(-\frac{xs}{2}\right)\left\{\cos\frac{xs\sqrt{3}}{2} + \sqrt{3}\sin\frac{xs\sqrt{3}}{2}\right\}\right].$$

5. $\varphi_4(x) = (\sinh x\alpha - \sin x\alpha)/2\alpha^3, \alpha = \lambda^{1/4}.$

EXERCISES 4b

1. (a) The first term of the series is an approximation with error not exceeding the second term.

   (b) $U(t) = \dfrac{1}{L}\displaystyle\int_0^L u(x,t)\,dx = \dfrac{8}{\pi^2}\sum_{m=0}^{\infty}\dfrac{e^{-\lambda_{2m+1}kt}}{(2m+1)^2}.$

3. $b_n = \dfrac{2}{n\pi}\left(\cos\dfrac{n\pi}{2} - \cos\dfrac{n\pi}{3}\right).$

4. $|b_n| \le 2M$, series solution dominated by convergent series $\displaystyle\sum_{n=1}^{\infty} 2Me^{-\lambda_n kt}$.

7. $\displaystyle\sum_{n=0}^{\infty}\frac{2L}{(n+\frac{1}{2})^2\pi^2}\exp\left\{-(n+\tfrac{1}{2})^2\frac{\pi^2 kt}{L^2}\right\}\cos(n+\tfrac{1}{2})\frac{\pi x}{L}.$

8. (b) $\sqrt{\lambda}\sin(\sqrt{\lambda}\log a) = 0$
   (c) $(1 - \cosh\lambda^{1/4}L\cos\lambda^{1/4}L)/2\lambda = 0.$

11. (b) $u = \displaystyle\sum_{n=0}^{\infty}\frac{4}{(2n+1)\pi}\exp\left(\frac{rt}{c\rho}\right)\exp\left\{-\left(\frac{(2n+1)\pi}{L}\right)^2 kt\right\}\sin\frac{n\pi x}{L}.$

EXERCISES 4c

2. (i) (a), (b); (ii) (a), (b).
4. (v) $Mg = (a_0 g)'''' - (a_1 g)''' + (a_2 g)'' - (a_3 g)' + a_4 g.$

EXERCISES 4d

1. If $\varphi(L) = 0$, then from the second B.C. we also get $\varphi'(L) = 0$. But the only solution of the D.E. which satisfies $\varphi(L) = \varphi'(L) = 0$ is the trivial solution, so φ is identically zero. This is a contradiction since an eigenfunction is non-trivial by definition.

2. $\int_a^b (Bf)g dx = (f'''g - f''g')|_a^b + \int_a^b f''g''\, dx$. If λ is an eigenvalue with eigenfunction φ, then

$$\lambda \int_0^L \varphi\bar{\varphi}\, dx = \int_0^L \varphi''\bar{\varphi}''\, dx\,.$$

6. Equation (4.7.6) is valid. λ_n is an eigenvalue of Problem 3 and $\varphi(x,\lambda_n)$ is a corresponding eigenfunction if and only if $\varphi(L,\lambda_n) = 0$, and then

$$\int_0^L [\varphi(x,\lambda_n)]^2\, dx = -\left[\varphi(L,\lambda) \frac{\partial}{\partial\lambda} \varphi'(L,\lambda)\right]_{\lambda=\lambda_n}.$$

7. If λ is an eigenvalue with eigenfunction φ, then

$$\lambda \int_0^L \varphi\bar{\varphi}\, dx = \int_0^L \varphi''\bar{\varphi}''\, dx + \int_0^L \varphi'\bar{\varphi}'\, dx\,.$$

EXERCISES 4e

1. (b) $h = -1/L$.

2. (a) $\cos L\sqrt{\lambda} + h \dfrac{\sin L\sqrt{\lambda}}{\sqrt{\lambda}} = 0$.

   (c) $[(n + \frac{1}{2})\pi/L]^2$.

3. (b) $h = 0,\ -2/L$.

   (c) $2h \cos L\sqrt{\lambda} + \left(\dfrac{h^2}{\sqrt{\lambda}} - \sqrt{\lambda}\right) \sin L\sqrt{\lambda} = 0$.

   (e) $(n\pi/L)^2$.

   (f) Two, if $h < -2/L$; one, if $-2/L < h < 0$.

7. $c_n = \dfrac{h}{\lambda_n\left[\dfrac{L}{2} + \dfrac{h(1 + Lh)}{2\lambda_n}\right]\cos L\sqrt{\lambda_n}}$.

EXERCISES 4f

1. (i) $\sqrt{\lambda} \sin L\sqrt{\lambda} - 2(1 - \cos L\sqrt{\lambda})/L = 0$.

   (iv) Use the result of part (i).

2. $\lambda_n = [(n + \frac{1}{2})(\pi/L)]^2$, $n = 0, 1, 2, \ldots$. Each eigenvalue is of multiplicity two. A pair of orthogonal eigenfunctions belonging to the eigenvalue λ_n is

$$\varphi_{n,1} = \cos\,(n + \tfrac{1}{2})\,\frac{\pi x}{L}, \quad \varphi_{n,2} = \sin\,(n + \tfrac{1}{2})\,\frac{\pi x}{L}.$$

4. $P_2(x) = 6x^2 - 6x + 1.$

7. $c_n = \int_a^b f(x)\overline{\varphi_n(x)} \left[\int_a^b \varphi_n(x)\overline{\varphi_n(x)}\, dx\right]^{-1}.$

EXERCISES 4g

3. (iii) $\log x = \sum_{n=1}^{\infty} \frac{2 \log b}{n\pi} (-1)^n \sin \frac{n\pi \log x}{\log b}.$

4. $u(x,t) = \sum_{n=1}^{\infty} c_n \exp\left\{-\left[\left(\frac{n\pi}{L}\right)^2 + \beta^2\right] t\right\} e^{-\beta x} \sin \frac{n\pi x}{L},$

$$c_n = \frac{2}{L} \int_0^L f(x) e^{\beta x} \sin \frac{n\pi x}{L}\, dx.$$

EXERCISES 5a

1. (iii) $u(x,t) = v(x,t) = A(t) + \frac{x}{L}[B(t) - A(t)],$

where $v(x,t)$ is the solution of

D.E. $\quad v_t - kv_{xx} = q(x,t) - A'(t) - \frac{x}{L}[B'(t) - A'(t)] \qquad 0 < x < L, \quad 0 < t$

B.C. $\quad \left.\begin{aligned} v(0,t) &= 0 \\ v(L,t) &= 0 \end{aligned}\right\} \qquad 0 < t$

I.C. $\quad v(x,0) = f(x) - A(0) - \frac{x}{L}[B(0) - A(0)] \qquad 0 < x < L.$

3. Set $u(x,t) = v(x,t) + \frac{x^2}{2L}[B(t) - A(t)] + xA(t).$

6.
$$u(x,t) = -\frac{Qx^2}{2k} + \left[\frac{B - A}{L} + \frac{QL}{2k}\right] x + A + \sum_{n=1}^{\infty} b_n \exp\left(-\frac{n^2\pi^2}{L^2} kt\right) \sin \frac{n\pi x}{L},$$

where

$$b_n = \frac{2QL^2}{kn^3\pi^3}[(-1)^n - 1] + \frac{2B(-1)^n}{n\pi} - \frac{2A}{n\pi}.$$

7. (ii) $\frac{A}{\omega} \sin \omega t + B$, where $B = \frac{1}{\pi} \int_0^{\pi} f(x)\, dx.$

EXERCISES 5b

1. (i) $u(x,t) = \sum_{n=0}^{\infty} \frac{4(-1)^n}{(2n+1)\pi} \frac{e^{-\lambda_n kt} - e^{-t}}{1 - \lambda_n k} \cos\left(n + \tfrac{1}{2}\right) \frac{\pi x}{L},$

where

$$\lambda_n = \left[(n + \tfrac{1}{2}) \frac{\pi}{L} \right]^2 .$$

(ii) $\left[\dfrac{\cos xk^{-1/2}}{\cos Lk^{-1/2}} - 1 \right] e^{-t}.$

3. $$\frac{Q}{\beta} \left\{ \left[1 - \frac{\cos\omega(x+L)\cosh\omega(x-L) + \cos\omega(x-L)\cosh\omega(x+L)}{\cos 2\omega L + \cosh 2\omega L} \right] \sin\beta t \right.$$
$$\left. + \left[\frac{\sin\omega(x+L)\sinh\omega(x-L) + \sin\omega(x-L)\sinh\omega(x+L)}{\cos 2\omega L + \cosh 2\omega L} \right] \cos\beta t \right\},$$

where $\omega = (\beta/2k)^{1/2}$.

5. (i) $$u(x,t) = \frac{b_0 + Lt}{2} + \sum_{n=1}^{\infty} \frac{2}{kL\lambda_n^2} [(-1)^n - 1] \cos \frac{n\pi x}{L}$$
$$+ \sum_{n=1}^{\infty} \left\{ b_n - \frac{2}{kL\lambda_n^2} [(-1)^n - 1] \right\} e^{-\lambda_n kt} \cos \frac{n\pi x}{L},$$

where

$$\lambda_n = \left(\frac{n\pi}{L} \right)^2$$

$$b_n = \frac{2}{L} \int_0^L f(x) \cos \frac{n\pi x}{L} \, dx , \qquad n = 0, 1, 2, \ldots .$$

(ii) The problem has an asymptotic solution of the form

$$U(x,t) = \frac{b_0 + Lt}{2} + V(x) .$$

7. (i) Use (5.3.12)–(5.3.13) and

$$\Phi(x,t;\tau) = \sum_{n=1}^{\infty} q_n(\tau) e^{\lambda_n (t-\tau)} \sin \frac{n\pi x}{L} ,$$

where $\lambda_n = (n\pi/L)^2$ and

$$q_n(\tau) = \frac{2}{L} \int_0^L q(x,\tau) \sin \frac{n\pi x}{L} \, dx , \qquad n = 1, 2, \ldots .$$

EXERCISES 5c

2. $$u(x,t) = \sum_{n=0}^{\infty} \frac{8QL}{(2n+1)^2 \pi^2} S_n(t) e^{-\lambda_n kt} \cos (n + \tfrac{1}{2}) \frac{\pi x}{L},$$

where $\lambda_n = \left[(n + \frac{1}{2}) \frac{\pi}{L} \right]^2$ and

$$S_n(t) = \begin{cases} 1 - e^{\lambda_n kt} & 0 < t \le 1 \\ 1 - e^{\lambda_n k} & 1 \le t \end{cases} \qquad n = 0, 1, 2, \ldots .$$

4. $$\Phi(x,t;\tau) = \sum_{n=1}^{\infty} \frac{2}{n\pi} [A(\tau) + (-1)^{n+1} B(\tau)][e^{-\lambda_n k(t-\tau)} - 1] \sin \frac{n\pi x}{L},$$

where $\lambda_n = (n\pi/L)^2$, $n = 1, 2, \ldots$.

6. (i) $u(x,t;\sigma) = \sum_{n=1}^{\infty} \frac{Q_n}{k\lambda_n} S_n(t;\sigma) e^{-\lambda_n kt} \sin \frac{n\pi x}{L}$

and

$$u(x,t) = \sum_{n=1}^{\infty} Q_n e^{-\lambda_n kt} \sin \frac{n\pi x}{L},$$

where for $n = 1, 2, \ldots, \lambda_n = (n\pi/L)^2$,

$$Q_n = \frac{2}{L} \int_0^L Q(x) \sin \frac{n\pi x}{L}\, dx,$$

and

$$S_n(t;\sigma) = \begin{cases} (e^{\lambda_n kt} - 1)/\sigma & 0 < t \leq \sigma \\ (e^{\lambda_n k\sigma} - 1)/\sigma & \sigma \leq t. \end{cases}$$

(ii) In part (i) replace $Q(x)$ by $q(x,\tau)$ and $u(x,t)$ by $u(x,t-\tau)$.

9. (ii) The Nth term of the series solution is

$$\left\{\frac{2}{L}\int_0^L f(x) \sin \frac{N\pi x}{L}\, dx + \frac{2Q}{N\pi}[1 - (-1)^N]t\right\} \sin \frac{N\pi x}{L}.$$

If N is odd, the problem has no equilibrium solution. Compare with Problem 8.

EXERCISES 6a

2. (ii) All a,b satisfying $2a + b = 23$.
   (iii) $a = 17$, $b = -11$.

4. (ii) $f(-L) = f(L), f'(-L) = f'(L)$.
   (iii) $f(-L) = f(L), f'(-L) = f'(L), f''(-L) = f''(L)$.

EXERCISES 6b

1. $f(x) \sim \sum_{k=0}^{\infty} \frac{4}{(2k+1)\pi} \sin \frac{(2k+1)\pi x}{3}$ $[-3,3]$.

2. $x^2 \sim \frac{\pi^2}{3} + \sum_{k=1}^{\infty} \frac{4(-1)^k}{k^2} \cos kx$ $[-\pi,\pi]$.

   $\frac{\pi^2}{2} = 1 - \frac{1}{2^2} + \frac{1}{3^2} - \cdots$.

3. $2x - 1 \sim -1 + \sum_{k=1}^{\infty} \frac{4(-1)^{k+1}}{k\pi} \sin k\pi x$ $[-1,1]$.

4. $f(x) \sim -\frac{2}{3} + \sum_{k=1}^{\infty} \frac{4(-1)^k}{k^2\pi^2} \cos k\pi x$.

EXERCISES 6c

1. $$2x + 1 \sim 2 - \frac{8}{\pi^2}\sum_{k=0}^{\infty}\frac{1}{(2k+1)^2}\cos(2k+1)\pi x \qquad [0,1]$$

$$2x + 1 \sim \frac{2}{\pi}\sum_{k=1}^{\infty}\frac{[1+3(-1)^{k+1}]}{k}\sin k\pi x \qquad [0,1].$$

2. $$ax^2 + bx \sim \frac{a}{3} + \frac{b}{2} + \sum_{k=1}^{\infty}\left\{\frac{4a(-1)^k + 2b[(-1)^k - 1]}{k^2\pi^2}\right\}\cos k\pi x \qquad [0,1]$$

$$\frac{\pi^2}{6} = \sum_{n=1}^{\infty}\frac{1}{n^2}.$$

6. (i) $f(0) = f''(0) = 0.$
   (ii) $f'(0) = 0.$

EXERCISES 6d

1. $$e^x \sim \sum_{k=-\infty}^{\infty}\frac{(-1)^k(3 + ik\pi)\sinh 3}{9 + k^2\pi^2}e^{ik\pi x/3} \qquad [-3,3].$$

2. $f(x) = \cos(\sin x)\exp(\cos x).$

EXERCISES 6e

3. A pointwise convergence theorem for the interval $0 < x < L$ is stated in Exercises 6c, Problem 4.

EXERCISES 6f

1. Use Equations (4.9.7)–(4.9.9) and (4.9.16)–(4.9.18).

EXERCISES 6g

1. (i) $$\frac{\pi^4}{96} = \sum_{k=0}^{\infty}\frac{1}{(2k+1)^4}.$$

   (ii) $$\frac{\pi^2}{8} = \sum_{k=0}^{\infty}\frac{1}{(2k+1)^2}.$$

2. $$\frac{\pi^4}{90} = \sum_{n=1}^{\infty}\frac{1}{n^4}.$$

6. (iii) Apply the result of part (ii) twice.

EXERCISES 7a

No answers or hints provided.

EXERCISES 7b

No answers or hints provided.

EXERCISES 8a

1. (i) $mu_{tt}(0,t) - T_0u_x(0,t) - au(0,t) - bu_t(0,t) = p(t)$.
   (ii) $mu_{tt}(L,t) + T_0u_x(L,t) - au(L,t) - bu_t(L,t) = p(t)$.

4. (i)

$$G(x,\xi) = \begin{cases} x(1-\xi) & 0 \le x \le \xi \\ (1-x)\xi & \xi \le x \le 1. \end{cases}$$

   (ii) $\int_0^1 G(x,\xi)q(\xi)\,d\xi = (1-x)\int_0^x \xi q(\xi)\,d\xi + x\int_x^1 (1-\xi)q(\xi)\,d\xi.$

6.
$$\rho_0 u_{tt} = \frac{\partial}{\partial s}\left[\rho_0 g(L-s)u_s\right] \qquad 0 < s < L, \quad 0 < t.$$

EXERCISES 8b

1. $\frac{LT_0}{4}\sum_{n=1}^{\infty}\left(\frac{n\pi}{L}\right)^2 A_n^2$, with A_n given by (8.4.9).

3. $u(x,t) = \sum_{n=1}^{\infty} B_n \frac{\sin(n\pi ct/L)}{(n\pi c/L)} \sin\frac{n\pi x}{L}$, where

$$B_n = \frac{2}{L}\int_0^L g(x)\sin\frac{n\pi x}{L}\,dx = \frac{2V}{n\pi}\left[\cos\frac{n\pi}{2} - \cos\frac{n\pi}{4}\right].$$

6. (i) Choosing $L = \pi$, $\rho_0 = T_0 = 1$ for convenience,

$$u(x,t) = \frac{g}{2}\left[x^2 - \pi x\right] + \frac{2g}{\pi}\sum_{n=1}^{\infty}\frac{[1-(-1)^n]}{n^3}$$
$$e^{-rt}\left[\cos t\sqrt{n^2-r^2} + r\frac{\sin t\sqrt{n^2-r^2}}{\sqrt{n^2-r^2}}\right]\sin nx,$$

   provided that $0 < r < 1$.

9. Use the result of Problem 8.

11.
$$\tan\frac{2\pi L}{3c_1}\nu + \frac{c_2}{c_1}\tan\frac{4\pi L}{3c_2}\nu = 0,$$

   where ν represents the value of a natural frequency and $c_1^2 = T/\rho_1$, $c_2^2 = T/\rho_2$.

EXERCISES 8c

No answers or hints provided.

EXERCISES 8d

5. (i)

$$u(x,t) = \begin{cases} -c\int_0^{t-x/c}\Phi(\xi)\,d\xi & 0 \le x \le ct \\ 0 & ct \le x. \end{cases}$$

6.

$$u(x,t) = \begin{cases} \dfrac{f(x+ct) + f(ct-x)}{2} - he^{h(x-ct)} \displaystyle\int_0^{ct-x} e^{h\xi} f(\xi)\, d\xi & 0 \leq x \leq ct \\ \dfrac{f(x+ct) + f(x-ct)}{2} & ct \leq x\,. \end{cases}$$

7.

$$u(x,t) = \begin{cases} \Phi\left(t - \dfrac{x}{c_1}\right) & x \leq -c_1 t \\ \Phi\left(t - \dfrac{x}{c_1}\right) + \dfrac{c_2 - c_1}{c_2 + c_1}\Phi\left(t + \dfrac{x}{c_1}\right) & -c_1 t \leq x \leq 0 \\ \dfrac{2c_2}{c_1 + c_2}\Phi\left(t - \dfrac{x}{c_2}\right) & 0 \leq x \leq c_2 t \\ 0 & c_2 t \leq x\,. \end{cases}$$

Note that $\Phi(\xi) = 0$ outside some interval $[a,b]$, where $0 < a < b$. Thus, for sufficiently large t and $x \leq 0$, $\Phi(t - x/c_1) = 0$.

9. Use the results of Exercises 6c, Problem 6 and Exercises 6a, Problem 4. Show that the B.C. are satisfied by the solution of the initial value problem because it is both odd and periodic with period $2L$.

EXERCISES 8e

3. Follow the argument of Section 8.5 using (8.7.6) instead of the energy equation.

EXERCISES 9a

2. From Equation (6.7.7)

$$s_n(0) = \frac{1}{\pi}\int_{-\pi}^{\pi} \frac{\sin\,(n+\frac{1}{2})t}{t}\,dt = \frac{1}{\pi}\int_{-(n+\frac{1}{2})\pi}^{(n+\frac{1}{2})\pi} \frac{\sin t}{t}\,dt\,,$$

and by the convergence theorem $\lim_{n\to\infty} s_n(0) = f(0) = 1$.

5. (a) (i) $f(x/a)/a$.
   (ii) $f(x - b)$.
   (iii) $\lambda F(\lambda)$.
   (iv) $ixf(x)$.
   (v) $(2 \sin \lambda a)/\lambda$.
   (vi) $2(1 - \cos \lambda a)/\lambda^2 a$.
   (vii) $a\pi^{-1}(a^2 + x^2)^{-1}$.
   (viii) $\frac{1}{2}e^{-a|x|}$.

   (b) (i) $\frac{1}{2}[f(x - b) + f(x + b)]$.
   (ii) $-f''(x)$.
   (iii) $f''''(x)$.
   (iv) $f(x) = \begin{cases} -\frac{1}{2}ie^{-ax}, & x > 0 \\ \frac{1}{2}ie^{ax}, & x < 0. \end{cases}$

EXERCISES 9b

1. (a) $h(x) = \begin{cases} 2a - |x|, & |x| < 2a \\ 0 & |x| \geq 2a. \end{cases}$

   (b) Use Exercise 9a, Problem 5, parts (v) and (vi).

   (c) π. [Use Fourier inversion formula to calculate $h(0)$.]

2. (a) $h(x) = \dfrac{1}{2\pi} \displaystyle\int_{-\pi}^{\pi} f(x - \xi)g(\xi)\, d\xi.$

   (b) $c_n = \displaystyle\sum_{m=-\infty}^{\infty} a_{n-m}b_m.$

EXERCISES 9c

1.
$$u(x,t) = \frac{1}{\sqrt{4\pi kt}} \int_0^\infty f(\xi)[e^{-(x-\xi)^2/4kt} - e^{-(x+\xi)^2/4kt}]\, d\xi\,.$$

4. (i) $v(x,t) = A \exp\left(-x\sqrt{\dfrac{\omega}{2k}}\right) \sin\left(\omega t - x\sqrt{\dfrac{\omega}{2k}}\right).$

   (ii) $4\sqrt{365}$ ft.

6. $\operatorname{erf} x = \dfrac{2}{\sqrt{\pi}} \displaystyle\sum_{n=0}^{\infty} \dfrac{(-1)^n x^{2n+1}}{n!(2n+1)}.$

8. $\operatorname{erf} x = 1 - \dfrac{2}{\sqrt{\pi}} \dfrac{e^{-x^2}}{2x} +$ error; the error is positive and does not exceed $\dfrac{2}{\sqrt{\pi}} \dfrac{e^{-x^2}}{4x^2}.$

10. $u(x,t) = u_0\left[\operatorname{erf} \dfrac{x}{\sqrt{4kt}} - \dfrac{1}{2} \operatorname{erf} \dfrac{x+L}{\sqrt{4kt}} - \dfrac{1}{2} \operatorname{erf} \dfrac{x-L}{\sqrt{4kt}}\right].$

    Use the identity
$$\frac{\sin \lambda x}{\lambda} = \int_0^x \cos \lambda\xi\, d\xi$$
    and Exercises 9a, Problem 5(ix).

11. See the hint for Problem 10.

13.
$$u(x,t) = \frac{x}{\sqrt{4\pi k}} \int_0^t U_0(t-\tau)\tau^{-3/2} \exp\left(\frac{-x^2}{4k\tau}\right) d\tau\,.$$

EXERCISES 9d

2.
$$u(x,t) = \frac{1}{2\pi} \int_{-\infty}^{\infty} F(\lambda) \cos t\sqrt{k + \lambda^2 c^2}\, e^{-i\lambda x}\, d\lambda + \frac{1}{2\pi} \int_{-\infty}^{\infty} G(\lambda) \frac{\sin t\sqrt{k + \lambda^2 c^2}}{\sqrt{k + \lambda^2 c^2}}\, e^{-i\lambda x}\, d\lambda.$$

4. $$u(x,t) = \frac{1}{2\pi} \int_{-\infty}^{\infty} e^{-i\lambda x} \int_0 Q(\lambda,\tau) \frac{\sin \lambda c(t-\tau)}{\lambda c} \, d\tau \, d\lambda,$$

$$Q(\lambda,\tau) = \int_{-\infty}^{\infty} q(x,\tau) e^{i\lambda x} \, dx.$$

5. $$u(x,t) = \begin{cases} -\int_0^{t-x} \Phi(\xi) \, d\xi & 0 < x < t \\ 0 & t < x. \end{cases}$$

6. $$u(x,t) = \begin{cases} e^{x-t} \int_0^{t-x} e^{\xi} \Phi(\xi) \, d\xi & 0 < x < t \\ 0 & t < x. \end{cases}$$

EXERCISES 9e

2. Assume $t \geq b > 0$ and $-a \leq s \leq a$. To prove that $u_x(x,t)$ exists and is given by the appropriate formula, show that

$$\left| \frac{\partial}{\partial x} f(\xi) \frac{1}{\sqrt{4\pi kt}} e^{-(x-\xi)^2/4kt} \right| \leq H_1(\xi) ,$$

where

$$H_1(\xi) = |f(\xi)| \frac{a + |\xi|}{\sqrt{\pi k b_0}} e^{-(a+|\xi|)^2/4kb_0} ,$$

and verify that $H_1(\xi)$ is absolutely integrable on $(-\infty,\infty)$.

4. Use the fact that for $t > 0$ the function

$$w(\xi; x,t) = \frac{1}{\sqrt{4\pi kt}} e^{-(x-\xi)^2/4kt}$$

is nonnegative and

$$\int_{-\infty}^{\infty} w(\xi; x,t) \, d\xi = 1 .$$

5. Let $f(x)$ be any bounded function and let $g(x)$ be a bounded function such that

$$\begin{aligned} g(x) &= f(x) , & -\infty < x \leq b \\ g(x) &> f(x) , & b < x < \infty . \end{aligned}$$

For example, let $g(x) = f(x) + \exp(b - x)$ for $b < x < \infty$. Show then that $v(x_0,t_0) > u(x_0,t_0)$.

EXERCISES 10a

1. $$u(x,y,t) = \tfrac{1}{2} \sum_{m=1}^{\infty} a_{m,0} e^{-\lambda_{m,0}kt} \varphi_{m,0}(x,y) + \sum_{m=1}^{\infty} \sum_{n=1}^{\infty} a_{m,n} e^{-\lambda_{m,n}kt} \varphi_{m,n}(x,y),$$

$$\lambda_{m,n} = (m + \tfrac{1}{2})^2 \frac{\pi^2}{a^2} + \frac{n^2\pi^2}{b^2}, \qquad m, n = 0, 1, \ldots$$

$$\varphi_{m,n}(x,y) = \sin (m + \tfrac{1}{2}) \frac{\pi x}{a} \cos \frac{n\pi y}{b},$$

$$a_{m,0} = \frac{(-1)^m 2a^2}{(m+\frac{1}{2})^2\pi^2} + \frac{2b}{(m+\frac{1}{2})\pi},$$

$$a_{m,n} = \frac{4}{(m+\frac{1}{2})\pi}\,\frac{b}{n^2\pi^2}\,[(-1)^n - 1], \qquad n \geq 1.$$

2. (b) $\lambda_{l,m,n} = \left(\frac{l\pi}{a}\right)^2 + \left(\frac{m\pi}{b}\right)^2 + \left(n+\frac{1}{2}\right)^2\left(\frac{\pi}{c}\right)^2,$

$$\varphi_{l,m,n}(x,y,z) = \cos\frac{l\pi x}{a}\sin\frac{m\pi y}{b}\sin\left(n+\frac{1}{2}\right)\frac{\pi z}{c},$$

$l = 0, 1, \ldots, m = 1, 2, \ldots, n = 0, 1, \ldots.$

4. (a) $\lambda_{39,52} = \lambda_{52,39} = \lambda_{25,60} = \lambda_{60,25}$.
   (c) 50.

6. (a) $\lambda_{m,n} = \left(\frac{m\pi}{\log 2}\right)^2 + n^2$, $\varphi_{m,n}(x,y) = \sin\frac{m\pi\log x}{\log 2}\cos ny,$

$m = 1, 2, \ldots, n = 0, 1, \ldots.$

   (b) $\frac{4}{\pi\log 2}\int_0^\pi\int_1^2 \varphi_{m,n}(x,y)\varphi_{p,q}(x,y)\,\frac{dx}{x}\,dy = \begin{cases} 0 & \text{if } m \neq p \text{ or } n \neq q \\ 1 & \text{if } m = p \text{ and } n = q. \end{cases}$

EXERCISES 10b

1. (i) Let λ be an eigenvalue with eigenfunction φ. In Green's first formula set $f = \varphi$, $g = \bar{\varphi}$. Since $\bar{\varphi} = 0$ on C, we get

$$-\lambda \iint_D |\varphi|^2\,dx\,dy = -\iint_D \left[\left|\frac{\partial\varphi}{\partial x}\right|^2 + \left|\frac{\partial\varphi}{\partial y}\right|^2\right] dx\,dy$$

and because

$$\iint_D |\varphi|^2\,dx\,dy > 0$$

it follows that λ is real and non-negative.

3. Proceeding as in Problem 1, we obtain

$$-\lambda \iint_D |\varphi|^2\,dx\,dy - \iint_D q|\varphi|^2\,dx\,dy = -\iint_D \left[\left|\frac{\partial\varphi}{\partial x}\right|^2 + \left|\frac{\partial\varphi}{\partial y}\right|^2\right] dx\,dy\,.$$

Hence λ is real and

$$0 \leq \lambda \iint_D |\varphi|^2\,dx\,dy + \iint_D q|\varphi|^2\,dx\,dy \leq (\lambda + M_2) \iint_D |\varphi|^2\,dx\,dy\,,$$

which shows that $\lambda + M_2 \geq 0$.

6. Compare with Exercises 8c, Problem 4.

EXERCISES 10c

2. Let $1 \leq m < n$ and $1 \leq p < q$. The function $\psi_{m,n}\psi_{p,q}$ is a product of two skew symmetric functions and is therefore symmetric, so that

$$\iint\limits_{D} \psi_{m,n}\psi_{p,q}\, dx\, dy = \tfrac{1}{2} \iint\limits_{D'} \psi_{m,n}\psi_{p,q}\, dx\, dy\,.$$

Note that $m < n$ and $p < q$ imply that not both of the equalities $m = q$ and $n = p$ can hold, and hence

$$\iint\limits_{D'} \varphi_{m,n}\varphi_{q,p}\, dx\, dy = \iint\limits_{D'} \varphi_{n,m}\varphi_{p,q}\, dx\, dy = 0\,.$$

Show that

$$\iint\limits_{D} \psi_{m,n}\psi_{p,q} = \begin{cases} 0 & \text{if } m \neq p \ \text{ or } \ n \neq q \\ \dfrac{ab}{4} & \text{if } m = p \text{ and } n = q\,. \end{cases}$$

4. $\lambda_{m,n} = \dfrac{\pi^2}{a^2}(m^2 + n^2), \qquad 0 \leq m \leq n < \infty,$

$$\psi_{m,n} = \cos\frac{m\pi x}{a}\cos\frac{n\pi y}{a} + \cos\frac{n\pi x}{a}\cos\frac{m\pi y}{a}.$$

EXERCISES 10d

2. (b) $v(x,y) = \displaystyle\sum_{m=1}^{\infty}\sum_{n=1}^{\infty} \frac{Q_{m,n}}{\lambda_{m,n}k}\varphi_{m,n}(x,y),$

$$Q_{m,n} = \frac{4}{ab}\iint\limits_{D} Q(x,y)\varphi_{m,n}(x,y)\, dx\, dy,$$

where $\lambda_{m,n}$ and $\varphi_{m,n}$ are given by (10.3.9) and (10.3.10). The problem is

D.E.	$k\Delta v + Q(x,y) = 0$	in D
B.C.	$v = 0$	on C.

The problem can be solved by assuming a series solution of the above form.

3. $$u(x,y) = B + \frac{1}{2}\sum_{n=1}^{\infty}\frac{a_{0,n}}{\lambda_{0,n}}\varphi_{0,n}(x,y) + \frac{1}{2}\sum_{m=1}^{\infty}\frac{a_{m,0}}{\lambda_{m,0}}\varphi_{m,0}(x,y) + \sum_{m=1}^{\infty}\sum_{n=1}^{\infty}\frac{a_{m,n}}{\lambda_{m,n}}\varphi_{m,n}(x,y),$$

where B is an arbitrary constant and

$$\lambda_{m,n} = \pi^2\left[\frac{m^2}{a^2} + \frac{n^2}{b^2}\right],$$

$$\varphi_{m,n}(x,y) = \cos\frac{m\pi x}{a}\cos\frac{n\pi y}{b},$$

$$a_{m,n} = \frac{4}{ab} \iint_D Q(x,y)\varphi_{m,n}(x,y)\, dx\, dy\, .$$

5. (a) $c_n = \dfrac{2}{b} \dfrac{n\pi/b}{\alpha^2 + (n\pi/b)^2}$.
   (b) $0 < y \leq b$.
   (c) Use (10.7.11) and Green's formula, as in part (a), to evaluate $C_{m,n}$.

7. (b) Let $g_0 = \dfrac{1}{a} \displaystyle\int_0^a g(\xi)\, d\xi$. Then

$$A = -kg_0/b.$$

   (c) D.E.

$$\begin{aligned} &\Delta v = 0 && \text{in } D, \\ &v_y(x,0) = g(x) - g_0, && 0 < x < a \\ &v_y(x,b) = 0, && 0 < x < a \\ &v_x(0,y) = v_x(a,y) = 0, && 0 < y < b. \end{aligned}$$

EXERCISES 10e

1. $$u(x,y,t) = \left(\frac{2}{\pi}\right)^2 \int_0^\infty \int_0^\infty \mu G(\lambda) \frac{1 - e^{-(\lambda^2+\mu^2)kt}}{\lambda^2 + \mu^2} \sin \lambda x \sin \mu y\, d\lambda\, d\mu,$$

   where $G(\lambda)$ is the Fourier sine transform of $g(x)$.

3. $$u(x,y,t) = \left(\frac{2}{\pi}\right)^2 \int_0^\infty \int_0^\infty U(\lambda,\mu,t) \sin \lambda x \cos \mu y\, d\lambda\, d\mu,$$

   where

$$U(\lambda,\mu,t) = F(\lambda,\mu) \frac{\sin (ct\sqrt{\lambda^2 + \mu^2})}{c\sqrt{\lambda^2 + \mu^2}} - [G(\lambda) + \lambda H(\mu)] \frac{1 - \cos (ct\sqrt{\lambda^2 + \mu^2})}{\lambda^2 + \mu^2},$$

   $F(\lambda,\mu) = \displaystyle\int_0^\infty \int_0^\infty f(x,y) \sin \lambda x \cos \mu y\, dx\, dy$, and

   $G(\lambda)$ is the Fourier sine transform of $g(x)$, and $H(\mu)$ is the Fourier cosine transform of $h(y)$.

EXERCISES 10f

1. (b) If $\varphi(a) \neq 0$ then, by the continuity of $\varphi(x)$, we can choose $\epsilon > 0$ so small that $\varphi(x)$ is of constant sign in the interval $a - \epsilon \leq x \leq a + \epsilon$.
   (e) Show that each integral is equal to

$$\frac{1}{(2\pi)^2} \int_{-\infty}^\infty \int_{-\infty}^\infty F(\lambda,\mu) H(-\lambda,-\mu)\, d\lambda\, d\mu\, .$$

EXERCISES 11a

4. (i) Let u_1 and u_2 be distinct solutions of the Dirichlet problem. Then $v = u_2 - u_1$ is non-trivial and is an eigenfunction of (11.2.9) with

eigenvalue γ. Conversely, if γ is an eigenvalue of (11.2.9) and v is a corresponding eigenfunction, then given any solution u_1 of the Dirichlet problem, $u_2 = u_1 + v$ is another solution.

(ii) The eigenvalues are non-negative.

5. (ii) The special case arises because $\gamma = 0$ is an eigenvalue of (11.2.10) with the constant eigenfunction $\varphi = 1$.

EXERCISES 11b

1. $$u(x,y,z) = \sum_{m=1}^{\infty}\sum_{n=1}^{\infty} f_{m,n} \frac{\sinh (c-z)\sqrt{\lambda_{m,n}}}{\sinh c\sqrt{\lambda_{m,n}}} \varphi_{m,n})(x,y),$$
where
$$\lambda_{m,n} = \pi^2\left[\frac{m^2}{a^2}+\frac{n^2}{b^2}\right], \quad \varphi_{m,n}(x,y) = \sin\frac{m\pi x}{a}\sin\frac{n\pi y}{b},$$
$$f_{m,n} = \frac{4}{ab}\int_0^b\int_0^a f(x,y)\varphi_{m,n}(x,y)\,dx\,dy .$$

4. (i) $$u(x,y) = B + \sum_{n=1}^{\infty}\frac{h_n\cosh(n\pi y/a) - g_n\cosh n\pi(b-y)/a}{(n\pi/a)\sinh(n\pi b/a)}\cos(n\pi x/a),$$
where B is an arbitrary constant.

5. $\alpha = (2ab)^{-1}\int_0^a (g-h)\,dx.$

6. (i) $$u(x,y) = B + \frac{1}{2}\sum_{m=1}^{\infty}\frac{Q_{m,0}}{\lambda_{m,0}}\varphi_{m,0} + \frac{1}{2}\sum_{n=1}^{\infty}\frac{Q_{0,n}}{\lambda_{0,n}}\varphi_{0,n} + \sum_{m=1}^{\infty}\sum_{n=1}^{\infty}\frac{Q_{m,n}}{\lambda_{m,n}}\varphi_{m,n}$$
where
$$\lambda_{m,n} = \pi^2\left[\frac{m^2}{a^2}+\frac{n^2}{b^2}\right], \quad \varphi_{m,n} = \cos\frac{m\pi x}{a}\cos\frac{n\pi y}{b},$$
$Q_{m,n}$ are the double Fourier cosine coefficients of $Q(x,y)$, and B is an arbitrary constant.

EXERCISES 11c

3. $$u(r,\theta) = \sum_{n=1}^{\infty} c_n\frac{\sinh(\alpha-\theta)\sqrt{\lambda_n}}{\sinh\alpha\sqrt{\lambda_n}}\varphi_n(r),$$
$$\lambda_n = (n\pi)^2(\log b - \log a)^{-2}, \ \varphi_n(r) = \sin\sqrt{\lambda_n}\,(\log r - \log a),$$
$$c_n = 2(\log b - \log a)^{-1}\int_a^b f(r)\varphi_n(r)\,\frac{dr}{r}.$$

6. $$u(r,\theta) = \sin\left[n\pi\frac{\log r - \log a}{\log b - \log a}\right]\exp\left[\frac{n\pi\theta}{\log b - \log a}\right], \qquad n = 1, 2, \ldots .$$

9. (ii) A solution of the related homogeneous problem is
$$u(r,\theta) = \left[\left(\frac{r}{b}\right)^{n\pi/\alpha} - \left(\frac{b}{r}\right)^{n\pi/\alpha}\right]\sin\frac{n\pi\theta}{\alpha},$$

where n is any positive integer.

10. $$u(r,\theta) = \frac{2}{\pi}\int_0^\infty \left[F(\lambda)\,\frac{\sinh \lambda(\alpha-\theta)}{\sinh \lambda\alpha} + G(\lambda)\,\frac{\sinh \lambda\theta}{\sinh \lambda\alpha}\right] \sin\left(\lambda \log \frac{b}{r}\right) d\lambda,$$

$$F(\lambda) = \int_0^b f(r) \sin\left(\lambda \log \frac{b}{r}\right) \frac{dr}{r},$$

$$G(\lambda) = \int_0^b g(r) \sin\left(\lambda \log \frac{b}{r}\right) \frac{dr}{r}.$$

EXERCISES 11d

1. $$u(r,\theta) = \frac{1}{2\alpha}$$

$$\int_0^\alpha f(\varphi)\left[\frac{1-\rho^2}{1-2\rho\cos(\theta-\varphi)+\rho^2} - \frac{1-\rho^2}{1-2\rho\cos(\theta+\varphi)+\rho^2}\right] d\varphi,$$

where $\rho = (r/b)^{\pi/\alpha}$.

2. (i) From (11.6.15) and the corresponding equation obtained by replacing $(b-r)^2$ by $(b+r)^2$, we have

$$(b-r)^2 \leq b^2 - 2rb\cos(\theta-\varphi) + r^2 \leq (b+r)^2\,,$$

and hence

$$\frac{1}{2\pi}\frac{b^2-r^2}{(b+r)^2} \leq \frac{1}{2\pi}\frac{b^2-r^2}{b^2-2rb\cos(\theta-\varphi)+r^2} \leq \frac{1}{2\pi}\frac{b^2-r^2}{(b-r)^2}\,.$$

Multiply this by the non-negative function $u(b,\varphi)$ and integrate over $-\pi \leq \varphi \leq \pi$.

(ii) Let $b \to \infty$ in Harnack's inequality.

EXERCISES 11e

2. Let D_b be a disc of radius b, C_b its circumference, and $L = 2\pi b$. Denote by w_0 and Δw_0 the values of w and Δw at the center of D_b. Apply the result of Problem 1 to the harmonic function Δw in a disc of radius r concentric with D_b, and then use (11.7.5). Multiply the resulting equation by r and integrate with respect to r from 0 to b.

The answer is

$$w_0 + \frac{b^2}{4}\Delta w_0 = \frac{1}{L}\int_{C_b} w ds\,.$$

3. If u is harmonic in a sphere of radius r, then its mean value over the surface S of the sphere is equal to its value at the center,

$$u(0,0,0) = \frac{1}{4\pi r^2}\iint_S u dS\,.$$

4. (ii) Since $I(r)$ satisfies the differential equation and is continuous at $r = 0$, we have

$$I(r) = CJ_0(r\sqrt{\gamma}),$$

where C is a constant. To find C, let $r \to 0$, and obtain

$$u(0,0)J_0(r\sqrt{\gamma}) = \frac{1}{2\pi r}\int_0^{2\pi} u(r,\theta)\, r d\theta .$$

EXERCISES 11f

2. (i) At a local maximum in D we have $u_x = u_y = 0$ and $\Delta u \leq 0$, and hence it follows from the differential equation that $c(x,y)u \geq 0$ at the local maximum. But $c(x,y) < 0$, so $u \leq 0$ at the local maximum. Since $-u$ satisfies the same equation and has a local maximum at any point where u has a local minimum, we have $-u \leq 0$ or $u \geq 0$ at a local minimum of u.
   (ii) If v is the difference of any two solutions then v satisfies D.E. and is zero on the boundary.

3. Let M be the maximum value of u on C and let K be the maximum value of $(x - x_0)^2$ on C. Since $u(x_0, y_0) > M$, we can choose ϵ positive but so small that $u(x_0, y_0) > M + \epsilon K$. The function

$$v(x,y) = u(x,y) + \epsilon(x - x_0)^2$$

satisfies Poisson's equation $\Delta v = -q$ with $q = -2\epsilon$. Since q is negative in D the function v cannot have a local maximum in D and hence attains its maximum on the boundary. But on the boundary v does not exceed $M + \epsilon K$, while $v(x_0, y_0) = u(x_0, y_0) > M + \epsilon K$. Thus the assumption that u does not attain is maximum on C leads to a contradiction.

EXERCISES 11g

1. $u(x,y) = \dfrac{2}{\pi}\displaystyle\int_0^\infty \cos \lambda x \left[\int_{-0}^{\infty} f(\xi) \cos \lambda\xi \, d\xi\right] e^{-\lambda y}\, d\lambda.$

3. $u(x,y) = \dfrac{2}{\pi}\displaystyle\int_0^\infty \frac{F(\mu) \sinh \mu(a - x) + K(\mu) \sinh \mu x}{\sinh \mu a} \sin \mu y \, d\mu,$

   where $F(\mu)$ and $K(\mu)$ are the Fourier sine transforms of $f(y)$ and $k(y)$, respectively.

5. $u(x,y) = \dfrac{1}{2\pi}\displaystyle\int_{-\infty}^{\infty} F(\mu) \frac{\sinh \mu x}{\sinh \mu a} e^{-i\mu y}\, d\mu,$

   where $F(\mu)$ is the Fourier transform of $f(y)$.

6. $u(x,y,z) = (2\pi)^{-2}\displaystyle\int_{-\infty}^{\infty}\int_{-\infty}^{\infty} F(\lambda,\mu)e^{-z\sqrt{\lambda^2+\mu^2}}e^{-i(\lambda x+\mu y)}\, d\lambda\, d\mu,$

   where $F(\lambda,\mu)$ is the double Fourier transform of $f(x,y)$.

8. $u(r,\theta) = \dfrac{1}{2\pi}\displaystyle\int_0^\infty F(\lambda)\,\frac{\sinh\lambda\theta}{\sinh\lambda\alpha}\,e^{i\lambda\log r}\,\frac{dr}{r},$

where

$$F(\lambda) = \int_0^\infty f(\rho)e^{-i\lambda\log\rho}\,\frac{d\rho}{\rho}\,.$$

12. Let the Fourier series of $f(\theta)$ and $g(\theta)$ be

$$f(\theta) \sim \tfrac{1}{2}\alpha_0 + \sum_{n=1}^{\infty}(\alpha_n\cos n\theta + \beta_n\sin n\theta)\,,$$

$$g(\theta) \sim \tfrac{1}{2}\gamma_0 + \sum_{n=1}^{\infty}(\gamma_n\cos n\theta + \delta_n\sin n\theta)\,.$$

Then

$$u(r,\theta) = \tfrac{1}{2}(A_0 + r^2A_0') + \sum_{n=1}^{\infty} r^n[(A_n + r^2A_n')\cos n\theta + (B_n + r^2B_n')\sin n\theta],$$

where

$$\begin{aligned} A_n &= [(n+2)\alpha_n - \gamma_n]/2\,, \quad & A_n' &= [\gamma_n - n\alpha_n]/2\,,\\ B_n &= [(n+2)\beta_n - \delta_n]/2\,, \quad & B_n' &= [\delta_n - n\beta_n]/2\,. \end{aligned}$$

13. $u(x,y) = \dfrac{1}{2\pi}\displaystyle\int_{-\infty}^{\infty}[F(\lambda) + y\{G(\lambda) + |\lambda|F(\lambda)\}]e^{-|\lambda|y}e^{-i\lambda x}\,d\lambda,$

where $F(\lambda)$ and $G(\lambda)$ are the Fourier transforms of $f(x)$ and $g(x)$, respectively.

15. $u(x,y) = \displaystyle\int_{-\infty}^{\infty}\frac{1}{\pi}\,\frac{\frac{1}{2}y\sqrt{3}}{\frac{3}{4}y^2 + (x - \frac{1}{2}y - \xi)^2}f(\xi)\,d\xi.$

EXERCISES 12a

1. Make a suitable change of the integration variable. Observe that the integrand is periodic with period 2π and the interval of integration has length 2π.

4. Show first that (12.2.12) can be rewritten as

$$J_0(\lambda) = \frac{1}{\pi}\int_{-1}^{1}(1 - x^2)^{-1/2}e^{i\lambda x}\,dx\,.$$

The answer is

$$\begin{cases} 2(a^2 - \lambda^2)^{-1/2} & |\lambda| < a\\ 0 & |\lambda| > a\,. \end{cases}$$

6.
$$\mu(r) = \exp\left[-\tfrac{1}{2}\int\alpha(r)\,dr\right]$$
$$N(r) = \beta(r) - \tfrac{1}{2}\alpha'(r) - \tfrac{1}{4}\alpha^2(r)\,.$$

9. Use the result of Problem 1.

11.
$$\varphi(r) = A\,\frac{\cos r\sqrt{\lambda}}{r} + B\,\frac{\sin r\sqrt{\lambda}}{r\sqrt{\lambda}},$$

A and B any constants.

EXERCISES 12b

3. (i) Let $\gamma_0, \gamma_1, \ldots, \gamma_m, \ldots$ be the roots of

$$\gamma J_0'(\gamma) = 0\,.$$

Then the eigenvalues and corresponding eigenfunctions are

$$\gamma_m = (\gamma_m/b)^2 \qquad m = 0, 1, 2, \ldots .$$
$$\varphi_m(r) = J_0\left(\frac{\gamma m}{b}\,r\right)$$

In particular, $\lambda_0 = 0$ and $\varphi_0(r) = 1$.

(iii)
$$\frac{b^2}{2}\,[J_0(\gamma_m)]^2\,, \qquad m = 0, 1, 2, \ldots .$$

6. Use the statement and results of Problem 3. The answer is

$$u(r,t) = \sum_{m=0}^{\infty} C_m \exp\left[-\left(\frac{\gamma_m}{b}\right)^2 kt\right] J_0\left(\frac{\gamma_m r}{b}\right),$$

where $C_0 = 1/4$ and

$$C_m = \frac{-J_0'(\gamma_m/2)}{\gamma_m[J_0(\gamma_m)]^2}, \qquad m = 1, 2, \ldots .$$

7. (ii)
$$U(r) = \frac{g}{4c^2}\,(r^2 - b^2)\,.$$

(iii) Use the theorem stated at the end of Section 3. The answer is

$$u(r,t) = U(r) + \sum_{m=1}^{\infty} A_m \cos\left(\frac{\beta_m ct}{b}\right) J_0\left(\frac{\beta_m r}{b}\right),$$

where

$$A_m = \frac{-2gb^2}{c^2\beta_m^3 J_0'(\beta_m)}, \qquad m = 1, 2, \ldots .$$

The calculation of the integral

$$\int_0^b U(r)\varphi_m(r)\,r dr\,,$$

where φ_m is given by (12.3.5), is effected by twice using the substitution

$$-\frac{1}{\lambda_m}\frac{d}{dr}\,(r\varphi_m') = r\varphi_m$$

and integration by parts, noting the B.C. satisfied by φ_m.

10.
$$u(r,z) = \sum_{m=1}^{\infty} B_m\,\frac{I_0(m\pi r/l)}{I_0(m\pi b/l)}\sin\frac{m\pi z}{l},$$

where

$$B_m = \frac{2}{l}\int_0^l h(z)\sin\frac{m\pi z}{l}\,dz\,, \qquad m = 1, 2, \ldots .$$

13. The equation whose roots are the appropriate eigenvalues for the problem is

$$\begin{vmatrix} \cos a\sqrt{\lambda} & \dfrac{\sin a\sqrt{\lambda}}{\sqrt{\lambda}} \\ \cos b\sqrt{\lambda} & \dfrac{\sin b\sqrt{\lambda}}{\sqrt{\lambda}} \end{vmatrix} = \frac{\sin(b-a)\sqrt{\lambda}}{\sqrt{\lambda}} = 0\,.$$

The eigenvalues and corresponding eigenfunctions are

$$\lambda_m = \left(\frac{m\pi}{b-a}\right)^2, \quad \varphi_m(r) = \frac{1}{r}\sin\left(m\pi\frac{r-a}{b-a}\right), \qquad m = 1, 2, \ldots .$$

The solution of the problem is

$$u(r,t) = \sum_{m=1}^{\infty} C_m e^{-\lambda_m kt}\varphi_m(r)\,,$$

where

$$C_m = \frac{2}{b-a}\int_a^b f(r)\varphi_m(r)r^2\,dr\,.$$

15. Set $\lambda_m = \beta_m^2/4L$, with β_m denoting the mth positive zero of J_0. Then

$$u(r,t) = \sum_{m=1}^{\infty}[A_m\cos t\sqrt{g\lambda_m} + B_m\sin t\sqrt{g\lambda_m}]J_0(\sqrt{4\lambda_m r})\,,$$

where

$$\begin{Bmatrix} A_m \\ B_m \end{Bmatrix} = \frac{1}{L[J_0'(\beta_m)]^2}\int_0^L \begin{Bmatrix} f(r) \\ g(r) \end{Bmatrix} J_0(\sqrt{4\lambda_m r})\,dr\,.$$

EXERCISES 12c

4. (i) Replace n by $n + 1$ in the second recurrence relation of Problem 3.
7. Writing

$$\psi(r) = \sum_{j=0}^{\infty} c_j r^{2j+\gamma}\,,$$

differentiating and substituting in (12.5.11), obtain the power series identity

$$(\gamma^2 - \nu^2)c_0 + \sum_{j=1}^{\infty}\{[(2j+\gamma)^2 - \nu^2]c_j + c_{j-1}\}r^{2j} = 0\,,$$

which can hold only if all the coefficients vanish. The vanishing of the first coefficient requires $\gamma = \pm\nu$, since $c_0 \neq 0$. Setting $\gamma = \nu$ in the remaining coefficients and equating them to zero, obtain the recurrence relation

$$c_j = \frac{-1}{2^2 j(j+\nu)}c_{j-1}\,, \qquad j = 1, 2, \ldots,$$

from which deduce

$$c_j = \frac{(-1)^j}{2^{2j} j!(j+\nu) \cdots (3+\nu)(2+\nu)(1+\nu)} c_0 .$$

The case $\gamma = -\nu$ is treated similarly.

9. Differentiate the second relation. In the first relation replace ν by $\nu - 1$ and use this relation to eliminate the term involving $Z'_{\nu-1}$ in the result of the differentiation. Finally, use the second relation to eliminate the term in $Z_{\nu-1}$ in the result.

13. (i) Let $\gamma_{\nu,1} < \gamma_{\nu,2} < \cdots < \gamma_{\nu,m} < \cdots$ be the positive roots of

$$J'_\nu(\gamma) = 0 .$$

Then the eigenvalues and corresponding eigenfunctions are

$$\left.\begin{aligned} \lambda_{\nu,m} &= \left(\frac{\gamma_{\nu,m}}{b}\right)^2 \\ \varphi_{\nu,m}(r) &= J_\nu\left(\frac{\gamma_{\nu,m}}{b} r\right) \end{aligned}\right\} \quad m = 1, 2, \ldots .$$

(iii)

$$\frac{b^2}{2}\left[1 - \left(\frac{\nu}{\gamma_{\nu,m}}\right)^2\right] [J_\nu(\gamma_{\nu,m})]^2 .$$

15. $$u(r,\theta,t) = \sum_{n=1}^{\infty} \sum_{m=1}^{\infty} \left[A_{n,m} \cos ct\sqrt{\lambda_{n,m}} + B_{n,m} \frac{\sin ct\sqrt{\lambda_{n,m}}}{c\sqrt{\lambda_{n,m}}} \right] \varphi_{n,m}(r,\theta),$$

where $\lambda_{n,m} = (\beta_{n,m}/b)^2$,

$$\varphi_{n,m}(r,\theta) = J_n(r\sqrt{\lambda_{n,m}}) \sin n\theta ,$$

$$\begin{Bmatrix} A_{n,m} \\ B_{n,m} \end{Bmatrix} = \frac{4}{\pi b^2 [J_{n+1}(\beta_{n,m})]^2} \int_0^b \int_0^\pi \begin{Bmatrix} f(r,\theta) \\ g(r,\theta) \end{Bmatrix} \varphi_{n,m}(r,\theta) r \, dr \, d\theta .$$

16. $$u(r,\theta,t) = \tfrac{1}{2} \sum_{m=1}^{\infty} A_{0,m} e^{-\lambda_{0,m} kt} \varphi_{0,m}(r,\theta) + \sum_{n=1}^{\infty} \sum_{m=1}^{\infty} A_{n,m} e^{-\lambda_{n,m} kt} \varphi_{n,m}(r,\theta) ,$$

where $\lambda_{n,m} = b^{-2} \beta^2_{n\pi/\omega,m}$,

$$\varphi_{n,m}(r,\theta) = J_{n\pi/\omega}(r\sqrt{\lambda_{n,m}}) \cos \frac{n\pi\theta}{\omega} ,$$

$$A_{n,m} = \frac{4}{\omega b^2 [J_{(n\pi/\omega)+1}(\beta_{n\pi/\omega,m})]^2} \int_0^\omega \int_0^b f(r,\theta) \varphi_{n,m}(r,\theta) r \, dr \, d\theta .$$

17. Use the results of Problem 13 and of Exercises 12b, Problem 3.

21. (i)

$$H_{2m}(x) = \sum_{k=0}^{m} \frac{(-1)^k (2m)!}{k!(2m-2k)!} (2x)^{2m-2k}$$

$$H_{2m+1}(x) = \sum_{k=0}^{m} \frac{(-1)^k (2m+1)!}{k!(2m-2k+1)!} (2x)^{2m-2k+1} .$$

(ii)

$$\left.\begin{aligned} H'_n(x) &= 2nH_{n-1}(x) \\ H_{n+1}(x) - 2xH_n(x) + 2nH_{n-1}(x) &= 0 \end{aligned}\right\} \quad n = 1, 2, \ldots .$$

Index